Sylvia G. Makowski

'1992'

ELEMENTARY
SURVEY
SAMPLING

ELEMENTARY SURVEY SAMPLING

THIRD EDITION

RICHARD L. SCHEAFFER
University of Florida

WILLIAM MENDENHALL
University of Florida, Emeritus

LYMAN OTT
Merrell Research Center

 PWS-KENT Publishing Company
Boston

PWS–KENT
Publishing Company

20 Park Plaza
Boston, Massachusetts 02116

PWS–KENT Publishing Company is a division of Wadsworth, Inc.

Library of Congress Cataloging in Publication Data

Scheaffer, Richard L.
  Elementary survey sampling.
  Includes index.
  1. Sampling (Statistics)   I. Mendenhall, William.
II. Ott, Lyman. III. Title.
QA276.6.S385 1986   519.5′2   85-7011
ISBN 0-87150-943-1

ISBN 0-87150-943-1

Printed in the United States of America.

88 89 90 — 10 9 8 7 6 5 4

Sponsoring Editor: *Michael Payne*
Production Coordinator/Interior Design: *Elise Kaiser*
Production: *Technical Texts, Inc.*
Interior Illustration: *Susan Linville*
Cover Design: *Elise Kaiser*
Composition: *J.W. Arrowsmith, Ltd.*
Cover Printer: *New England Book Components, Inc.*
Text Printer/Binder: *Maple-Vail Book Manufacturing Group*

Cover art "Visitation" by Michael Lasuchin; a screenprint used with the permission of the artist.

PREFACE

Elementary Survey Sampling is an introductory text on the design and analysis of sample surveys intended for students of business, the social sciences, or natural resource management. The only prerequisite is an elementary course in statistics. The numerous examples, with solutions, also make it suitable for use as a supplemental text for higher-level courses.

Since it is written to appeal to students of limited mathematical background, the text emphasizes the practical aspects of survey problems. Each major chapter introduces a sample survey design or a possible estimation procedure by describing a pertinent practical problem and then explaining the suitability of the methodology proposed. This introduction is followed by the appropriate estimation procedures and a compact presentation of the formulas; then a practical example is worked out. The text is not entirely cookbook in nature. Explanations that appeal to the students' intuition are supplied to justify many of the formulas and to support the choice of particular sample survey designs. Examples and exercises have been selected from many fields of application. Answers, given for selected exercises, may be subject to small rounding errors because of the complexity of some of the formulas.

The "Experiences with Real Data" sections found at the end of most chapters include suggestions on how the student can become involved with real sampling problems. These problems may be large or small projects, with some requiring computations to be handled by a computer, but we have found such projects to be valuable learning experiences for students taking a sampling course. Working on a real project forces students to think about every aspect of the survey and causes them to realize that some ideas that sound simple in the textbook are not so easily carried out in practice.

The text includes a review of elementary concepts (Chapters 1 and 2) and a description of terms pertinent to survey sampling, along with a discussion of the design of questionnaires and methods of data collection (Chapter 3). Chapters 4, 5, 7, and 8 present the four most common sample survey designs—namely, simple random sampling, stratified random sampling, systematic sampling, and cluster sampling, respectively. Chapter 6 discusses ratio and regression estimation. The remaining chapters deal with two-stage cluster sampling, sampling of animal populations, and other specialized problems that occur in survey sampling.

The third edition of this text emphasizes the practical aspects of conducting sample surveys, with new sections on sources of errors in surveys, methods of data collection, designing questionnaires, and guidelines for planning surveys. Most chapters now contain more examples of how the various survey designs are actually used in practice. Many practical applications of surveys employ sampling with probabilities proportional to size, and discussions of this method have been added to three chapters.

Other topics new to the third edition include stratification after selection of the sample, sample size determination in two-stage cluster sampling, and a broader discussion of estimating population sizes.

New exercises have been added to most chapters, and Chapter 12 now has a set of exercises that may require some careful thought in the selection of the appropriate analysis. These exercises can serve as a review of the major methods presented in the book. A large, real data set is included in the Appendix, with numerous exercises referring to it. A solutions manual is available from the publisher.

The Appendix also includes the mathematical derivations of many of the main results in the text. The understanding of many of these derivations requires a working knowledge of elementary probability theory.

We wish to express our sincere appreciation to the many people who have helped in the preparation of this text. Particular thanks are due to the reviewers for their helpful comments during the preparation of this manuscript. Thanks are also due to Professor A. Hald for his kind permission to use the table of normal curve areas reprinted in the Appendix. We are also deeply indebted to the typists who have given much of their time in preparing this text: Judith Donnelley, Mary Jackson, Catherine Kennedy, and Shirley Morley. Finally, we thank our families for assistance and encouragement throughout the duration of this project.

Richard L. Scheaffer
William Mendenhall
Lyman Ott

CONTENTS

1
INTRODUCTION

Introductory courses stress that modern statistics is a theory of information with inference as its objective. The target of our curiosity is a set of measurements, a *population*, that exists in fact or may be generated by repeated experimentation. The medium of inference is the *sample*, which is a subset of measurements selected from the population. We wish to make an inference about the population on the basis of characteristics of the sample—or, equivalently, the information contained in the sample.

For example, suppose that a chain of department stores maintains customer charge accounts. The amount of money owed the company will vary from day to day as new charges are made and some accounts are paid. Indeed, the set of amounts due the company on a given day represents a population of measurements of considerable interest to the management. The population characteristic of interest is the total of all measurements in the population or, equivalently, the daily total credit load.

Keeping track of the daily total credit associated with charge accounts may seem to be a simple task for an electronic computer. However, the data must be updated daily, and updating takes time. A simpler method for determining the total credit load associated with the charge accounts is to randomly sample the population of accounts on a given day, estimate the average amount owed per account, and multiply by the number of accounts. In other words, we employ a statistical estimator to make an inference about the population total. Elementary statistics tells us that this estimate can be made as accurate as we wish simply by increasing the sample size. The resulting estimate either is accompanied by a bound on the error of estimation (Mendenhall, 1983, Chapter 8) or is expressed as a confidence interval. Thus information in the sample is used to make an inference about the population.

Information from sample surveys affects almost every facet of our daily lives. Such information determines government policies on, for example, the control of the economy and the promotion of social programs. Opinion polls are the basis of much of the news reported by the various news media.

1

Ratings of television shows determine which shows are to be available for viewing in the future.

One usually thinks of the U.S. Census as contacting every household in the country. Actually, in the 1980 census only 14 questions were asked of all households. Information on an additional 42 questions was obtained from only a sample of households. The resulting information is used by the federal government in determining allocations of funds to states and cities. It is used by business to forecast sales, to manage personnel, and to establish future site locations. It is used by urban and regional planners to plan land use, transportation networks, and energy consumption. It is used by social scientists to study economic conditions, racial balance, and other aspects of the quality of life.

The U.S. Bureau of Labor Statistics routinely conducts over twenty surveys. Some of the best known and most widely used are the surveys that establish the consumer price index (CPI). The CPI is a measure of price change for a fixed market basket of goods and services over time. It is used as a measure of inflation and serves as an economic indicator for government policies. Businesses have wage rates and pension plans tied to the CPI. Federal health and welfare programs, as well as many state and local programs, tie their bases of eligibility to the CPI. Escalator clauses in rents and mortgages are based on the CPI. So we can see that this one index, determined on the basis of sample surveys, plays a fundamental role in our society.

Many other surveys at the Bureau of Labor Statistics (BLS) are crucial to society. The monthly Current Population Survey establishes basic information on the labor force, employment, and unemployment. The consumer expenditure surveys collect data on family expenditures for goods and services used in day-to-day living. The Establishment Survey collects information on employment hours and earnings for nonagricultural business establishments. The survey on occupational outlook provides information on future employment opportunities for a variety of occupations, projecting to approximately ten years ahead. Other activities of the BLS are addressed in the *BLS Handbook of Methods* (1982).

Opinion polls are constantly in the news, and the names of Gallup and Harris have become well known to everyone. These polls, or sample surveys, reflect the attitudes and opinions of citizens on everything from politics and religion to sports and entertainment. The Nielsen ratings determine the success or failure of TV shows.

Businesses conduct sample surveys for their internal operations, in addition to using government surveys for crucial management decisions. Auditors estimate account balances and check on compliance with operating rules by sampling accounts. Quality control of manufacturing processes relies heavily on sampling techniques.

One particular area of business activity that depends on detailed sampling activities is marketing. Decisions on which products to market, where to market them, and how to advertise them are often made on the basis of sample survey data. The data may come from surveys conducted by the firm that manufactures the product or may be purchased from survey firms that specialize in marketing data. The activities of three such firms are outlined next.

The Nielsen retail index is less famous than the Nielsen television ratings, but it is very important to firms marketing products for retail sale. This index furnishes continuous sales data on foods, cosmetics, pharmaceuticals, beverages, and many other classes of products. It can provide estimates of total sales for a product class, sales for a client's particular brand, sales for a competing brand, retail and wholesale price data, and the percentage of stores stocking a particular item. The data comes from auditing inventories and sales in 1600 stores across the United States every 60 days.

Selling Areas—Marketing, Inc. (SAMI), collects information on the movement of products from warehouses and wholesalers. Data is obtained in 36 major television market areas, containing 74% of national food sales, and covers 425 product categories.

The Market Research Corporation of America provides many types of marketing data through the use of surveys, but some of the more interesting results come from its National Menu Census. This survey samples families and observes their eating patterns for two weeks. As many as four thousand families may participate during a year. Data are obtained on the number of times a particular food item is served, how it is served, how many persons eat the item, and many other details, including what happens to the leftovers. Such details are important for product development and advertising.

Many interesting examples of the practical uses of statistics in general and sampling in particular can be found in *Statistics: A Guide to the Unknown* (see the references in the Appendix). In this book you might want to look at some of the methods and uses of opinion polling discussed in the articles "Opinion Polling in a Democracy" by George Gallup and "Election Night on Television" by R. F. Link. Those interested in wildlife ecology should read "The Plight of the Whales" by D. G. Chapman. Find out how interrailroad and interairline billing is handled economically through sampling by reading "How Accountants Save Money by Sampling" by John Neter.

Since the objective of modern statistics is inference, you may question what particular aspect of statistics will be covered in a course on sample survey design. The answer to this question is twofold. First, we will focus on the economics of purchasing a specific quantity of information. More specifically, how can we design sampling procedures that reduce the cost of a fixed quantity of information? Although introductory courses in statistics acknowledge the importance of this subject, they place major emphasis on basic concepts and on how to make inferences in specific situations *after* the data have been collected. The second distinguishing feature of our topic is that it is aimed at the particular types of sampling situations and inferential problems most frequently encountered in business, the social sciences, and natural resource management (timber, wildlife, and recreation) rather than in the physical sciences.

Even the terminology of the social scientist differs from that of the physical scientist. Social scientists conduct *surveys* to collect a sample, while physical scientists perform *experiments*. Thus we acknowledge that differences exist from one field of science to another in the nature of the populations and the manner in which a sample can be drawn. For example, populations of voters, financial accounts, or animals of a particular species may contain only a small number of elements. In contrast, the conceptual population of responses generated by measuring the yield of a chemical process is very large indeed. (You may recall that the properties of estimators and test statistics covered in most introductory courses assume that the population of interest is large relative to the sample.) Limitations placed on the sampling procedure also vary from one area of science to another. Sampling in the biological and physical sciences can frequently be performed under controlled experimental conditions. Such control is frequently impossible in the social sciences, business, and natural resource management. For example, a medical researcher might compare the growth of rats subjected to two different drugs. For this experiment the initial weights of the rats and the daily intake of food could be controlled to reduce unwanted variation in the experiment. In contrast, very few variables can be controlled in comparing the effect of two different television advertisements on sales for a given product; no control is possible when studying the effect of environmental conditions on the number of seals in the North Pacific Ocean.

In summary, this text is concerned with the peculiarities of sampling and inference commonly encountered in business, the social sciences, and natural resource management. Specifically, we will consider methods for actually selecting the sample from an existing population and ways of circumventing various difficulties that arise. Methods for designing surveys that capitalize on characteristics of the population will be presented along with associated estimators to reduce the cost for acquiring an estimate of specified accuracy.

Chapter 2 reviews some of the basic concepts encountered in introductory statistics, including the fundamental role that probability plays in making inferences. Chapter 3 presents some of the basic terminology of sampling, as well as a discussion of problems arising in sample survey design. Simple random sampling, familiar to the beginning student, is carefully presented in Chapter 4; it includes physical procedures for actually selecting the sample. Following chapters cover economical methods for selecting a sample and associated methods for estimating population parameters.

In reading this text, keep in mind that the ultimate objective of each chapter is *inference*. Identify the sampling procedure associated with each chapter, the population parameters of interest, their estimators, and the associated bounds on the errors of estimation. Develop an intuitive understanding of and appreciation for the benefits to be derived from specialized sampling procedures. Focus on the broad concepts, and do not become hypnotized by the formulas for estimators and variances that sometimes are unavoidably complicated. In short, focus on the forest rather than the trees. Work some exercises, and the details will fall into place.

2

A REVIEW
OF SOME
BASIC CONCEPTS

2.1
INTRODUCTION

Knowledge of the basic concepts of statistics is a prerequisite for a study of sample survey design. Thus in this chapter we will review some of these basic concepts.

The ultimate objective of statistics is to make inferences about a population from information contained in a sample. The target of our inference, the population, is a set of measurements, finite or infinite, existing or conceptual. Hence the first step in statistics is to find a way to phrase an inference about a population or, equivalently, to describe a set of measurements. Thus frequency distributions and numerical descriptive measures are the first topics of our review.

The second step in statistics is to consider how inferences can be made about the population from information contained in a sample. For this step we must consider probability distributions of sample quantities, or sampling distributions. Knowledge of probability distributions associated with the sample allows us to choose proper inference-making procedures and to attach measures of goodness to such inferences.

The method of inference primarily employed in business and the social sciences is estimation. We may wish to estimate the total assets of a corporation, the fraction of voters favoring candidate Jones, or the number of campers using a state park during a given period of time. Hence we must understand the basic concepts underlying the selection of an estimator of a population parameter, the method for evaluating its goodness, and the concepts involved in interval estimation. Because the bias and the variance of estimators determine their goodness, we need to review the basic ideas concerned with the expectation of a random variable and the notions of variance and covariance.

5

The subsequent sections follow the outline just given. We begin with a review of the primary problem, namely, how to describe a set of measurements. We then rapidly review the probabilistic model for the repetition of an experiment. We explain how the model can be used to infer the characteristics of a population and discuss random variables, probability distributions, and expectations. Finally, we present the basic concepts associated with point and interval estimation.

2.2
SUMMARIZING INFORMATION IN POPULATIONS AND SAMPLES

Since grasping the essential characteristics of a large set of measurements by looking at a listing of the numbers is not easy, we usually must summarize the measurements through the use of graphical or numerical techniques. Even though all the measurements in a population for a study are generally not available, we may still be able to assume some reasonable graphical shape for the relative frequency distribution of this population. Of course, we can always construct a frequency or relative frequency histogram for a sample, since the sample measurements are known, and use it to make an empirical assessment of the shape of the population.

Once a relative frequency distribution is established for a population, we can, by using probability arguments, calculate summarizing numerical measures like the mean, variance, and standard deviation. Similar quantities can be calculated directly from sample measurements.

For purposes of illustration, let's assume that a population consists of a large number of integers, 0, 1, 2, . . . , 9, in equal proportions. We may think of these integers as written on slips of paper and mixed up in a box, as stored on a table (like a random number table), or as generated in a computer file. Since all integers occur in equal proportions, the *relative frequency histogram*,

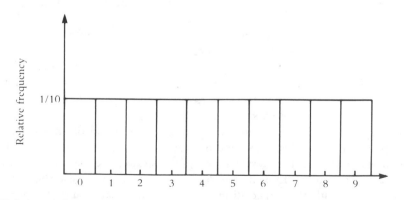

FIGURE 2.1 Distribution of population containing integers 0 through 9 with equal frequency

which shows the distribution of the population measurements, is as shown in Figure 2.1.

These relative frequencies can be thought of in probabilistic terms. If one number is selected *at random* (for example, if someone reaches into the box and blindly pulls out one piece of paper), then the *probability* that the selected number will be a 4 is $\frac{1}{10}$. Suppose one number is to be selected at random from the population under study, and let its value be denoted by y. Then the possible values for y $(0, 1, 2, \ldots, 9$ in this case) and the probabilities associated with those values ($\frac{1}{10}$ for each in this case) constitute the *probability distribution* for the *random variable* y. The probability associated with y is sometimes denoted by $p(y)$. Thus for this population

$$p(0) = p(1) = \cdots = p(8) = p(9) = \frac{1}{10}$$

The numerical measures used to summarize the characteristics of a population are defined as *expected values* of y or a function of y. By definition, the *expected value* of y, $E(y)$, is given by

$$E(y) = \sum_y yp(y)$$

where the summation is over all values of y for which $p(y) > 0$.

For the population and random variable y under study,

$$
\begin{aligned}
E(y) &= \sum_y yp(y) \\
&= 0p(0) + 1p(1) + 2p(2) + \cdots + 8p(8) + 9p(9) \\
&= \tfrac{1}{10}(45) = 4.5
\end{aligned}
$$

One can see that $E(y)$ is equal to the average value, or mean value, of all the measurements in our conceptual population. In general, a population mean will be denoted by μ, and it follows that

$$\mu = E(y)$$

where y is the value of a single measurement chosen at random from the population.

The variability of measurements in a population can be measured by the *variance*, which is defined as the expected value, or average value, of the square of the deviation between a randomly selected measurement y and its mean value μ. Thus the variance of y, $V(y)$, is given by

$$V(y) = E(y - \mu)^2 = \sum_y (y - \mu)^2 p(y)$$

For the population used for illustration in this section,

$$
\begin{aligned}
V(y) &= E(y - \mu)^2 = \sum_y (y - \mu)^2 p(y) \\
&= (0 - 4.5)^2(\tfrac{1}{10}) + (1 - 4.5)^2(\tfrac{1}{10}) + \cdots + (9 - 4.5)^2(\tfrac{1}{10}) \\
&= \tfrac{1}{10}[(0 - 4.5)^2 + (1 - 4.5)^2 + \cdots + (9 - 4.5)^2] \\
&= \tfrac{1}{10}(82.5) = 8.25
\end{aligned}
$$

Variance $V(y)$ is commonly denoted by σ^2.

The *standard deviation* is defined to be the square root of the variance, and it is denoted by $\sigma = \sqrt{\sigma^2}$. For the specific population under discussion,

$$\sigma = \sqrt{8.25} = 2.9$$

In statistical studies the population of interest consists of *unknown* measurements; hence we can only speculate about the nature of the relative frequency histogram or the size of μ and σ. To gain some information about the population, we select a sample of n measurements and study the properties of this sample. We then *infer* characteristics of the population from what we observe in the sample. The sample measurements, in general, will be denoted by $y_1, y_2, \ldots, y_n$.

Following the pattern set for summarizing the information in a population, we can calculate the mean, the variance, and the standard deviation of a sample. These numerical descriptive measures are given, respectively, by

$$\bar{y} = \frac{1}{n} \sum_{i=1}^{n} y_i$$

$$s^2 = \frac{\sum_{i=1}^{n} (y_i - \bar{y})^2}{n - 1}$$

and

$$s = \sqrt{s^2}$$

Note that s^2 has a divisor of $n - 1$ instead of n.

For the population of integers $0, 1, \ldots, 9$ in equal proportion, a sample of $n = 10$ measurements was selected. Each of the 10 measurements was selected at random. (Think of drawing 10 slips of paper out of a box containing a large number of slips, each marked with an integer between 0 and 9.) The sample measurements were

$$6, 9, 3, 8, 1, 7, 8, 8, 4, 0$$

For this sample

$$\bar{y} = \frac{1}{n} \sum_{i=1}^{n} y_i$$

$$= \tfrac{1}{10}(6 + 9 + 3 + \cdots + 4 + 0)$$

$$= \tfrac{1}{10}(54) = 5.4$$

$$s^2 = \frac{1}{n - 1} \sum_{i=1}^{n} (y_i - \bar{y})^2$$

$$= \tfrac{1}{9}[(6 - 5.4)^2 + (9 - 5.4)^2 + \cdots + (0 - 5.4)^2]$$

$$= \tfrac{1}{9}(92.4) = 10.27$$

and

$$s = \sqrt{s^2} = \sqrt{10.17} = 3.2$$

Uses for these sample quantities will be discussed in the next two sections, but we can see that $\bar{y}$ might form a reasonable approximation to μ if μ were unknown. Likewise, s^2 might form a reasonable approximation to σ^2 if σ^2 were unknown, and s might form a reasonable approximation to σ.

2.3
SAMPLING DISTRIBUTIONS

In the following chapters sample quantities like $\bar{y}$ will be used extensively for making inferences about unknown population quantities; hence we must study the properties of certain functions of sample observations. This study begins with a numerical illustration, which will then be generalized to cover a wide variety of sampling situations.

Consider the population discussed in Section 2.2 in which the integers $0, 1, \ldots, 9$ were represented in equal proportions. Fifty samples each of size $n = 10$ were selected from this population, each sample chosen in a manner similar to the method used for the one sample selected at the end of Section 2.2. The sample means $\bar{y}$ for these 50 samples are listed in increasing numerical order in Table 2.1.

TABLE 2.1 Sample means for 50 samples, each of size $n = 10$

2.3	3.6	4.1	4.3	4.8
2.6	3.7	4.1	4.3	4.8
2.6	3.7	4.1	4.4	4.8
3.2	3.7	4.1	4.5	4.9
3.3	3.8	4.2	4.7	5.0
3.4	3.9	4.3	4.7	5.1
3.5	4.0	4.3	4.7	5.3
3.5	4.1	4.3	4.8	5.5
3.6	4.1	4.3	4.8	6.0
3.6	4.1	4.3	4.8	6.6

A frequency histogram for the 50 sample means is shown in Figure 2.2. This distribution is an approximation to the *theoretical sampling distribution* of $\bar{y}$, since it shows how the $\bar{y}$'s tend to be distributed when repeated samples are taken. The sampling distribution can be thought of as a probability distribution for $\bar{y}$. Note that the distribution of $\bar{y}$'s tends to center close to the population mean of $\mu = 4.5$, has much less spread (or variability) than the original population measurements, and has a mound shape rather than the flat shape of the population distribution.

From an elementary statistics course (see Mendenhall, 1983) we know that the sampling distribution of $\bar{y}$ should have a mean of μ, a standard deviation of $\sigma/\sqrt{n}$, and a shape like that of a *normal curve* (a symmetric, bell-shaped curve). This display of 50 sample means has an average of 4.22 (which is close to $\mu = 4.5$) and a standard deviation of .79 (which is close to $\sigma/\sqrt{n} = 2.9/\sqrt{10} = .92$). Also, the frequency histogram has an approxi-

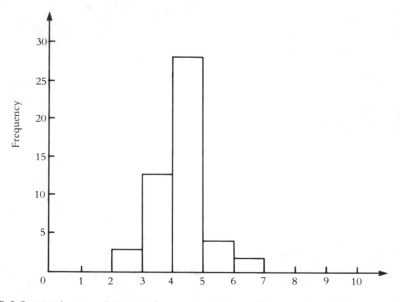

FIGURE 2.2 Distribution of 50 sample means with $n = 10$ for each sample

mate bell shape, although it is not quite symmetric. These facts concerning the behavior of sample means will be important in the development of inference procedures.

From known properties of the normal curve it follows that approximately 68% of the values of $\bar{y}$, in repeated sampling, should fall within one standard deviation of the mean of the sampling distribution of $\bar{y}$'s. Approximately 95% of the values of $\bar{y}$, in repeated sampling, should fall within two standard deviations of the mean. Checking these statements for the observed sample of 50 $\bar{y}$'s, we see that

$$4.11 \pm .79 \quad \text{or} \quad (3.43, 5.01)$$

contains 39 out of 50 (78%) of the $\bar{y}$ values in the sample, and

$$4.22 \pm 2(.79) \quad \text{or} \quad (2.64, 5.80)$$

contains 45 out of 50 (90%) of the $\bar{y}$'s. These percentages are reasonably close to the theoretical values of 68% and 95%. (Remember, we've only seen an approximation, based on 50 samples, to the true sampling distribution of $\bar{y}$.)

If the sampling distribution of some sample quantity does not follow a normal distribution, at least approximately, then relative frequency interpretations can still be obtained from Tchebysheff's theorem. This theorem states that for any $k \geq 1$ *at least* $(1 - 1/k^2)$ of the measurements in any set must lie within k standard deviations of their mean. For example, setting $k = 2$ yields that at least $(1 - \frac{1}{2^2}) = (1 - \frac{1}{4}) = \frac{3}{4}$ of any set of measurements must lie within two standard deviations of their mean. Usually, this fraction is much greater than $\frac{3}{4}$.

The high percentage of measurements falling within two standard deviations of the mean, from either the normal distribution or Tchebysheff's theorem, suggests that the range of any set of measurements usually encompasses a little more than four standard deviations. Put another way, the standard deviation of a set of measurements can be approximated as one-fourth of the range of that set of measurements.

The discussion of sampling distributions thus far has assumed that the population from which the samples were selected was essentially infinite. But we may want to work with populations of N measurements, where N may be relatively small. Does the approximate normality of the sampling distribution of $\bar{y}$ still hold? Results of two empirical investigations into this question follow.

A population of $N = 100$ measurements was generated by computer and yielded a distribution like the one shown in Figure 2.3. A sample of

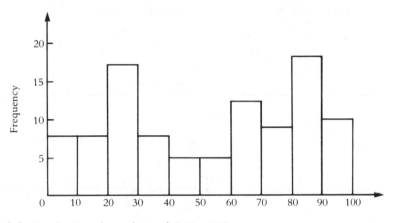

FIGURE 2.3 Distribution of population for $N - 100$

$n = 20$ measurements was drawn from this population in such a way that every possible sample of size 20 had an equal chance of being selected (see Chapter 4). This process was repeated until 50 such samples were selected. The sample mean $\bar{y}$ was calculated for each sample, and the distribution of these sample means is shown in Figure 2.4. Notice the tendency of this histogram to be bell-shaped, although not perfectly symmetric.

A population of $N = 20$ measurements was then generated, yielding the distribution shown in Figure 2.5. The methodology outlined above was used to select 50 samples, each of size $n = 15$. The distribution of the 50 sample means is shown in Figure 2.6. Note once again a tendency toward a bell-shaped, somewhat symmetric distribution. However, this distribution differs considerably from that of Figure 2.4. Here the sample means are grouped very tightly around the population mean, as should be expected since the sample size of 15 is nearly equal to the population size of 20. One might suggest that the distribution in Figure 2.6 does not resemble the normal distribution nearly as much as does the one in Figure 2.4.

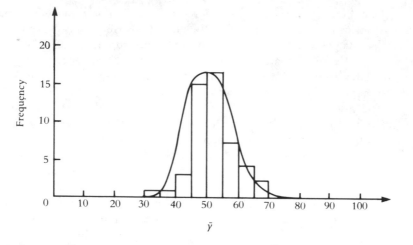

FIGURE 2.4 Distribution of sample means for $N = 100$ and $n = 20$

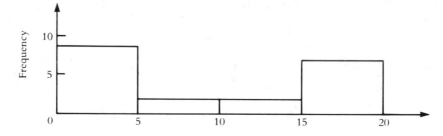

FIGURE 2.5 Distribution of population for $N = 20$

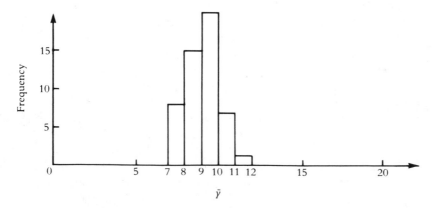

FIGURE 2.6 Distribution of sample means for $N = 20$ and $n = 15$

In summary, the sampling distribution of $\bar{y}$ should closely approximate a normal distribution if n is no more than 20% of N. This observation is especially true if $n \geq 30$. When n is a very high percentage of N, the sampling distribution of $\bar{y}$ is grouped even more tightly around the population mean than one might expect.

2.4
COVARIANCE AND CORRELATION

Often an experiment yields more than one random variable of interest. For example, the psychologist measures more than one characteristic per individual in a study of human behavior. Typical variables might be a measure of intelligence, y_1, a personality measure, y_2, and other variables representing test scores or measures of physical characteristics. Often we are interested in the simple dependence of pairs of variables, such as the relationship between personality and intelligence, or between college achievement and college board scores. Particularly, we ask whether data representing paired observations of y_1 and y_2 on a number of people imply a dependence between the two variables. If so, how strong is the dependence?

Intuitively, we think of dependence of two random variables, y_1 and y_2, as implying that one, say y_1, either increases or decreases as y_2 changes. We will confine our attention to two measures of dependence, the *covariance* and the *simple coefficient of linear correlation*, and will utilize Figures 2.7(a) and 2.7(b) to justify choosing them as measures of dependence. These figures represent plotted points for two (random) samples of $n = 10$ experimental units drawn from a population. Measurements of y_1 and y_2 were made on each experimental unit. If all of the points lie on a straight line, as indicated in Figure 2.7(a), y_1 and y_2 are obviously dependent. In contrast, Figure 2.7(b) indicates little or no dependence between y_1 and y_2.

Suppose we actually know μ_1 and μ_2, the means of y_1 and y_2, respectively, and locate this point on the graphs, Figure 2.7. Now locate a plotted point on Figure 2.7(a) and measure the deviations, $(y_1 - \mu_1)$ and $(y_2 - \mu_2)$. Note that both deviations assume the same algebraic sign for a particular

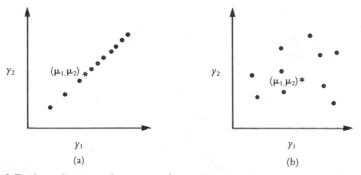

FIGURE 2.7 Plotted points of two samples

point; hence their product, $(y_1 - \mu_1)(y_2 - \mu_2)$, is positive. This result will be true for all plotted points on Figure 2.7(a). Points to the right of (μ_1, μ_2) will yield pairs of positive deviations, points to the left will produce pairs of negative deviations, and the average of the product of the deviations $(y_1 - \mu_1)(y_2 - \mu_2)$ will be "large" and positive. If the linear relation indicated in Figure 2.7(a) has sloped downward to the right, all corresponding pairs of deviations would have been of the opposite sign, and the average value of $(y_1 - \mu_1)(y_2 - \mu_2)$ would have been a large negative number.

The situation just described will not occur for Figure 2.7(b), where little or no dependence exists between y_1 and y_2. Corresponding deviations, $(y_1 - \mu_1)$ and $(y_2 - \mu_2)$, will assume the same algebraic sign for some points and opposite signs for others. Thus the product $(y_1 - \mu_1)(y_2 - \mu_2)$ will be positive for some points, negative for others, and will average to some value near zero.

Clearly, then, the expected (average) value of $(y_1 - \mu_1)(y_2 - \mu_2)$ provides a measure of the linear dependence of y_1 and y_2. This quantity, defined over the two corresponding populations associated with y_1 and y_2, is called the *covariance* of y_1 and y_2. We denote the covariance between y_1 and y_2 thus:

$$\text{Cov}(y_1, y_2) = E[(y_1 - \mu_1)(y_2 - \mu_2)]$$

The larger the absolute value of the covariance of y_1 and y_2, the greater will be the linear dependence between y_1 and y_2. Positive values indicate that y_1 increases as y_2 increases; negative values indicate that y_1 decreases as y_2 increases. A zero value of the covariance indicates no linear dependence between y_1 and y_2.

Unfortunately, to use the covariance as an absolute measure of dependence is difficult, because its value depends on the scale of measurement. Consequently, to determine, at first glance, whether a particular covariance is "large" is difficult. We can eliminate this difficulty by standardizing its value, using the simple coefficient of linear correlation. Thus the population linear coefficient of correlation,

$$\rho = \frac{\text{Cov}(y_1, y_2)}{\sigma_1 \sigma_2}$$

(where σ_1 and σ_2 are the standard deviations of y_1 and y_2, respectively) is related to the covariance and can assume values in the interval $-1 \le \rho \le 1$. The sample coefficient of correlation is used as an estimator of ρ and is discussed in most introductory courses. Further information on this subject can be found in Mendenhall (1983, Chapter 10).

2.5
ESTIMATION

The objective of any sample survey is to make inferences about a population of interest on the basis of information obtained in a sample from that population. Inferences in sample surveys are usually aimed at the *estimation* of certain numerical characteristics of the population, such as the mean, total, or variance. These numerical descriptive measures of the population are called *parameters*.

An *estimator* is a function of observable random variables, and perhaps other known constants, used to estimate a parameter. For example, the sample mean $\bar{y}$ can be used as an estimator of the population mean μ. Mean $\bar{y}$ is an estimator since it is a function of sample observations. Note that $\bar{y}$ is, however, a random variable and has a probability distribution, or sampling distribution, that depends on the sampling mechanism, as pointed out in Section 2.3. Some of the possible values that $\bar{y}$ can take on will be close to μ, and others may be quite far from μ on either the positive or the negative side. If we are to take a sample and calculate a specific value as our best estimate of μ, we would like to know that on the average $\bar{y}$ generates values that center about μ and are in general quite close to μ. Thus we will want to select a sampling plan that ensures us that $E(\bar{y}) = \mu$ and $V(\bar{y})$ is "small."

In general, suppose that $\hat{\theta}$ is an estimator of the parameter θ. Two properties that we would like $\hat{\theta}$ to possess are as follows:

1. $E(\hat{\theta}) = \theta$.
2. $V(\hat{\theta}) = \sigma_{\hat{\theta}}^2$ is small.

An estimator possessing property 1 is said to be *unbiased*. As for property 2, we will not discuss minimum–variance unbiased estimators in this text, but we will compare unbiased estimators on the basis of their variances. If two unbiased estimators are available for θ, we will generally give preference to the one with the smaller variance.

To summarize, this text will investigate a number of combinations of sampling plans and estimators that give rise to unbiased estimators with small variance.

Although the probability distribution of $\bar{y}$, a common estimator, will depend on the sampling mechanism and the sizes of the sample and the population, in many instances the sample mean tends to have a bell–shaped symmetric distribution known as the *normal distribution*. This observation is especially true if n is large, say $n \geq 30$.

Once we know which estimator $\hat{\theta}$ we are using in a situation and something about its probability distribution, we can assess the magnitude of the error of estimation. We define the *error of estimation* to be $|\hat{\theta} - \theta|$. How good will a single estimate be? We cannot state that an observed estimate will be within a specified distance of θ, but we can, at least approximately, find a bound B such that

$$P(|\hat{\theta} - \theta| \leq B) = 1 - \alpha$$

for any desired probability $1 - \alpha$, where $0 < \alpha < 1$. If $\hat{\theta}$ has a normal distribution, then $B = z_{\alpha/2}\sigma_{\hat{\theta}}$, where $z_{\alpha/2}$ is the value cutting off an area of $(\alpha/2)$ in the right-hand tail of the standard normal distribution. The table values of $z_{\alpha/2}$ are given in Table 1 of the Appendix. If $1 - \alpha = .95$, then $z_{.025} = 1.96$, or approximately 2. Since many estimators we use throughout the text will not be precisely normally distributed for many values of n and N, and since Tchebysheff's theorem states that at least 75% of the observations for *any* probability distribution will be within two standard deviations of their mean, we will use $2\sigma_{\hat{\theta}}$ as a bound on the error of estimation. This value gives $P(|\hat{\theta} - \theta| \le B) \doteq .95$ for the approximately normal cases and $P(|\hat{\theta} - \theta| \le B) \ge .75$ in any case. Note that for a normally distributed $\hat{\theta}$, a bound to satisfy any desired probability $(1 - \alpha)$ can be found through use of Table 1.

If $P(|\hat{\theta} - \theta| \le B) = 1 - \alpha$, then $P(\hat{\theta} - B \le \theta \le \hat{\theta} + B) = 1 - \alpha$. In this form $(\hat{\theta} - B, \hat{\theta} + B)$ is called a *confidence interval* for θ with *confidence coefficient* $(1 - \alpha)$. The quantity $\hat{\theta} - B$ is called the lower confidence limit (LCL), and $\hat{\theta} + B$ is called the upper confidence limit (UCL).

2.6
SUMMARY

Chapter 2 presents a capsule review of the basic concepts of statistics. Making an inference about a population requires a method for describing a set of measurements and, consequently, requires a discussion of frequency histograms and numerical descriptive measures. Two very useful numerical descriptive measures are the mean and the standard deviation. Although the mean is an easily understood measure of central tendency, the standard deviation acquires meaning as a measure of variation only when interpreted by using Tchebysheff's theorem or some specific distribution such as the normal.

Another important concept is the role that probability plays in making inferences about a population. The probabilist reasons from a known population to a sample. In contrast, the statistician uses probability as the vehicle to make inferences about a population from information contained in a sample. Although a good background in probability is desirable, knowledge of the basic concepts of probability and the use of probability in inference making provides a sufficient background for understanding this text.

Random variables and their probability distributions are presented to provide a background for describing the properties of estimators of population parameters. The notions of expectations, covariance, and correlation assist in evaluating the properties of estimators.

The estimation of population parameters is the primary method of inference making used in sample survey methods. The concept of a point estimator with its corresponding measure of goodness (bound on the error of estimation) is presented and is used as the method of inference in all subsequent chapters.

EXERCISES

2.1   What is the objective of statistics?

2.2   How does a course on sample survey design differ from the standard introductory course on statistics?

2.3   Why is knowing how to describe a set of measurements essential?

2.4   How can you describe a set of measurements?

2.5   What is a parameter?

2.6   State Tchebysheff's theorem.

2.7   Show that the sample variance s^2, given in Section 2.2, is equivalent to

$$s^2 = \frac{1}{n-1}\left[\sum_{i=1}^{n} y_i^2 - n\bar{y}^2\right]$$

The latter form is usually easier for numerical calculation.

2.8   Given $n = 20$ sample measurements: 1, 2, 0, 2, 2, 4, 0, 3, 1, 2, 3, 2, 0, 1, 2, 2, 4, 2, 1, 3.

    (a)   Calculate the sample mean $\bar{y}$.

    (b)   Calculate s^2.

    (c)   What fraction of the measurements lie within one standard deviation of the mean? Two? Three? How do these fractions agree with those given by Tchebysheff's theorem? (This exercise illustrates the effectiveness of the standard deviation as a measure of the variability of a set of measurements.)

2.9   Given $n = 10$ sample measurements: 5, 2, 4, 4, 3, 4, 1, 3, 5, 4.

    (a)   Calculate the sample mean.

    (b)   Calculate the sample variance.

    (c)   Find the fraction of measurements lying within one standard deviation of the mean. Compare these with the corresponding figures given by Tchebysheff's theorem and the normal distribution.

2.10 Explain what is meant by the term *sampling distribution* for the random variable $\bar{y}$.

2.11 In the sampling distribution of $\bar{y}$, how should the mean and the variance relate to the mean and the variance of the population from which the sample is selected?

2.12 What is an estimator?

2.13 How does one evaluate the goodness of an estimator?

2.14 Describe two desirable properties for an estimator.

2.15 What is an unbiased estimator?

2.16 What is the error of estimation?

2.17 What is a reasonable bound on the error of estimation?

2.18 Of what value is Tchebysheff's theorem in making statements about the error of estimation?

2.19 Suppose a population consists of measurements denoted by $u_1, u_2, \ldots, u_N$. A single observation y is chosen at random from this population. Show that

$$\sigma^2 = V(y) = \frac{1}{N}\sum_{i=1}^{N}(u_i - \mu)^2$$

where μ is the population mean.

2.20 Generate an approximate sampling distribution of $\bar{y}$ by selecting 25 samples of the same size from a population of measurements of interest to you. (Replace the first sample before the second is drawn so that each sample comes from the same population.) Do the sample means appear to agree with the results discussed in this chapter for sampling distributions?

3

ELEMENTS OF THE SAMPLING PROBLEM

3.1

INTRODUCTION

You will recall that the objective of statistics is to make inferences about a population from information contained in a sample. This same objective motivates our discussion of the sampling problem. We will consider the particular problem of sampling from a finite collection of measurements (population). We will refer occasionally to populations composed of an infinite number of measurements. In most cases the inference will be in the form of an estimate of a population parameter, such as a mean, total, or proportion, with a bound on the error of estimation. For those more interested in methodology than theory, intuitive arguments will be given whenever possible to justify the use of estimators.

The first part of our discussion of the sampling problem introduces certain technical terms common to sample surveys. Next, we discuss how to select a sample from the population.

Each observation, or item, taken from the population contains a certain amount of information about the population parameter or parameters of interest. Since information costs money, the experimenter must determine how much information he or she should buy. Too little information prevents the experimenter from making good estimates, while too much information results in a waste of money. The quantity of information obtained in the sample depends on the number of items sampled and on the amount of variation in the data. This latter factor can be controlled by the method of selecting the sample, called the *design of the sample survey*. The design of the survey and the sample size determine the quantity of information in the sample pertinent to a population parameter, provided that accurate measurements are obtained on each sampled element. Several sample survey designs are introduced in Section 3.3.

If accurate measurements are not obtained on each element of the survey, then other errors are introduced. These errors are discussed in Section 3.4. Accuracy of measurements can be enhanced by proper methods of data collection, discussed in Section 3.5, and by good questionnaire construction, discussed in Section 3.6. Section 3.7 presents the main elements one should be careful to check when planning a survey.

3.2
TECHNICAL TERMS

Technical terminology is kept to a minimum in this text; however, certain common terms must be defined. Let us introduce these terms by way of an example. In a certain community an opinion poll was conducted to determine public sentiment toward a bond issue in an upcoming election. The objective of the survey was to estimate the proportion of voters in the community who favored the bond issue.

DEFINITION 3.1   An *element* is an object on which a measurement is taken.

In our example an element is a registered voter in the community. The measurement taken on an element is the voter's preference on the bond issue. Since measurements are usually considered to be numbers, the experimenter can obtain numerical data by recording a 1 for a voter in favor of the bond issue and a 0 for a voter not in favor.

DEFINITION 3.2   A *population* is a collection of elements about which we wish to make an inference.

The population in our example is the collection of voters in the community. The characteristic (numerical measurement) of interest for each member of this population is his or her preference on the bond issue.

An important task for the investigator is to carefully and completely define the population *before* collecting a sample. The definition must include a description of the elements to be included and a specification of the measurements to be taken, since these two components are interrelated. For example, if the population in the bond issue study consists of registered voters, then one may want to collect information on whether or not each sampled person plans to vote in the upcoming election. Sampling the desired target population is not always possible, and the investigator may have to collect additional information so that answers can be provided for questions of interest. If in our example the only population available for sampling is a list of residents of the community, then information about whether each sampled person is, in fact, a registered voter should be collected.

DEFINITION 3.3   *Sampling units* are nonoverlapping collections of elements from the population that cover the entire population.

In the bond issue example a sampling unit may be a registered voter in the community. However, a more efficient process may be to sample households, which are collections of elements, in order to obtain information on voter preferences. If households are the sampling units, they must be defined so that no voter in the population can be sampled more than once and so that each voter has a chance of being selected in the sample.

As the definition states, sampling units should be nonoverlapping. However, situations do arise in which the nonoverlapping condition is virtually impossible to achieve. Field plot samples taken, for example, in studies of animal habitat are often circular. The circular pattern is a convenient one to lay out and has advantages in terms of the amount of walking necessary to study the plot. Obviously, circular plots cannot cover a field without some overlap. The intent here is to suggest that the overlap should be as small as possible for efficient sampling.

If each sampling unit contains one and only one element of the population, then a sampling unit and an element from the population are identical. This situation arises if we sample individual voters rather than households within the community.

DEFINITION 3.4   A *frame* is a list of sampling units.

If we specify the individual voter as the sampling unit, a list of all registered voters may serve as a frame for a public opinion poll. Note that this frame will not include all the elements in the population, because updating the list daily is impossible. If we take the household as the sampling unit, then a telephone directory, a city directory, or a list of household heads obtained from census data can serve as a frame.

All these frames have some inadequacies. The lists will not be up to date. They will contain many names of unregistered household heads, and hence a sample drawn from the lists will contain many units that are not in the population. Also, some registered voters may not appear on any of these lists. It is hoped, however, that the gap between the frame and the population is small enough to permit inferences to be made about the population on the basis of a sample drawn from the frame.

Some sampling schemes may involve multiple frames. In sampling voters, one could start by sampling housing units or city blocks and then sampling voters within the selected housing units or blocks. One frame, then, is a list of housing units or a list of city blocks, and the second frame is a list of voters within those larger units. The second frame may not be available until the housing units or blocks are selected and studied in some detail. As another example, estimation of crop yields in a state may involve sampling from a list of growers to be interviewed and a list of fields to be objectively measured.

DEFINITION 3.5   A *sample* is a collection of sampling units drawn from a frame or frames.

Data are obtained from the elements of the sample and used in describing the population. Let the individual voter be our sampling unit and the list of registered voters be our frame. In the public opinion poll a number of voters (the sample) will be contacted to determine their preference for the upcoming bond issue. We then can use the information obtained from these voters to make an inference about voter preference throughout the community.

3.3
HOW TO SELECT THE SAMPLE: THE DESIGN OF THE SAMPLE SURVEY

The objective of sampling is to estimate population parameters, such as the mean or the total, from information contained in a sample. As stated previously, the experimenter controls the quantity of information contained in the sample by the number of sampling units he or she includes in the sample and by the method used to select the sample data. How do we determine which procedure to use and the number of observations (sampling units) to include in the sample? The answer depends on how much information we want to buy. If θ is the parameter of interest and $\hat{\theta}$ is an estimator of θ, we should specify a bound on our error of estimation; that is, we should specify that θ and $\hat{\theta}$ differ in absolute value by less than some value B. Stated symbolically,

$$\text{error of estimation} = |\theta - \hat{\theta}| < B$$

We also must state a probability, $(1 - \alpha)$, that specifies the fraction of times in repeated sampling we require the error of estimation to be less than B. This condition can be stated as

$$P[\text{error of estimation} < B] = 1 - \alpha$$

We will usually select $B = 2\sigma_{\hat{\theta}}$, and hence $(1 - \alpha)$ will be approximately .95 for bell-shaped distributions. Most estimators used in this book will exhibit bell-shaped distributions for reasonably large sample sizes, even when the parent population is skewed.

After we obtain a specified bound with its associated probability $(1 - \alpha)$, we can compare different designs (methods of selecting the sample) to determine which procedure yields the desired precision at minimum cost. The problem of selecting the sample size to achieve a certain bound on error is discussed in Mendenhall (1983, Chapter 8).

The basic design (*simple random sampling*) consists of selecting a group of n sampling units in such a way that each sample of size n has the same chance of being selected. Thus we can obtain a random sample of n eligible voters in the bond issue poll by drawing names from the list of registered voters in such a way that each sample of size n has the same probability of selection. The details of simple random sampling are discussed in Chapter 4. At this point we merely state that a simple random sample will contain as much information on the community preference as any other sample survey design, provided all voters in the community have similar socioeconomic backgrounds.

Suppose, however, that the community consists of people in two distinct income brackets, high and low. Voters in the high bracket may have opinions on the bond issue that are quite different from the opinions of voters in the low bracket. Therefore, to obtain accurate information about the population, we want to sample voters from each bracket. We can divide the population elements into two groups, or strata, according to income and select a simple random sample from each group. The resulting sample is called a *stratified random sample*.

Note that stratification is accomplished by using knowledge of an auxiliary variable, namely, personal income. By stratifying on high and low values of income, we increase the accuracy of our estimator. *Ratio estimation* is a second method for using the information contained in an auxiliary variable. Ratio estimators not only use measurements on the response of interest but also incorporate measurements on an auxiliary variable. Ratio estimation can also be used with stratified random sampling.

Although individual preferences are desired in the survey, a more economical procedure, especially in urban areas, may be to sample specific families, apartment buildings, or city blocks rather than individual voters. Individual preferences can then be obtained from each eligible voter within the unit sampled. This technique is called *cluster sampling*. Although we divide the population into groups for both cluster sampling and stratified random sampling, the techniques differ. In stratified random sampling we take a simple random sample within each group, while in cluster sampling we take a simple random sample of groups and then sample all items within the selected groups (clusters).

Sometimes, the names of persons in the population of interest are available in a list, such as a registration list, or on file cards stored in a drawer. For this situation an economical technique is to draw the sample by selecting one name near the beginning of the list and then selecting every tenth or fifteenth name thereafter. If the sampling is conducted in this manner, we obtain a *systematic sample*. As you might expect, systematic sampling offers a convenient means of obtaining sample information; unfortunately, we do not necessarily obtain the most information for a specified amount of money.

We know that observations cost money. Note that the cost of an observation may vary from design to design, and even within a design, depending on the method of data collection. The experimenter should choose the design that gives the desired bound on error with the smallest number of observations (assuming the same cost per observation). However, if the cost per observation varies from design to design, the experimenter should choose the design that gives the desired bound on the error of estimation at a minimum cost.

3.4
SOURCES OF ERRORS IN SURVEYS

The error of estimation discussed in Section 3.3 arises because a sample cannot give complete information on a population. This kind of error is called *sampling error*. Sampling error can be controlled by carefully designing the survey, a topic discussed in the remaining chapters in this book.

Other types of errors can subtly creep into a survey, however, and these errors are more difficult to control. These errors, called *nonsampling errors*, are mainly due to nonresponse, inaccurate response, and selection bias.

The first cause of error, nonresponse, is important because nonresponse to a question put to an individual selected to be included in the sample can introduce a bias into the sample data. Those in the sample who do respond may not represent the population about which we wish to make inferences. For example, in a survey to determine employee acceptance of a monthly parking fee, quite likely, only those people violently opposed to the fee will respond to a mailed questionnaire. If we consider the percentage of respondents favoring the fee, we will probably obtain a distorted estimate of the true percentages for the entire population.

The second problem is that respondents or measuring equipment frequently give false information. For example, if a person is asked in an interview whether he cheated on his income tax, for fear of discovery he will probably respond negatively whether he had or had not. The same person, however, might give a truthful statement to the same question posed on a mailed questionnaire. As another example, forest areas measured on aerial photographs may always read either high or low because of an improperly calibrated planimeter.

Inaccurate responses are sometimes caused by errors of definition in survey questions. For example, in an employment survey, what does the term *unemployed* mean? Should the unemployed include those who gave up looking for work, teenagers who cannot find summer jobs, those who lost part-time jobs, and so on? Even simpler terms like *number of years of education* are easy to misinterpret. Does education include only formal schooling at the elementary, secondary, and college levels, or does it also include technical training, on-the-job classes, and summer institutes? Items to be measured in a survey must be precisely defined and able to be measured unambiguously.

The third problem concerns arbitrary changes in the sampled elements. Data must be obtained from the exact sampling units that were selected in accordance with a sampling design. An interviewer must not substitute a next-door neighbor for a person whose name was sampled. Theoretically, samples selected according to a design have known probabilities associated with them. These known probabilities allow us to calculate the expected values and the variances of estimators, such as the sample mean, and thus to determine the goodness of these estimators. If haphazard substitutions are made in the sample, this probability structure is altered and the goodness of the estimator is uncertain.

Practically speaking, haphazard substitutions may bias the results. For example, suppose next-door neighbors are substituted for families not at home. This substitution may lead to a sample that contains an unduly high proportion of families with children, because these families will more frequently be found at home. If the response depends on the number of children in the family, the resulting estimate will be biased.

One of the classical errors in opinion-polling history came about because of nonsampling errors related to nonresponse and poor selection of a frame. The *Literary Digest* attempted to predict the outcome of the 1936 presidential election by sending postcard questionnaires to 10 million people selected, evidently, from subscribers to the *Digest*, telephone directories, and automobile owners. The 2,376,523 returned cards showed Landon a winner over Roosevelt by 57% to 43%. However, Roosevelt won the election by 62.5% to 37.5%. The large error may have been partly due to the frame being weighted toward higher-income people, but, certainly, the high non-response rate was a significant factor.

According to the account in his book *The Sophisticated Poll Watcher's Guide*, George Gallup polled a subsample of 3000 of the 10 million polled by the *Digest*, and he *predicted* that the *Digest* poll would show 56% for Landon and 44% for Roosevelt. Furthermore, another Gallup poll predicted that Roosevelt would win with 56% of the vote. The error in the Gallup poll was still sizable but much smaller than the error in the *Digest* poll, even though the *Digest* had many more respondents. For further details on this interesting case, read the article by Maurice C. Bryson listed in the references in the Appendix.

Nonsampling errors must be controlled by careful attention to the construction of the questionnaire and to the details of the fieldwork. These errors in surveys can be minimized by adhering to the points presented in the following subsections.

CALLBACKS

Nonresponse can be minimized by having a carefully prepared plan for callbacks on sampled elements. A fixed number of callbacks should be required for each sampled element, and these callbacks should be on different days of the week and at different hours of the day. A specific method for determining an appropriate number of callbacks will be given in Chapter 11. That some responses be obtained on at least a subset of the original nonrespondents is important so that large biasing factors can be eliminated. For example, in surveying opinions on gun control legislation, one would want to make sure that all the nonrespondents were not people who favor gun control but not strongly enough to bother responding to a questionnaire.

REWARDS AND INCENTIVES

Sometimes, an appropriate tactic for encouraging responses is to offer an award for responding. This reward may be a cash payment to a person who agrees to participate in a study. Or in studies of consumer products a participant may be given a supply of the product. The rewards should be offered to potential participants in a study only *after* they have been selected for the sample by some objective procedure. To take as a sample those who respond to an advertised reward is usually not appropriate, since those who respond under such inducements may not be representative of the target population.

Incentives to respond are particularly helpful for samples from groups that have a particular interest in the problem under study. Insured motorists may be more willing to respond to a questionnaire on automobile insurance if a cover letter from the state insurance commissioner's office states that the results may help to promote lower rates. Hunters will respond to a questionnaire on game management practices if they are assured that the results may improve hunting conditions. Many similar examples can be given, but the important point is that people are more likely to respond to a survey if they see some potential benefit coming from the results.

TRAINED INTERVIEWERS

The skill of the interviewer is directly related to the quantity and quality of data resulting from a survey, whether the interview is in person or over the telephone. Good interviewers can ask questions in such a way as to encourage honest responses and can tell the difference between those who really don't know the answer and those who are simply reluctant to answer. Newly recruited interviewers should practice on typical respondents like those they might meet in the field. These practice sessions should be under the watchful eye of experienced interviewers, who can then evaluate the interview and suggest improvements in interview technique.

DATA CHECKS

Completed questionnaires should be scrutinized carefully by someone other than the interviewer to see that the form is filled out correctly. At this stage, and again later if data is entered into a computer, a predesigned system of data checks should be made to spot obvious errors in information.

The ranges of measurements can be checked to sort out the cases in which, say, the age of a person is listed as 1040, or a married adult is listed as 9 years old, or a family is reported to have 53 children under the age of 12. Data can be cross-checked in a well-designed questionnaire to see, for example, whether the respondent's reported age agrees with the reported year of birth. Simple arithmetic facts—for instance, proportions must be between 0 and 1 and the hours per day assigned to different work tasks cannot sum to more than 24—can be included in these data checks.

Checking data quickly, so that questionable responses can be corrected while the respondent is still available, is very important to the success of a sample survey.

After all the responses are collected and the data is being analyzed, additional data checks can be employed. The survey results should be representative of the population, and, sometimes, sample data can be checked against known facts for the population to see whether potential problem areas exist. For example, if the population is 50% female but the sample is only 10% female, there may be serious errors in summary measurements that average over males and females. If average income for survey respondents is well below the reported average from other sources for the target population, then large errors may show up in summary measurements on variables related to income. Some of these potential problems may be solved by augmenting the sample or by changing the form of analysis, but even if they cannot be solved, any inconsistencies should be pointed out in the final analysis.

QUESTIONNAIRE CONSTRUCTION

After sample selection, the most important component of a well-run, informative, and accurate sample survey is a properly designed questionnaire. This subject is the topic of Section 3.6.

3.5
METHODS OF DATA COLLECTION

The most commonly used methods of data collection in sample surveys are personal interviews and telephone interviews. These methods, with appropriately trained interviewers and carefully planned callbacks, commonly achieve response rates of 60% to 75%, and, sometimes, these rates can be even higher. A mailed questionnaire sent to a specific group of interested persons can achieve good results, but, generally, response rates for this type of data collection are so low that all reported results are suspect. Frequently, objective information can be found from direct observation rather than from an interview or mailed questionnaire. These four types of data collection are discussed in the following subsections.

PERSONAL INTERVIEWS

Data are frequently obtained by *personal interviews*. For example, we can use personal interviews with eligible voters to obtain a sample of the public sentiments toward a community bond issue. The procedure usually requires the interviewer to ask prepared questions and to record the respondent's answers. The primary advantage of these interviews is that people will usually respond when confronted in person. In addition, the interviewer can note specific reactions and eliminate misunderstandings about the questions asked. The major limitations of the personal interview (aside from the cost involved) concern the interviewers. If they are not thoroughly trained, they may deviate from the required protocol, thus introducing a bias into the sample data. Any movement, facial expression, or statement by the interviewer can affect the response obtained. For example, a leading question such as "Are you also in favor of the bond issue?" may tend to elicit a positive response. Finally, errors in recording the responses can also lead to erroneous results.

TELEPHONE INTERVIEWS

Information can also be obtained from persons in the sample through *telephone interviews*. With the advent of wide-area telephone service lines (WATS lines), an interviewer can place any number of calls to specified areas of the country for a fixed monthly rate. Surveys conducted through telephone interviews are frequently less expensive than personal interviews, owing to the elimination of travel expenses. The investigator can also monitor the interviews to be certain the specified interview procedure is being followed.

A major problem with telephone surveys is the establishment of a frame that closely corresponds to the population. Telephone directories have many numbers that do not belong to households, and many households have unlisted numbers. A few households have no phone service, although lack of phone service is now only a minor problem for most surveys in the United States. A technique that avoids the problem of unlisted numbers is random-digit dialing. In this method a telephone exchange number (the first three digits of the seven-digit number) is selected, and then the last four digits are dialed randomly until a fixed number of households of a specified type are reached. This technique seems to produce unbiased samples of households in selected target populations and avoids many of the problems of trying to sample a telephone directory.

Telephone interviews generally must be kept shorter than personal interviews because respondents tend to get impatient more easily when talking over the telephone. With appropriately designed questionnaires and trained interviewers, telephone interviews can be as successful as personal interviews. [See Schuman and Presser (1981) for more details.]

SELF-ADMINISTERED QUESTIONNAIRES

Another useful method of data collection is the *self-administered questionnaire*, to be completed by the respondent. These questionnaires usually are mailed to the individuals included in the sample, although other distribution methods can be used. The questionnaire must be carefully constructed if it is to encourage participation by the respondents.

The self-administered questionnaire does not require interviewers, and thus its use results in a savings in the survey cost. This savings in cost is usually bought at the expense of a lower response rate. Nonresponse can be a problem in any form of data collection, but since we have the least contact with respondents in a mailed questionnaire, we frequently have the lowest rate of response. The low response rate can introduce a bias into the sample because the people who answer questionnaires may not be representative of the population of interest. To eliminate some of this bias, investigators frequently contact the nonrespondents through follow-up letters, telephone interviews, or personal interviews.

DIRECT OBSERVATION

The fourth method for collecting data is *direct observation*. For example, if we were interested in estimating the number of trucks that use a particular road during the 4–6 P.M. rush hours, we could assign a person to count the number of trucks passing a specified point during this period. Possibly, electronic-counting equipment could also be used. The disadvantage in using an observer is the possibility of errors in observation.

Direct observation is used in many surveys that do not involve measurements on people. The U.S. Department of Agriculture, for instance, measures certain variables on crops in sections of fields in order to produce estimates of crop yields. Wildlife biologists may count animals, animal tracks, eggs, or nests in order to estimate the size of animal populations.

A closely related notion is that of getting data from objective sources that are not affected by the respondents themselves. Health information can sometimes be obtained from hospital records, income information from employer's records (especially for state and federal government workers). This approach may take more time but may yield large rewards in important surveys.

3.6
DESIGNING A QUESTIONNAIRE

As stated earlier, one objective of any survey design is to minimize the nonsampling errors that may occur. If a survey is to obtain information from people, then many potential nonsampling errors should be considered and, it is hoped, controlled by the careful design of the questionnaire. We briefly discuss questionnaire construction in this section, but it is a very important topic and should be investigated further by those attempting to design complex questionnaires for surveys. An excellent reference, and the one on which we rely extensively for the discussion that follows, is Schuman and Presser (1981). Some major concerns in questionnaire construction are outlined in the following subsections.

QUESTION ORDERING

Respondents to questionnaires generally try to be consistent in their responses to questions. Respondent consistency may cause the ordering of the questions to affect the responses, sometimes in ways that seem unpredictable to the inexperienced investigator. An example discussed in Schuman and Presser (1981) illustrates the point.

An experiment was conducted with the following two questions:

A. Do you think the United States should let Communist newspaper reporters from other countries come in here and send back to their papers the news as they see it?

B. Do you think a Communist country like Russia should let American newspaper reporters come in and send back to America the news as they see it?

For surveys in 1980 in which the questions appeared in the order (A, B), 54.7% of the respondents answered yes to A and 63.7% answered yes to B. For surveys in which the questions appeared in the order (B, A), 74.6% answered yes to A and 81.9% answered yes to B. So the evidence suggests that asking question B first puts the respondents in a more lenient frame of mind toward allowing Communist reporters into the United States. In other words, those who answered yes to B, when it was asked first, tried to be consistent and also answer yes to a similar question, A. Thus the context in which a question is asked is very important and should be understood and explained in the analysis of questionnaire data.

Order is also important in the relative positioning of specific versus general questions. Respondents may be asked the following questions:

A. Will you support an increase in state taxes for education?

B. Will you support an increase in state taxes?

It would not be unusual to find more people supporting B if asked in the order (B, A) than if asked in the order (A, B). If question A is asked first, persons who support taxes for education and answer A affirmatively may think that B implies an increase in taxes *not* necessarily going to education, and they may then say no to this question. If B is presented first, the same people who support more taxes for education may answer affirmatively since they have not yet seen a specific question on taxes for education.

Attitude toward a question in a survey is very often set, or changed, by preceding questions that bear on the same topic. Schuman and Presser report that more crime victimization was reported by respondents when the question on victimization occurred *after* a series of questions on crime than when it occurred by itself. Evidently, the questions on crime helped the person responding to remember small incidents when he or she was a victim of crime, incidents that might otherwise have been forgotten. Attitudes toward government can be quite negative after a series of questions emphasizing government waste and inefficiency, and they can be much more positive after a series of questions emphasizing the necessary and timely functions government performs.

In a series of questions involving ratings, the first question is often considered in a different light from those that follow, and it tends to receive more extreme ratings. For example, suppose a person is to rate a number of possible vacation sites, with each one receiving a numerical rating from 1 to 10, 10 being very good. If the first site looks good to the respondent, it will tend to be rated close to 10 and the others will tend to be rated lower. If the first site looks unattractive, it will tend to be rated close to 1 and the others will tend to be rated higher. Thus, among the group of good sites, each will tend to receive its highest ratings when it appears first on the list. Similarly, each bad site will tend to receive its lowest rating when it appears first on the list. Evidently, the first item on the list is used as a reference point, and other items are rated up or down relative to the first item.

For many survey questions the order of the possible responses (or choices) to a particular question is as important as the position of the question on the questionnaire. If a person being interviewed is presented with a long list of possible choices, or if each possible choice is wordy or difficult to interpret, then the person is likely to respond with the most recent choice (the last one on the list). If a respondent must choose items from a long written list, then the items appearing toward the top of the list have a selection advantage. For example, consider the election of candidates for office from a long slate: Those toward the top of the list tend to get elected. In a list of simple choices, such as strongly agree, agree, disagree, and strongly disagree in an attitude survey, alternatives tend to receive their highest frequency of response when listed first. That is, the proportion who strongly agree will tend to be higher when that option is a first choice rather than a fourth choice.

Researchers attempting to design a questionnaire should be aware of the common ordering problems for question and response. They should attempt to counter potential difficulties by considering the following techniques:

1. Printing questionnaires with different orderings for different subsets of the sample.

2. Using show cards or repeating the question as often as necessary in an interview so that the question and possible answers are clearly understood.

3. Carefully explaining the context in which a question was asked in the analysis of the survey data.

Open versus Closed Questions

Since questionnaires today are often designed to be electronically scored after completion, with the data in a form for computer handling, most questions will be *closed questions*. That is, each question will have either a single numerical answer (like age of the respondent) or a fixed number of predetermined choices, one of which is to be selected by the respondent.

Even though closed questions allow for easy data coding and analysis, some thought should be given to *open questions*, in which the respondent is allowed to freely state an unstructured answer. The open question allows the respondent to express some depth and shades of meaning in the answer.

But it can cause great difficulties in analysis because answers may not be easily quantified and may be nearly impossible to compare across questionnaires. In contrast, the closed question may not always provide the appropriate alternatives, and the alternatives listed may, themselves, influence the opinion of the person responding. Once a questionnaire is completed, however, the data handling is fairly routine, and valid statistical summaries of reported answers are easily constructed.

A typical open question, similar to ones actually used in Gallup polls, is as follows:

> What is the most important problem facing the United States today?

This question can provide meaningful results as it is, since many people will choose similar problems as being most important. However, their choices could be forced into predetermined categories by the following closed question:

> The most important problem facing the United States today is (check one):
>
> a. national security.
> b. crime.
> c. inflation.
> d. unemployment.
> e. budget deficits.

One can see that any closed form of this question will limit the alternatives and may force a respondent into an answer that would not necessarily be a first choice.

A good plan for designing a closed question with appropriate alternatives is to use a similar open question on a pretest; then choose as the fixed alternatives those that most nearly represent the choices expressed in the open answers. To come up with a short list of alternatives from the open-ended answers will not always be easy, but this approach will provide more realistic alternatives than could be obtained from mere speculation.

RESPONSE OPTIONS

On almost any question that can be posed, someone being interviewed will want to say that he doesn't know or has no opinion. Since such responses give no useful information about the question and essentially reduce the sample size, typical survey practice is to avoid using these options. The respondent is forced to make a choice from among the listed informative answers, unless the interviewer decides that such a choice simply cannot be made.

However, to force people to make decisions on questions they know nothing about seems inappropriate. Thus a good questionnaire will provide screening questions to determine whether the respondent has enough information to form an opinion on a certain issue. If so, the main question is asked without a "no opinion" option. If not, the question may be skipped.

In other words, questions about which nearly everyone has enough information to form some opinion, such as questions on stricter enforcement of speed limit laws for drivers, should be stated without a "no opinion" option. Questions of a specific, narrow, or detailed nature, such as questions on a recently passed city ordinance, should be prefaced by screening questions to see whether the respondent has any information on the subject.

Even after the "no opinion" option is eliminated from a question, there remains the problem of deciding how many options to allow. Frequently, questionnaires attempt to polarize opinion on one side or the other, as in the following question:

> Do you think the enforcement of traffic laws in our city is too strict or too lenient?

Here no middle ground is offered. One reason for not allowing a middle choice, such as "just right the way things are," is that respondents may take this choice far too often as an easy way out. The two-choice option forces the person responding to think about the direction of the response, but the interviewer should explain that various degrees of strictness or leniency can be taken into account. "Which pole am I closest to?" is the point that the respondent is urged to consider. Of course, if one wants to categorize the degrees of strictness or leniency in this question, then more than two options can be presented. However, questionnaire designers usually wish to keep the number of options as small as possible.

WORDING OF QUESTIONS

Even for questions in which the number of response options is clearly determined, the designer should be concerned about the phrasing of the main body of the question. Yes–no questions like

> Do you favor the use of capital punishment?

should be asked in a more balanced form, such as

> Do you favor or oppose the use of capital punishment?

Some questions have strong arguments and counterarguments woven into them. Schuman and Presser (1981, p. 186) show results for a comparison of the following questions:

A. If there is a union at a particular company or business, do you think that all the workers there should be required to be union members, or are you opposed to this?

B. If there is a union at a particular company or business, do you think that all the workers there should be required to be union members, or should it be left to the individual to decide whether or not he wants to be in the union?

Among persons presented with question A, 32.1% responded that workers should be required to be union members; but among those presented with question B, only 23.0% responded in this way. Question B has a stronger counterargument in the second phase of the question. People with no strong feelings either way are particularly susceptible to strong arguments or counterarguments within the body of the question. Again, questions should be asked in a balanced form, with little argument or counterargument within the text of the question.

Sometimes, the respondent may tend to show agreement with the interviewer when the question is a leading question. For instance, the question

Do you agree that courts are too lenient with criminals?

will receive many more yes responses than it should simply because that response seems to agree with the interviewer's notion of the correct response. Leading questions should be rephrased in a balanced form, as discussed earlier in this subsection.

Responses to many questions can be drastically altered just by an appropriate, or inappropriate, choice of words. Schuman and Presser (1981, p. 277) report on studies of the following questions:

A. Do you think the United States should forbid public speeches against democracy?
B. Do you think the United States should allow public speeches against democracy?

In one study those presented with question A gave 21.4% yes responses, while those presented with question B gave 47.8% no responses. People are somewhat reluctant to *forbid* public speeches against democracy, but they are much more willing to *not allow* such speeches. *Forbid* is a strong word and elicits a negative feeling that many cannot favor. *Allow* is a much milder word and doesn't elicit strong feelings. The important point to remember is that the tone of the question, set by the words employed, can have a significant impact on the responses.

Questions also must be stated in clearly defined terms in order to minimize response errors. A question like

How much water do you drink?

is unnecessarily vague. It may be reworded as follows:

Here is an eight-ounce glass. (Hold one up.) How many eight-ounce glasses of water do you drink each day?

If total water intake is important, the interviewer must remind the person that coffee, tea, and other drinks are mostly water.

Similarly, a question like

How many children are in your family?

is too ambiguous. It may be restated as follows:

> How many persons under the age of 21 live in your household and receive more than one-half of their financial support from you?

Again, the question must be specific, with all components well defined.

Many more items could be discussed on the topic of questionnaire construction. But the items presented here are the most important ones, and each should be considered very carefully before sampling is begun.

3.7
PLANNING A SURVEY

We will now review and extend some of the ideas presented in previous sections in the form of a checklist. Each item on the checklist should be carefully considered in the planning of any survey.

1. **STATEMENT OF OBJECTIVES.** State the objectives of the survey clearly and concisely, and refer to these objectives regularly as the design and the implementation of the survey progress. Keep the objectives simple enough to be understood by those working on the survey and to be met successfully when the survey is completed.

2. **TARGET POPULATION.** Carefully define the population to be sampled. If adults are to be sampled, then define what is meant by *adult* (all those over the age of 18, for example), and state what group of adults are included (all permanent residents of a city, for example). Keep in mind that a sample must be selected from this population, and define the population so that sample selection is possible.

3. **THE FRAME.** Select the frame (or frames) so that the list of sampling units and the target population show close agreement. Keep in mind that multiple frames may make the sampling more efficient. For example, residents of a city can be sampled from a list of city blocks coupled with a list of residents within blocks.

4. **SAMPLE DESIGN.** Choose the design of the sample, including the number of sample elements, so that the sample provides sufficient information for the objectives of the survey. Many surveys have produced little or no useful information because they were not properly designed.

5. **METHOD OF MEASUREMENT.** Decide on the method of measurement, usually one or more of the following methods: personal interviews, telephone interviews, mailed questionnaires, or direct observations.

6. **MEASUREMENT INSTRUMENT.** In conjunction with step 5, carefully specify how and what measurements are to be obtained. If a questionnaire is to be used, plan the questions so that they minimize nonresponse and incorrect response bias.

7. **SELECTION AND TRAINING OF FIELD-WORKERS.** Carefully select and train the field-workers. After the sampling plan is clearly and

completely set up, someone must collect the data. Those collecting data, the field-workers, must be carefully taught what measurements to make and how to make them. Training is especially important if interviews, either personal or telephone, are used, because the rate of response and the accuracy of responses are affected by the interviewer's personal style and tone of voice.

8. **THE PRETEST.** Select a small sample for a pretest. The pretest is crucial, since it allows you to field-test the questionnaire or other measurement device, to screen interviewers, and to check on the management of field operations. The results of the pretest usually suggest that some modifications must be made before a full-scale sampling is undertaken.

9. **ORGANIZATION OF FIELDWORK.** Plan the fieldwork in detail. Any large-scale survey involves numerous people working as interviewers, coordinators, or data managers. The various jobs should be carefully organized and lines of authority clearly established before the survey is begun.

10. **ORGANIZATION OF DATA MANAGEMENT.** Outline how each piece of data is to be handled for all stages of the survey. Large surveys generate huge amounts of data. Hence a well-prepared data management plan is of the utmost importance. This plan should include the steps for processing data from the time a measurement is taken in the field until the final analysis is completed. A quality control scheme should also be included in the plan in order to check for agreement between processed data and data gathered in the field.

11. **DATA ANALYSIS.** Outline the analyses that are to be completed. Closely related to step 10, this step involves the detailed specification of what analyses are to be performed. It may also list the topics to be included in the final report. If you think about the final report before a survey is run, you may be more careful in selecting items to be measured in the survey.

If these steps are followed diligently, the survey is off to a good start and should provide useful information for the investigator.

3.8
SUMMARY

The objective of a sample survey is to make inferences about the population of interest from information contained in a sample. The population consists of the body of data about which we wish to make an inference and is composed of elements or bits of information. Nonoverlapping collections of elements from the population are called sampling units. The frame is a list of sampling units that we use to represent the population. The sample is a collection of sampling units drawn from the frame. Using the sample data, we will estimate certain population parameters and place bounds on our error of estimation.

The quantity of information obtained from the sample can be controlled by the number of sampling units drawn and the sample design or method of data collection used. Some of the designs introduced were simple random sampling, stratified random sampling, cluster sampling, and systematic sampling. Each is discussed in detail in a later chapter. The best design for a given problem is the one that provides the necessary precision in terms of a bound on the error of estimation for a minimum cost.

After the design has been selected, there are various methods of collecting the sample data. Personal interviews, telephone interviews, direct observations, and questionnaires were discussed and assessed as means of collecting the sample data. Each method has its advantages and limitations.

In Section 3.6 we discussed the actual construction of questionnaires. Again, we emphasize the importance of obtaining information in the sample that is representative of the population. This problem is of prime significance when we consider methods of data collection.

EXERCISES

3.1   An experimenter wants to estimate the average water consumption per family in a city. Discuss the relative merits of choosing individual families, dwelling units (single-family houses, apartment buildings, etc.), and city blocks as sampling units. What would you use as a frame in each case?

3.2   A forester wants to estimate the total number of trees on a tree farm that possess diameters exceeding 12 inches. A map of the farm is available. Discuss the problem of choosing appropriate sampling units and an appropriate frame.

3.3   A safety expert is interested in estimating the proportion of automobile tires with unsafe tread. Should he use individual cars or collections of cars, such as those in parking lots, as sampling units? What could he use as a frame?

3.4   An industry is composed of many small plants located throughout the United States. An executive wants to survey the opinions of the employees on the vacation policy of the industry. What would you suggest she use as sampling units? What could she use as a frame?

3.5   A state department of agriculture desires to estimate the number of acres planted in corn within the state. Suggest possible sampling units and frames.

3.6   A political scientist wants to estimate the proportion of adult residents of a state who favor a unicameral legislature. Discuss possible units and frames. Also, discuss the relative merits of personal interviews, telephone interviews, and mailed questionnaires as methods of data collection.

3.7   Discuss the relative merits of using personal interviews, telephone interviews, and mailed questionnaires as methods of data collection for each of the following situations:

(a)   A television executive wants to estimate the proportion of viewers in the country who are watching her network at a certain hour.

(b) A newspaper editor wants to survey the attitudes of the public toward the type of news coverage offered by his paper.

(c) A city commissioner is interested in determining how homeowners feel about a proposed zoning change.

(d) A county health department wants to estimate the proportion of dogs that have had rabies shots within the last year.

3.8 Discuss problems associated with question ordering. Give a list of two or three questions for which you think order is important, and explain why.

3.9 Discuss the use of open versus closed questions. Give an example of an appropriate open question. Give an example of how a similar question could be closed. What are the advantages of closed questions?

3.10 Give an example of a question that contains a weak counterargument. Give an example of a question that contains a strong counterargument.

3.11 Discuss the use of a no-opinion option in a closed question.

3.12 Give an example of a question that could force a response in a certain direction because of its strong wording.

3.13 Discuss the importance of proper data management techniques and quality control in a survey.

3.14 Discuss the importance of having a pretest.

3.15 Why is the response rate an important consideration in surveys? Discuss methods for reducing the nonresponse rate.

3.16 Respondents commonly receive telephone calls from people taking surveys during the evening dinner hour. Those planning the survey probably think that many potential respondents will be home at that time. Discuss the pros and cons of this approach.

3.17 You are hired to estimate the proportion of registered Republicans in your county who favor an increase in the number of nuclear weapons owned by the United States. How would you plan the survey? (Go through the eleven steps of Section 3.7, realizing that steps 4 and 11 cannot be answered completely at this time.)

3.18 A Yankelovich, Skelly and White poll taken in the fall of 1984 showed that a fifth of the 2207 people surveyed admitted to having cheated on their federal income taxes. Do you think that this fraction is close to the actual proportion who cheated? Why? (Discuss the difficulties of obtaining accurate information on a question of this type.)

3.19 In a Gallup youth survey (*Gainesville Sun*, February 13, 1985), 414 high school juniors and seniors were asked:

> What course or subject that you have studied in high school has been the best for preparing you for your future education or career?

In their responses to this question 25% of the students chose mathematics and 25% chose English. Do you think this question is a good question with informative results?

3.20 A survey by Group Attitudes, Inc., was said to measure American attitudes toward college (*Gainesville Sun*, September 9, 1982). The polling firm mailed questionnaires to 4200 people across the United States and received 1188 responses. About 55% of those polled said they had major concerns about being able to pay for their child's college education. Would you regard this figure as highly reliable and representative of the true proportion of Americans with this concern? (What groups of people are likely to respond to such a question?)

CASE STUDY

ALMOST every daily newspaper contains a discussion of at least one opinion poll. Alert readers can use knowledge of basic statistical procedures to decide whether the claims made in the articles are supported by the data.

A Yankelovich, Skelly and White poll, taken in late 1984, surveyed the opinions of 2207 United States residents on questions related to the federal income tax (*Gainesville Sun*, January 15, 1985). The results showed that 54% of those surveyed considered the Internal Revenue Service tax forms too complicated for their tax situations. Is the headline stating "Most Find Income Tax Complicated" justified? Statistical techniques explained in Chapter 4 will help in answering this question. The analysis for this case will be presented at the end of the chapter.

SIMPLE
RANDOM
SAMPLING

4.1
INTRODUCTION

The objective of a sample survey is to make an inference about the population from information contained in a sample. Two factors affect the quantity of information contained in the sample and hence affect the precision of our inference-making procedure. The first is the size of the sample selected from the population. The second is the amount of variation in the data; variation can frequently be controlled by the method of selecting the sample. The procedure for selecting the sample is called the *sample survey design*. For a fixed sample size n we will consider various designs, or *sampling procedures*, for obtaining the n observations in the sample. Since observations cost money, a design that provides a precise estimator of the parameter for a fixed sample size yields a savings in cost to the experimenter. The basic design, or sampling technique, called *simple random sampling* is discussed in this chapter.

DEFINITION 4.1   If a sample of size n is drawn from a population of size N in such a way that every possible sample of size n has the same chance of being selected the sampling procedure is called *simple random sampling*. The sample thus obtained is called a *simple random sample*.

We will use simple random sampling to obtain estimators for population means, totals, and proportions.

Consider the following problem. A federal auditor is to examine the accounts for a city hospital. The hospital records obtained from a computer show a particular accounts receivable total, and the auditor must verify this total. If there are 28,000 open accounts in the hospital, the auditor cannot afford the time to examine every patient record to obtain a total accounts receivable figure. Hence the auditor must choose some sampling scheme for obtaining a representative sample of patient records. After examining the patient accounts in the sample, the auditor can then estimate the accounts receivable total for the entire hospital. If the computer figure lies within a specified distance of the auditor's estimate, the computer figure is accepted as valid. Otherwise, more hospital records must be examined for possible discrepancies between the computer figure and the sample data.

Suppose that all $N = 28,000$ patient records are recorded on computer cards and a sample size $n = 100$ is to be drawn. The sample is called a simple random sample if every possible sample of $n = 100$ records has the same chance of being selected.

Simple random sampling forms the basis of most of the sampling designs discussed in this book, and it forms the basis of most scientific surveys done in practice. The Nielsen Television Index (NTI) is the most widely used audience measurement service in existence. It is based on a random sample of approximately twelve hundred households that have a storage instantaneous audimeter connected to the television set. This meter records whether or not the television set is on, what channel is being viewed, and channel changes. In an additional random sample of families each family keeps a diary on who watches various shows. The NTI reports the number of households in the audience, the type of audience, and the amount of television viewing for various time periods.

The Gallup poll actually begins with a random sample of approximately 300 election districts, sampled from the 200,000 election districts in the United States. Then households for interviewing are selected from each district by another randomization device. The sampling is in two stages, but simple random sampling plays a key role at each stage.

Auditors study simple random samples of accounts in order to check for compliance with audit controls set up by the firm or to verify the actual dollar value of the accounts. Thus they may wish to estimate the proportion of accounts not in compliance with controls or the total value of, say, accounts receivable.

Marketing research often involves a simple random sample of potential users of a product. The researcher may want to estimate the proportion of potential buyers who prefer a certain color of car or flavor of food.

A forester may estimate the volume of timber or proportion of diseased trees in a forest by selecting geographic points in the area covered by the forest and then attaching a plot of fixed size and shape (such as a circle of 10-meter radius) to that point. All the trees within the sample plots may be studied, but, again, the basic design is a simple random sample.

Two problems now face the experimenter: (1) how does he or she draw the simple random sample, and (2) how can he or she estimate the various population parameters of interest? These topics are discussed in the following sections.

4.2
How to Draw a Simple Random Sample

To draw a simple random sample from the population of interest is not as trivial as it may first appear. How can we draw a sample from a population in such a way that every possible sample of size n has the same chance of being selected? We might use our own judgment to "randomly" select the sample. This technique is frequently called haphazard sampling. A second technique, representative sampling, involves choosing a sample that we consider to be typical or representative of the population. Both haphazard and representative sampling are subject to investigator bias, and, more importantly, they lead to estimators whose properties cannot be evaluated. Thus neither of these techniques leads to a simple random sample.

Simple random samples can be selected by using tables of random numbers. A table of random numbers is shown in Table 2 of the Appendix.

A random number table is set of integers generated so that in the long run the table will contain all ten integers $(0, 1, \ldots, 9)$ in approximately equal proportions, with no trends in the pattern in which the digits were generated. Thus if one number is selected from a random point in the table, it is equally likely to be any of the digits 0 through 9.

Choosing numbers from the table is analogous to drawing numbers out of a hat containing those numbers on thoroughly mixed pieces of paper. Suppose we want a simple random sample of three persons to be selected from seven. We could number the people from 1 to 7, put slips of paper containing these numbers (one number to a slip) into a hat, mix them, and draw out three, *without replacing* drawn numbers. Analogously, we could drop a pencil point on a random starting point in Table 2 of the Appendix. Suppose the point falls on the 15th line of column 9 and we decide to use the right-most digit (a 5, in this case). This procedure is like drawing a 5 from the hat. We may now proceed in any direction to obtain the remaining numbers in the sample. Suppose we decide before starting to proceed down the page. The number immediately below the 5 is a 2, so our second sampled person is number 2. Proceeding, we next come to an 8, but there are only seven people in our population; hence the 8 must be ignored. Two more 5s then appear, but both must be ignored since person 5 has already been selected. (The 5 has been removed from the hat.) Finally, we come to a 1, and our sample of three is completed with persons numbered 5, 2, and 1.

Note that any starting point can be used and one can move in any predetermined direction. If more than one sample is to be used in any problem, each should have its own unique starting point.

A more realistic illustration is given in Example 4.1.

Example 4.1

For simplicity, assume there are $N = 1000$ patient records from which a simple random sample of $n = 20$ is to be drawn. We know that a simple random sample will be obtained if every possible example of $n = 20$ records has the same chance of being selected. The digits in Table 2 of the Appendix, and in any other table of random numbers, are generated to satisfy the conditions of simple random sampling. Determine which records are to be included in a sample of size $n = 20$.

SOLUTION

We can think of the accounts as being numbers $001, 002, \ldots, 999, 000$. That is, we have 1000 three-digit numbers, where 001 represents the first record, 999 the 999th patient record, and 000 the 1000th.

Refer to Table 2 of the Appendix and use the first column; if we drop the last two digits of each number, we see that the first three-digit number formed is 104, the second is 223, the third is 241, and so on. Taking a random sample of 20 digits, we obtain the numbers shown in Table 4.1.

TABLE 4.1 Patient records to be included in the sample

104	779	289	510
223	995	635	023
241	963	094	010
421	895	103	521
375	854	071	070

If the records are actually numbered, we merely choose the records with the corresponding numbers, and these records represent a simple random sample of $n = 20$ from $N = 1000$. If the patient accounts are not numbered, we can refer to a list of the accounts and count from the 1st to the 10th, 23rd, 70th, and so on, until the desired numbers are reached. If a random number occurs twice, the second occurrence is omitted, and another number is selected as its replacement.

4.3
ESTIMATION OF A POPULATION MEAN AND TOTAL

We stated previously that the objective of survey sampling is to draw inferences about a population from information contained in a sample. One way to make inferences is to estimate certain population parameters by utilizing the sample information. The objective of a sample survey is often to estimate a population mean, denoted by μ, or a population total, denoted by τ. Thus the auditor of Example 4.1 might be interested in the mean dollar value for the accounts receivable or the total dollar amount in these accounts. Hence we consider estimation of the two population parameters, μ and τ, in this section.

Suppose that a simple random sample of n accounts is drawn, and we are to estimate the mean value per account for the total population of hospital records. Intuitively, we would employ the sample average,

$$\bar{y} = \frac{\sum\limits_{i=1}^{n} y_i}{n}$$

to estimate μ.

Of course, a single value of $\bar{y}$ tells us very little about the population mean μ, unless we are able to evaluate the goodness of our estimator. Hence in addition to estimating μ, we would like to place a bound on the error of estimation. It can be shown that $\bar{y}$ possesses many desirable properties for estimating μ. In particular, $\bar{y}$ is an unbiased estimator of μ and has a variance that decreases as the sample size n increases. More precisely for a simple random sample chosen without replacement from a population of size N,

$$E(\bar{y}) = \mu$$

and

$$V(\bar{y}) = \frac{\sigma^2}{n}\left(\frac{N-n}{N-1}\right) \tag{4.1}$$

These properties are formally derived in the Appendix, but at this point we will show that they hold for a particular case. Suppose we have a population of $N = 4$ measurements given by $\{1, 2, 3, 4\}$. If a single observation y is selected at random from this population, then y can take on any of the four possible values, each with probability $\frac{1}{4}$. Thus

$$\mu = E(y) = \sum yp(y) = 1(\tfrac{1}{4}) + 2(\tfrac{1}{4}) + 3(\tfrac{1}{4}) + 4(\tfrac{1}{4})$$
$$= (\tfrac{1}{4})(1 + 2 + 3 + 4) = \tfrac{10}{4} = 2.50$$

and

$$\sigma^2 = V(y) = E(y - \mu)^2 = \sum (y - \mu)^2 p(y)$$
$$= (1 - 2.5)^2(\tfrac{1}{4}) + (2 - 2.5)^2(\tfrac{1}{4}) + (3 - 2.5)^2(\tfrac{1}{4}) + (4 - 2.5)^2(\tfrac{1}{4})$$
$$= \tfrac{5}{4}$$

Now suppose we are to select a random sample of size $n = 2$ (without replacement) from this population of four measurements. There are six possible samples, listed as follows:

$$\{1, 2\}, \{1, 3\}, \{1, 4\}, \{2, 3\}, \{2, 4\}, \{3, 4\}$$

All of these samples should be equally likely, and hence a probability of $\frac{1}{6}$ should be attached to the occurrence of any one sample. The six sample means, $\bar{y}$, are, respectively,

$$1.5, 2.0, 2.5, 2.5, 3.0, 3.5$$

Since each of these sample means can occur with probability $\frac{1}{6}$, we know the sampling distribution of $\bar{y}$ and can compute $E(\bar{y})$ and $V(\bar{y})$. From our definition of expected value,

$$E(\bar{y}) = \sum \bar{y} p(\bar{y}) \qquad \text{(summed over all values of } \bar{y})$$
$$= (1.5)(\tfrac{1}{6}) + (2.0)(\tfrac{1}{6}) + (2.5)(\tfrac{1}{6}) + (2.5)(\tfrac{1}{6}) + (3.0)(\tfrac{1}{6}) + (3.5)(\tfrac{1}{6})$$
$$= 2.50 = \mu$$

Also, $V(\bar{y}) = E(\bar{y} - \mu)^2 = \sum (\bar{y} - \mu)^2 p(\bar{y})$
$$= (1.5 - 2.5)^2(\tfrac{1}{6}) + (2.0 - 2.5)^2(\tfrac{1}{6}) + (2.5 - 2.5)^2(\tfrac{1}{6})$$
$$+ (2.5 - 2.5)^2(\tfrac{1}{6}) + (3.0 - 2.5)^2(\tfrac{1}{6}) + (3.5 - 2.5)^2(\tfrac{1}{6})$$
$$= (2.5)(\tfrac{1}{6}) = \tfrac{5}{12}$$

Recalling that for this example $\sigma^2 = \tfrac{5}{4}$, $N = 4$, and $n = 2$, we have

$$\frac{\sigma^2}{n}\left(\frac{N-n}{N-1}\right) = \frac{(\tfrac{5}{4})}{2}\left(\frac{4-2}{4-1}\right)$$
$$= \tfrac{5}{8}(\tfrac{2}{3}) = \tfrac{5}{12}$$

Thus we have demonstrated that

$$E(\bar{y}) = \mu$$

and
$$V(\bar{y}) = \frac{\sigma^2}{n}\left(\frac{N-n}{N-1}\right)$$

It will also be shown in the Appendix that

$$E(s^2) = \frac{N}{N-1}\sigma^2$$

so that $V(\bar{y})$ can be unbiasedly estimated from the sample by

$$\hat{V}(\bar{y}) = \frac{s^2}{n}\left(\frac{N-n}{N}\right)$$

where
$$s^2 = \frac{1}{n-1}\sum_{i=1}^{n}(y_i - \bar{y})^2$$

The variance of the estimator $\bar{y}$ is the same as that given in an introductory course except that it is multiplied by a correction factor to adjust for sampling from a finite population. The correction factor takes into account the fact that an estimate based on a sample $n = 10$ from a population of $N = 20$ items contains more information about the population than a sample of $n = 10$ from a population of $N = 20,000$.

Estimator of the population mean μ:

$$\hat{\mu} = \bar{y} = \frac{\sum_{i=1}^{n} y_i}{n} \tag{4.2}$$

Estimated variance of $\bar{y}$:

$$\hat{V}(\bar{y}) = \frac{s^2}{n}\left(\frac{N-n}{N}\right) \tag{4.3}$$

where
$$s^2 = \frac{\displaystyle\sum_{i=1}^{n}(y_i - \bar{y})^2}{n-1} = \frac{\displaystyle\sum_{i=1}^{n} y_i^2 - n\bar{y}^2}{n-1}$$

Bound on the error of estimation:

$$2\sqrt{\hat{V}(\bar{y})} = 2\sqrt{\frac{s^2}{n}\left(\frac{N-n}{N}\right)} \tag{4.4}$$

The quantity $(N-n)/N$ is called the finite population correction (fpc). Note that this correction factor differs slightly from the one encountered in the true variance of $\bar{y}$. When n remains small relative to the population size N, the fpc is close to unity. Practically speaking, the fpc can be ignored if $(N-n)/N \geq .95$, or, equivalently, $n \leq (\frac{1}{20})N$. In that case the estimated variance of $\bar{y}$ is the more familiar quantity s^2/n.

In many cases the population size is not clearly defined or is unknown. Suppose very small laboratory specimens are selected from a large bulk tank of raw sugar in order to measure pure sugar content. How N will be determined is unclear, but it can generally be assumed to be quite large. Hence the fpc can be ignored. If a sample of voters is selected from the population of a state, to obtain a precise N for that point in time is generally impossible. Again, N is assumed large and the fpc is ignored.

To show the behavior of confidence intervals for the mean, we selected 50 random samples of size $n = 20$ from the population of $N = 100$ elements graphed in Figure 2.3. An approximate 95% confidence interval was constructed for each sample, with the results shown in Table 4.2. Note that four (or 8%) of the observed intervals fail to cover the true population mean. This result is quite close to the nominal value of 5%.

Table 4.3 shows the results of a similar experiment drawn from the data of Figure 2.5. Here only two (4%) of the intervals fail to cover the true mean.

Notice that in both cases the confidence intervals change in both location and length as we move from sample to sample. Remember, also that the intervals are *random*. In repeated sampling roughly 95% of the intervals include μ, but any one interval may or may not include μ.

EXAMPLE 4.2

Refer to the hospital audit of Example 4.1 and suppose that a random sample of $n = 200$ accounts is selected from the total of $N = 1000$. The sample

TABLE 4.2 Confidence intervals for $N = 100$ and $n = 20$

$\bar{y}$	s^2	LCL	UCL	$\mu = 52.575$
56.020	1047.629	43.332	68.708	
53.650	973.679	41.418	65.882	
60.052	1044.769	47.381	72.722	
49.350	606.324	39.697	59.002	
49.082	994.433	36.721	61.444	
49.038	1058.878	36.282	61.794	
42.857	937.009	30.858	54.856	
46.682	901.619	34.911	58.453	
42.694	677.978	32.487	52.901	
52.922	1086.781	39.999	65.844	
47.778	926.727	35.845	59.712	
48.950	705.443	38.539	59.362	
52.200	1227.258	38.467	65.933	
50.395	714.205	39.919	60.871	
54.384	845.914	42.982	65.785	
49.296	968.221	37.099	61.494	
50.167	957.080	38.040	62.295	
50.082	948.243	38.010	62.153	
58.146	840.061	46.785	69.508	
51.010	1144.449	37.749	64.271	
54.947	1021.469	42.418	67.476	
51.596	907.564	39.787	63.405	
60.053	612.693	50.350	69.756	
61.360	730.304	50.767	71.954	
37.612	642.730	27.674	47.550	
45.641	788.646	34.632	56.640	
47.266	678.076	37.059	57.474	
51.645	815.394	40.452	62.839	
48.601	760.584	37.790	59.412	
49.368	1003.110	36.953	61.784	
52.723	874.174	41.133	64.313	
43.005	622.081	33.228	52.782	
33.760	586.996	24.262	43.257	
57.683	656.446	47.639	67.726	
68.100	750.229	57.363	78.837	
59.298	695.199	48.962	69.634	
47.474	1021.986	34.942	60.006	
47.749	962.295	35.588	59.909	
50.098	785.590	39.111	61.085	
51.697	893.741	39.978	63.416	
45.989	731.062	35.390	56.588	
54.382	735.614	42.392	66.373	
56.294	898.002	44.547	68.041	
52.548	1333.015	38.236	66.860	
53.236	1147.398	39.958	66.514	
57.694	766.730	46.840	68.548	
63.771	860.750	52.270	75.271	
48.835	875.848	37.234	60.437	
66.375	645.377	56.416	76.333	
56.731	1070.385	43.906	69.556	

TABLE 4.3 Confidence intervals for $N = 20$ and $n = 15$

$\bar{y}$	s^2	LCL	UCL	$\mu = 9.035$
10.172	62.698	8.168	12.175	
10.312	53.296	8.465	12.160	
10.435	58.390	8.501	12.368	
8.198	15.953	6.340	10.057	
7.410	46.677	5.681	9.139	
10.455	49.063	8.682	12.227	
9.133	64.951	7.094	11.172	
9.255	51.192	7.445	11.066	
9.392	54.933	7.516	11.267	
10.386	59.018	8.442	12.330	
8.700	62.707	6.696	10.703	
8.126	46.516	6.401	9.852	
8.869	53.483	7.018	10.719	
9.074	58.610	7.137	11.012	
7.719	52.275	5.889	9.548	
7.815	49.723	6.031	9.600	
8.794	54.153	6.932	10.656	
8.778	53.764	6.923	10.634	
11.350	50.345	9.554	13.145	
8.205	55.182	6.326	10.085	
8.371	56.787	6.464	10.278	
9.556	57.116	7.644	11.468	
9.442	58.971	7.499	11.385	
7.865	53.681	6.011	9.719	
9.323	67.261	7.248	11.398	
9.300	48.507	7.538	11.062	
9.400	47.895	7.667	11.133	
9.738	56.042	7.844	11.632	
10.100	50.812	8.297	11.904	
9.540	58.265	7.608	11.471	
9.204	55.947	7.311	11.096	
9.057	55.205	7.177	10.937	
9.514	48.569	7.750	11.277	
10.533	50.279	8.738	12.327	
9.076	51.918	7.252	10.899	
8.159	52.775	6.321	9.997	
7.453	55.096	5.575	9.331	
8.323	58.580	6.386	10.260	
8.704	57.380	6.787	10.620	
9.146	57.078	7.234	11.057	
9.301	53.757	7.446	11.156	
8.908	59.789	6.951	10.864	
7.418	55.680	5.530	9.306	
7.335	49.494	5.555	9.115	
9.601	58.002	7.674	11.528	
8.175	55.629	6.288	10.063	
8.634	52.570	6.799	10.468	
9.200	55.837	7.310	11.091	
7.136	41.977	5.496	8.775	
8.032	50.013	6.243	9.822	

mean of the accounts is found to be $\bar{y} = \$94.22$, and the sample variance is $s^2 = 445.21$. Estimate μ, the average due for all 1000 hospital accounts, and place a bound on the error of estimation.

SOLUTION

We use $\bar{y} = \$94.22$ to estimate μ. A bound on the error of estimation can be found by using Equation (4.4).

$$2\sqrt{\hat{V}(\bar{y})} = 2\sqrt{\frac{s^2}{n}\left(\frac{N-n}{N}\right)} = 2\sqrt{\frac{445.21}{200}\left(\frac{1000-200}{1000}\right)}$$
$$= 2\sqrt{1.7808} = \$2.67$$

Thus we estimate the mean value per account, μ, to be $\bar{y} = \$94.22$. Since n is large, the sample mean should possess approximately a normal distribution, so that $\$94.22 \pm \$2.67$ is approximately a 95% confidence interval for the population mean.

EXAMPLE 4.3

A simple random sample of $n = 9$ hospital records is drawn to estimate the average amount of money due on $N = 484$ open accounts. The sample values for these nine records are listed in Table 4.4. Estimate μ, the average amount outstanding, and place a bound on your error of estimation.

TABLE 4.4 Amount of money owed

y_1	33.50
y_2	32.00
y_3	52.00
y_4	43.00
y_5	40.00
y_6	41.00
y_7	45.00
y_8	42.50
y_9	39.00

SOLUTION

Displaying the sample data and computations as indicated in Table 4.5 is convenient.

Summing the entries in the y column, we get

$$\sum_{i=1}^{9} y_i = 368.00$$

Using the y^2 column, we have

$$\sum_{i=1}^{9} y_i^2 = 15,332.50$$

TABLE 4.5 Data and computations for Example 4.3

y	y^2
33.50	1,122.25
32.00	1,024,00
52.00	2,704.00
43.00	1,849.00
40.00	1,600.00
41.00	1.681.00
45.00	2,025.00
42.50	1,806.25
39.00	1,521.00
$\sum y_i = 368.00$	$\sum y_i^2 = 15{,}332.50$

We need both of these quantities to calculate $\bar{y}$ and s^2. Our estimate of μ is

$$\bar{y} = \frac{\sum\limits_{i=1}^{9} y_i}{9} = \frac{368.00}{9} = \$40.89$$

To find a bound on the error of estimation, we must compute

$$s^2 = \frac{\sum\limits_{i=1}^{n} (y_i - \bar{y})^2}{n-1} = \frac{\sum\limits_{i=1}^{9} y_i^2 - \left(\sum\limits_{i=1}^{9} y_i\right)^2 \Big/ 9}{8}$$

$$= \frac{1}{8}\left[15{,}332.50 - \frac{(368)^2}{9}\right] = \frac{1}{8}[15{,}332.50 - 15{,}047.11]$$

$$= 35.67$$

Utilizing Equation (4.4), we obtain the bound on the error of estimation,

$$2\sqrt{\hat{V}(\bar{y})} = 2\sqrt{\frac{s^2}{n}\left(\frac{N-n}{N}\right)} = 2\sqrt{\frac{35.67}{9}\left(\frac{484-9}{484}\right)}$$

$$= 2\sqrt{3.890} = 3.944 = \$3.94$$

To summarize, the estimate of the mean amount of money owed per account, μ, is $\bar{y} = \$40.89$. Although we cannot be certain how close $\bar{y}$ is to μ, we are reasonably confident that the error of estimation is less than $3.94.

Many sample surveys are conducted to obtain information about a population total. The federal auditor of Example 4.1 would probably be interested in verifying the computer figure for the total accounts receivable (in dollars) for the $N = 1000$ open accounts.

You recall that the mean for a population of size N is the sum of all observations in the population divided by N. The population total—that is, the sum of all observations in the population—is denoted by the symbol τ. Hence

$$N\mu = \tau$$

Intuitively, we expect the estimator of τ to be N times the estimator of μ, which it is.

Estimator of the population total τ:

$$\hat{\tau} = N\bar{y} = \frac{N \sum\limits_{i=1}^{n} y_i}{n} \tag{4.5}$$

Estimated variance of $\hat{\tau}$:

$$\hat{V}(\hat{\tau}) = \hat{V}(N\bar{y}) = N^2 \left(\frac{s^2}{n}\right)\left(\frac{N-n}{N}\right) \tag{4.6}$$

where

$$s^2 = \frac{\sum\limits_{i=1}^{n} (y_i - \bar{y})^2}{n-1}$$

Bound on the error of estimation:

$$2\sqrt{\hat{V}(N\bar{y})} = 2\sqrt{N^2\left(\frac{s^2}{n}\right)\left(\frac{N-n}{N}\right)} \tag{4.7}$$

Note that the estimated variance of $\hat{\tau} = N\bar{y}$ in Equation (4.6) is N^2 times the estimated variance of $\bar{y}$ given in Equation (4.3).

EXAMPLE 4.4

An industrial firm is concerned about the time per week spent by scientists on certain trivial tasks. The time log sheets of a simple random sample of $n = 50$ employees show the average amount of time spent on these tasks is 10.31 hours with a sample variance $s^2 = 2.25$. The company employs $N = 750$ scientists. Estimate the total number of man-hours lost per week on trivial tasks, and place a bound on the error of estimation.

SOLUTION

We know the population consists of $N = 750$ employees from which a random sample of $n = 50$ time log sheets was obtained. The average amount of time lost for the 50 employees was $\bar{y} = 10.31$ hours per week. Therefore the estimate of τ is

$$\hat{\tau} = N\bar{y} = 750(10.31) = 7732.5 \text{ hours}$$

To place a bound on the error of estimation, we apply Equation (4.7) to obtain

$$2\sqrt{\hat{V}(\hat{\tau})} = 2\sqrt{(750)^2\left(\frac{2.25}{50}\right)\left(\frac{750-50}{750}\right)}$$

$$= 2\sqrt{23,625} = 307.4 \text{ hours}$$

Thus the estimate of total time lost is $\hat{\tau} = 7732.5$ hours. We are reasonably confident that the error of estimation is less than 307.4 hours.

4.4
SELECTING THE SAMPLE SIZE FOR ESTIMATING POPULATION MEANS AND TOTALS

At some point in the design of the survey, someone must make a decision about the size of the sample to be selected from the population. So far we have discussed a sampling procedure (simple random sampling) but have said nothing about the number of observations to be included in the sample. The implications of such a decision are obvious. Observations cost money. Hence if the sample is too large, time and talent are wasted. Conversely, if the number of observations included in the sample is too small, we have bought inadequate information for the time and effort expended and have again been wasteful.

The number of observations needed to estimate a population mean μ with a bound on the error of estimation of magnitude B is found by setting two standard deviations of the estimator, $\bar{y}$, equal to B and solving this expression for n. That is, we must solve

$$2\sqrt{V(\bar{y})} = B \tag{4.8}$$

for n.

You will recall that the estimated variance of $\bar{y}$, $\hat{V}(\bar{y})$, is given by

$$\hat{V}(\bar{y}) = \frac{s^2}{n}\left(\frac{N-n}{N}\right) \tag{4.9}$$

Also,

$$V(\bar{y}) = \frac{\sigma^2}{n}\left(\frac{N-n}{N-1}\right) \tag{4.10}$$

You will recognize Equation (4.10) from an introductory course as the familiar variance of $\bar{y}$, that is, σ^2/n, multiplied by the factor

$$(N-n)/(N-1)$$

The required sample size can now be found by solving the following equation for n:

$$2\sqrt{V(\bar{y})} = 2\sqrt{\frac{\sigma^2}{n}\left(\frac{N-n}{N-1}\right)} = B \tag{4.11}$$

The solution is given in Equation (4.12).

Sample size required to estimate μ with a bound on the error of estimation B:

$$n = \frac{N\sigma^2}{(N-1)D + \sigma^2} \tag{4.12}$$

where

$$D = \frac{B^2}{4}$$

Solving for n in a practical situation presents a problem because the population variance σ^2 is unknown. Since a sample variance s^2 is frequently available from prior experimentation, we can obtain an approximate sample size by replacing σ^2 with s^2 in Equation (4.12). We will illustrate a method for guessing a value of σ^2 when very little prior information is available. If N is large, as it usually is, the $(N-1)$ can be replaced by N in the denominator of Equation (4.12).

EXAMPLE 4.5

The average amount of money μ for a hospital's accounts receivable must be estimated. Although no prior data is available to estimate the population variance σ^2, that most accounts lie within a $100 range is known. There are $N = 1000$ open accounts. Find the sample size needed to estimate μ with a bound on the error of estimation $B = \$3$.

SOLUTION

We need an estimate of σ^2, the population variance. Since the range is often approximately equal to four standard deviations (4σ), one-fourth of the range will provide an approximate value of σ. Hence

$$\sigma \approx \frac{\text{range}}{4} = \frac{100}{4} = 25$$

and

$$\sigma^2 \approx (25)^2 = 625$$

Using Equation (4.12), we obtain

$$n = \frac{N\sigma^2}{(N-1)D + \sigma^2}$$

where
$$D = \frac{B^2}{4} = \frac{(3)^2}{4} = 2.25$$

So
$$n = \frac{1000(625)}{999(2.25) + 625} = 217.56$$

That is, we need approximately 218 observations to estimate μ, the mean accounts receivable, with a bound on the error of estimation of $3.00.

In like manner, we can determine the number of observations needed to estimate a population total τ, with a bound on the error of estimation of magnitude B. The required sample size is found by setting two standard deviations of the estimator equal to B and solving this expression for n. That is, we must solve

$$2\sqrt{V(N\bar{y})} = B$$

or, equivalently,

$$2N\sqrt{V(\bar{y})} = B \tag{4.13}$$

[The reason for this equivalence is given directly after Equation (4.7).]

Sample size required to estimate τ with a bound on error B:

$$n = \frac{N\sigma^2}{(N-1)D + \sigma^2} \tag{4.14}$$

where
$$D = \frac{B^2}{4N^2}$$

EXAMPLE 4.6

An investigator is interested in estimating the total weight gain in 0 to 4 weeks for $N = 1000$ chicks fed on a new ration. Obviously, to weigh each bird would be time-consuming and tedious. Therefore, determine the number of chicks to be sampled in this study in order to estimate τ with a bound on the error of estimation equal to 1000 grams. Many similar studies on chick nutrition have been run in the past. Using data from these studies, the investigator found that σ^2, the population variance, was approximately equal to 36.00 (grams)^2. Determine the required sample size.

SOLUTION

We can obtain an approximate sample size using Equation (4.14) with σ^2 equal to 36.00 and

$$D = \frac{B^2}{4N^2} = \frac{(1000)^2}{4(1000)^2} = .25$$

That is,

$$n = \frac{N\sigma^2}{(N-1)D + \sigma^2} = \frac{1000(36.00)}{999(0.25) + 36.00} = 125.98$$

The investigator, therefore, needs to weight $n = 126$ chicks to estimate τ, the total weight gain for $N = 1000$ chickens in 0 to 4 weeks, with a bound on the error of estimation equal to 1000 grams.

4.5
ESTIMATION OF A POPULATION PROPORTION

The investigator conducting a sample survey is frequently interested in estimating the proportion of the population that possesses a specified characteristic. For example, a congressional leader investigating the merits of an 18-year-old voting age may want to estimate the proportion of the potential voters in the district between the ages of 18 and 21. A marketing research group may be interested in the proportion of the total sales market in diet preparations that is attributable to a particular product. That is, what percentage of sales is accounted for by a particular product? A forest manager may be interested in the proportion of trees with a diameter of 12 inches or more. Television ratings are often determined by estimating the proportion of the viewing public that watches a particular program.

You will recognize that all these examples exhibit a characteristic of the binomial experiment, that is, an observation either does belong or does not belong to the category of interest. For example, one can estimate the proportion of eligible voters in a particular district by examining population census data for several of the precincts within the district. An estimate of the proportion of voters between 18 and 21 years of age for the entire district will be the fraction of potential voters from the precincts sampled that fell into this age range.

In subsequent discussion we denote the population proportion and its estimator by the symbols p and $\hat{p}$, respectively. The properties of $\hat{p}$ for simple random sampling parallel those of the sample mean $\bar{y}$ if the response measurements are defined as follows: Let $y_i = 0$ if the ith element sampled does not possess the specified characteristic and $y_i = 1$ if it does. Then the total number of elements in a sample of size n possessing a specified characteristic is

$$\sum_{i=1}^{n} y_i$$

If we draw a simple random sample of size n, the sample proportion $\hat{p}$ is the fraction of the elements in the sample that possess the characteristic of interest. For example, the estimate $\hat{p}$ of the proportion of eligible voters between the ages of 18 and 21 in a certain district is

$$\hat{p} = \frac{\text{number of voters sampled between the ages of 18 and 21}}{\text{number of voters sampled}}$$

or $\qquad \hat{p} = \dfrac{\displaystyle\sum_{i=1}^{n} y_i}{n} = \bar{y}$

In other words, $\hat{p}$ is the average of the 0 and 1 values from the sample. Similarly, we can think of the population proportion as the average of the 0 and 1 values for the entire population (that is, $p = \mu$).

Estimator of the population proportion p:

$$\hat{p} = \bar{y} = \frac{\displaystyle\sum_{i=1}^{n} y_i}{n} \tag{4.15}$$

Estimated variance of $\hat{p}$:

$$\hat{V}(\hat{p}) = \frac{\hat{p}\hat{q}}{n-1}\left(\frac{N-n}{N}\right) \tag{4.16}$$

where $\qquad \hat{q} = 1 - \hat{p}$

Bound on the error of estimation:

$$2\sqrt{\hat{V}(\hat{p})} = 2\sqrt{\frac{\hat{p}\hat{q}}{n-1}\left(\frac{N-n}{N}\right)} \tag{4.17}$$

EXAMPLE 4.7

A simple random sample of $n = 100$ college seniors was selected to estimate (1) the fraction of $N = 300$ seniors going on to graduate school and (2) the fraction of students that have held part-time jobs during college. Let y_i and x_i ($i = 1, 2, \ldots, 100$) denote the responses of the ith student sampled. We will set $y_i = 0$ if the ith student does not plan to attend graduate school and $y_i = 1$ if he does. Similarly, let $x_i = 0$ if he has not held a part-time job sometime during college and $x_i = 1$ if he has. Using the sample data presented in the accompanying table, estimate p_1, the proportion of seniors planning to attend graduate school, and p_2, the proportion of seniors who have had a part-time job sometime during their college careers (summers included).

Student	y	x
1	1	0
2	0	1
3	0	1
4	1	1
5	0	0
6	0	0
7	0	1
⋮	⋮	⋮
96	0	1
97	1	0
98	0	1
99	0	1
100	1	1

$$\sum_{i=1}^{100} y_i = 15 \qquad \sum_{i=1}^{100} x_i = 65$$

SOLUTION

The sample proportions from Equation (4.15) are given by

$$\hat{p}_1 = \frac{\sum\limits_{i=1}^{n} y_i}{n} = \frac{15}{100} = .15$$

and

$$\hat{p}_2 = \frac{\sum\limits_{i=1}^{n} x_i}{n} = \frac{65}{100} = .65$$

The bounds on the errors of estimation of p_1 and p_2 are, respectively,

$$2\sqrt{\hat{V}(\hat{p}_1)} = 2\sqrt{\frac{\hat{p}_1 \hat{q}_1}{n-1}\left(\frac{N-n}{N}\right)}$$

$$= 2\sqrt{\frac{(.15)(.85)}{99}\left(\frac{300-100}{300}\right)} = 2(0.0293) = .059$$

and

$$2\sqrt{\hat{V}(\hat{p}_2)} = 2\sqrt{\frac{\hat{p}_2 \hat{q}_2}{n-1}\left(\frac{N-n}{N}\right)}$$

$$= 2\sqrt{\frac{(.65)(.35)}{99}\left(\frac{300-100}{300}\right)} = 2(.0391) = .078$$

Thus we estimate that .15 (15%) of the seniors plan to attend graduate school, with a bound on the error of estimation equal to .059 (5.9%). We estimate that .65 (65%) of the seniors have held a part-time job during college, with a bound on the error of estimation equal to .078 (7.8%).

We have shown that the population proportion p can be regarded as the average (μ) of the 0 and 1 values for the entire population. Hence the problem of determining the sample size required to estimate p to within B units should be analogous to determining a sample size for estimating μ with a bound on the error of estimation B. You will recall that the required sample size for estimating μ is given by

$$n = \frac{N\sigma^2}{(N-1)D + \sigma^2} \qquad (4.18)$$

where $D = B^2/4$ [see Equation (4.12)]. The corresponding sample size needed to estimate p can be found by replacing σ^2 in Equation (4.18) with the quantity pq.

Sample size required to estimate p with a bound on the error of estimation B:

$$n = \frac{Npq}{(N-1)D + pq} \qquad (4.19)$$

where $\qquad q = 1 - p \qquad$ and $\qquad D = \dfrac{B^2}{4}$

In a practical situation we do not know p. An approximate sample size can be found by replacing p with an estimated value. Frequently, such an estimate can be obtained from similar past surveys. However, if no such prior information is available, we can substitute $p = .5$ into Equation (4.19) to obtain a conservative sample size (one that is likely to be larger than required).

EXAMPLE 4.8

Student government leaders at a college want to conduct a survey to determine the proportion of students that favors a proposed honor code. Since interviewing $N = 2000$ students in a reasonable length of time is almost impossible, determine the sample size (number of students to be interviewed) needed to estimate p with a bound on the error of estimation of magnitude $B = .05$. Assume that no prior information is available to estimate p.

SOLUTION

We can approximate the required sample sizes when no prior information is available by setting $p = .5$ in Equation (4.19). We have

$$D = \frac{B^2}{4} = \frac{(.05)^2}{4} = .000625$$

Hence
$$n = \frac{Npq}{(N-1)D + pq}$$

$$= \frac{(2000)(.5)(.5)}{(1999)(.000625) + (.5)(.5)} = \frac{500}{1.499}$$

$$= 333.56$$

That is, 334 students must be interviewed to estimate the proportion of students that favors the proposed honor code with a bound on the error of estimation of $B = .05$.

EXAMPLE 4.9

Referring to Example 4.8, suppose that in addition to estimating the proportion of students that favors the proposed honor code, student government leaders also want to estimate the number of students who feel the student union building adequately serves their needs. Determine the combined sample size required for a survey to estimate p_1, the proportion that favors the proposed honor code, and p_2, the proportion that believes the student union adequately serves its needs, with bounds on the errors of estimation of magnitude $B_1 = .05$ and $B_2 = .07$. Although no prior information is available to estimate p_2, approximately 60% of the students believed the union adequately met their needs in a similar survey run the previous year.

SOLUTION

In this example we must determine a sample size n that will allow us to estimate p_1 with a bound $B_1 = .05$ and p_2 with a bound $B_2 = .07$. First, we determine the sample sizes that satisfy each objective separately. The larger of the two will then be the combined sample size for a survey to meet both objectives. From Example 4.8 the sample size required to estimate p_1 with a bound on the error of estimation of $B_1 = .05$ was $n = 334$ students. We can use data from the survey of the previous year to determine the sample size needed to estimate p_2. We have

$$D = \frac{B^2}{4} = \frac{(.07)^2}{4} = .001225$$

and hence, with $p_2 = .60$,

$$n = \frac{Npq}{(N-1)D + pq}$$

$$= \frac{(2000)(.6)(.4)}{(1999)(.001225) + (.6)(.4)} = \frac{480}{2.68877}$$

$$= 178.52$$

That is, 179 students must be interviewed to estimate p_2, the proportion of the $N = 2000$ students that believes the student union meets its needs, with a bound on the error of estimation equal to .07.

The sample size required to achieve both objectives in one survey is 334, the larger of the two sample sizes.

4.6
SAMPLING WITH PROBABILITIES PROPORTIONAL TO SIZE

Previous work in this chapter has depended on the sample being a simple random sample, according to Definition 4.1. We will now show that varying the probabilities with which different sampling units are selected is sometimes advantageous. Suppose, for example, we wish to estimate the number of job openings in a city by sampling industrial firms within the city. Typically, many such firms will be quite small and employ few workers, while some firms will be very large. In a simple random sample, size of firm is not taken into account, and a typical sample will contain mostly small firms. But the information desired (number of job openings) is heavily influenced by the large firms. Thus we should be able to improve on the simple random sample by giving the large firms a greater chance to appear in the sample. A method for accomplishing this sampling is called *sampling with probabilities proportional to size*, or *pps sampling*.

For a sample $y_1, y_2, \ldots, y_n$ from a population of size N, let

$$\pi_i = \text{probability that } y_i \text{ appears in the sample}$$

Unbiased estimators of τ and μ, along with their estimated variances and bounds on the error of estimation, are as follows:

Estimator of the population total τ:

$$\hat{\tau}_{\text{pps}} = \frac{1}{n} \sum_{i=1}^{n} \left(\frac{y_i}{\pi_i} \right) \tag{4.20}$$

Estimated variance of $\hat{\tau}_{\text{pps}}$:

$$\hat{V}(\hat{\tau}_{\text{pps}}) = \frac{1}{n(n-1)} \sum_{i=1}^{n} \left(\frac{y_i}{\pi_i} - \hat{\tau}_{\text{pps}} \right)^2 \tag{4.21}$$

Bound on the error of estimation:

$$2\sqrt{\hat{V}(\hat{\tau}_{\text{pps}})} = 2\sqrt{\frac{1}{n(n-1)} \sum_{i=1}^{n} \left(\frac{y_i}{\pi_i} - \hat{\tau}_{\text{pps}} \right)^2} \tag{4.22}$$

Estimator of the population mean μ:

$$\hat{\mu}_{pps} = \frac{1}{N} \hat{\tau}_{pps} = \frac{1}{Nn} \sum_{i=1}^{n} \left(\frac{y_i}{\pi_i} \right) \tag{4.23}$$

Estimated variance of $\hat{\mu}_{pps}$:

$$\hat{V}(\hat{\mu}_{pps}) = \frac{1}{N^2 n(n-1)} \sum_{i=1}^{n} \left(\frac{y_i}{\pi_i} - \hat{\tau}_{pps} \right)^2 \tag{4.24}$$

Bound on the error of estimation:

$$2\sqrt{\hat{V}(\hat{\mu}_{pps})} = 2\sqrt{\frac{1}{N^2 n(n-1)} \sum_{i=1}^{n} \left(\frac{y_i}{\pi_i} - \hat{\tau}_{pps} \right)^2} \tag{4.25}$$

The estimators $\hat{\tau}_{pps}$ and $\hat{\mu}_{pps}$ are unbiased for any choices of π_i, but it is clearly in the best interest of the experimenter to choose these π_i's so that the variances of the estimators are as small as possible. How should this choice be made? Suppose, for the moment, that the value of y_i is known for each of the N units in the population. Thus the population total τ is also known. Under these conditions we can select each unit for the sample with probability proportional to its actual measured value y_i, assuming all measurements are positive. That is, we can make $\pi_i = y_i/\tau$.

With $\pi_i = y_i/\tau$ for each sampled item, $\hat{\tau}_{pps}$ becomes

$$\hat{\tau}_{pps} = \frac{1}{n} \sum_{i=1}^{n} \frac{y_i}{\pi_i} = \frac{1}{n} \sum_{i=1}^{n} \frac{y_i}{(y_i/\tau)}$$

$$= \frac{1}{n} \sum_{i=1}^{n} \tau = \frac{1}{n}(n\tau) = \tau$$

Thus every $\hat{\tau}_{pps}$ estimates τ exactly. In addition,

$$\hat{V}(\hat{\tau}_{pps}) = \frac{1}{n(n-1)} \sum_{i=1}^{n} \left(\frac{y_i}{\pi_i} - \hat{\tau}_{pps} \right)^2$$

$$= \frac{1}{n(n-1)} \sum_{i=1}^{n} \left(\frac{y_i}{y_i/\tau} - \tau \right)^2$$

$$= \frac{1}{n(n-1)} \sum_{i=1}^{n} (\tau - \tau)^2 = 0$$

which again shows that we have a precise estimator.

Now to know the values y_i for every unit in the population before sampling is impossible. (If they were known, no sampling would be necessary.) Hence the choice $\pi_i = y_i/\tau$ is not possible, but it does provide a criterion for selecting π_i's that can be used in sampling. The best practical way to choose the π_i's is to choose them proportional to a known measurement that is highly correlated with y_i. In the problem of estimating total number of job openings, firms can be sampled with probabilities proportional to their total work force, which should be known fairly accurately before the sample is selected. The number of job openings per firm is not known before sampling, but it should be highly correlated with the total number of workers in the firm. We will see a very useful application of pps sampling in Chapter 8.

　　To show how pps sampling works numerically, we will return to the population of $N = 4$ elements, $\{1, 2, 3, 4\}$. Recall that for simple random samples of size $n = 2$, $E(\bar{y}) = 2.5$ and $V(\bar{y}) = \frac{5}{12} = .417$. Suppose that we decide to sample $n = 2$ elements with varying probabilities and choose $\pi_1 = .1$, $\pi_2 = .1$, $\pi_3 = .4$, and $\pi_4 = .4$. To accomplish this sampling, we can choose a random digit from the random number table and take our first sampled element to be

　　1　if the random digit is 0,

　　2　if the random digit is 1,

　　3　if the random digit is 2, 3, 4, or 5,

　　4　if the random digit is 6, 7, 8, or 9.

The process is then repeated for the second sampled element. (Note that the same element can be selected twice, which is not possible in simple random sampling.) This choice for π_1, π_2, π_3, and π_4 gives the smaller values in the population, (1, 2), the smaller chance of getting selected in the sample, and the larger values, (3, 4), have an appreciably larger chance of getting selected. Note that these probabilities are not exactly proportional to size, but they do tend in that direction.

　　Table 4.6 contains a listing of the ten possible samples, the probability of obtaining each sample, and the $\hat{\tau}_{pps}$ estimate produced from each sample. The sample $\{1, 2\}$ will result if our first random digit is a 0 and our second

TABLE 4.6 Sampling with varying probabilities (samples of size $n = 2$ from $\{1, 2, 3, 4\}$ with $\pi_1 = \pi_2 = .1$, $\pi_3 = \pi_4 = .4$)

Sample	Probability of Obtaining Sample	$\hat{\tau}_{pps}$
$\{1, 2\}$	.02	15
$\{1, 3\}$	.08	$\frac{35}{4}$
$\{1, 4\}$	.08	10
$\{2, 3\}$	.08	$\frac{55}{4}$
$\{2, 4\}$	.08	15
$\{3, 4\}$	.32	$\frac{35}{4}$
$\{1, 1\}$	.01	10
$\{2, 2\}$	.01	20
$\{3, 3\}$	.16	$\frac{15}{2}$
$\{4, 4\}$	.16	10
	1.00	

random digit is a 1. The $\hat{\tau}_{\text{pps}}$ value then comes from Equation (4.20):

$$\hat{\tau}_{\text{pps}} = \frac{1}{n} \sum_{i=1}^{n} \left(\frac{y_i}{\pi_i} \right) = \frac{1}{2} \left(\frac{1}{.1} + \frac{2}{.1} \right)$$
$$= \tfrac{1}{2}(10 + 20) = 15$$

From Table 4.6 we see that

$$E(\hat{\tau}_{\text{pps}}) = 15(.02) + \tfrac{35}{4}(.08) + \cdots + 10(.16) = 10$$

And so for this sample $\hat{\tau}_{\text{pps}}$ is demonstrated to be an unbiased estimator of τ. Also,

$$V(\hat{\tau}_{\text{pps}}) = (15 - 10)^2(.02) + (\tfrac{35}{4} - 10)^2(.08) + \cdots + (10 - 10)^2(.16)$$
$$= 6.250$$

If we use simple random sampling with $n = 2$, our estimate $N\bar{y}$ of τ will have a variance computed as follows:

$$V(N\bar{y}) = N^2 V(\bar{y}) = (4)^2(.417) = 6.672$$

which is larger than $V(\hat{\tau}_{\text{pps}})$. The proportional reduction in variance by using a pps estimator would be greater if the population measurements had more variability and if the π_i's were even closer to being truly proportional to the size of the population measurements.

Sampling with varying probabilities will lower the variance of an estimator, thus allowing for more precise estimates, if the probabilities are proportional, or approximately proportional, to the size of the sampled measurements. If, however, the probabilities are improperly chosen, then $\hat{\tau}_{\text{pps}}$ and $\hat{\mu}_{\text{pps}}$ can have *larger* variance than the corresponding $\hat{\tau}$ and $\hat{\mu}$ from simple random sampling. Suppose, for example, in samples of size $n = 2$ from $\{1, 2, 3, 4\}$, we choose $\pi_1 = \pi_2 = .4$ and $\pi_3 = \pi_4 = .1$. Then $\hat{\tau}_{\text{pps}}$ is still an unbiased estimator of τ, but $V(\hat{\tau}_{\text{pps}}) = 81.25$. In this case $\hat{\tau}_{\text{pps}}$ is a much poorer estimator than $\hat{\tau}$ from simple random sampling.

In summary, pps sampling involves sampling *with* replacement, which means that a sampled item is not removed from the population after it is selected for the sample. Thus a particular sampling unit can be selected more than once. A repeated selection is usually undesirable, but it will not happen often if n/N is small. Moreover, this undesirable feature is often more than offset by the reduction in variance that can occur. The pps estimators of τ and μ only produce smaller variances if the probabilities π_i are proportional, or approximately proportional, to the size of the y_i's under investigation.

We illustrate the practical use of pps estimators in Example 4.10.

EXAMPLE 4.10

An investigator wishes to estimate the average number of defects per board on boards of electronic components manufactured for installation in computers. The boards contain varying numbers of components, and the investigator feels that the number of defects should be positively correlated with the number of components on a board. Thus pps sampling is used, with the probability of selecting any one board for the sample being proportional to the number of components on that board. A sample of $n = 4$ boards is to be selected from the $N = 10$ boards of one day's production. The number of components on the 10 boards are, respectively,

$$10, 12, 22, 8, 16, 24, 9, 10, 8, 31$$

Show how to select $n = 4$ boards with probabilities proportional to size.

SOLUTION

We list the number of components (our measure of size) in a column and list the *cumulative ranges* and desirable π_i's in adjacent columns, as follows:

Board	Number of Components	Cumulative Range	π_i
1	10	1–10	10/150
2	12	11–22	12/150
3	22	23–44	22/150
4	8	45–52	8/150
5	16	53–68	16/150
6	24	69–92	24/150
7	9	93–101	9/150
8	10	102–111	10/150
9	8	112–119	8/150
10	31	120–150	31/150

There are 150 components in the population to be sampled. We can think of these components as being numbered from 1 to 150. The cumulative range column keeps track of the interval of numbered components on each board. Board number 1 has the first 10 components, board number 2 has components 11 through 22, and so on.

The π's are simply the number of components per board divided by the total number of components. The boards having greater numbers of components have larger probabilities of selection.

To choose the sample of $n = 4$ boards, we enter the random number table and select four random numbers between 1 and 150. The numbers we selected were 14, 56, 94, and 25. We locate these numbers in the cumulative range column. The boards corresponding to those range intervals constitute the sample.

Since 14 lies in the range of board 2, that board enters the sample. Similarly, 56 lies in the range of board 5, 94 lies in the range of board 7, and 25 lies in the range of board 3. Thus the sample consists of boards 2, 3, 5, and 7. These boards have been selected with probabilities proportional to their numbers of components. Note that with this method we could have sampled a particular board more than once.

EXAMPLE 4.11

After the sampling of Example 4.10 was completed, the number of defects found on boards 2, 3, 5, and 7 were, respectively, 1, 3, 2, and 1. Estimate the average number of defects per board, and place a bound on the error of estimation.

SOLUTION

From Equation (4.23) the mean μ is estimated by

$$\hat{\mu}_{pps} = \frac{1}{Nn} \sum_{i=1}^{n} \left(\frac{y_i}{\pi_i} \right)$$

where $n = 4$, $N = 10$,

$$y_1 = 1, \qquad y_2 = 3, \qquad y_3 = 2, \qquad y_4 = 1$$

and $\qquad \pi_1 = \dfrac{12}{150}, \qquad \pi_2 = \dfrac{22}{150}, \qquad \pi_3 = \dfrac{16}{150}, \qquad \pi_4 = \dfrac{9}{150}$

Hence

$$\hat{\mu}_{pps} = \frac{1}{10(4)} \left[1 \left(\frac{150}{12} \right) + 3 \left(\frac{150}{22} \right) + 2 \left(\frac{150}{16} \right) + 1 \left(\frac{150}{9} \right) \right]$$

$$= \tfrac{1}{40}(68.37) = 1.71$$

Also, from (4.24)

$$\hat{V}(\hat{\mu}_{pps}) = \frac{1}{N^2 n(n-1)} \sum_{i=1}^{n} \left(\frac{y_i}{\pi_i} - \hat{\tau}_{pps} \right)^2$$

$$= \frac{1}{(10)^2(4)(3)} \left\{ \left(\frac{150}{12} - 17.10 \right)^2 + \left[\frac{3(150)}{22} - 17.10 \right]^2 \right.$$

$$\left. + \left[\frac{2(150)}{16} - 17.10 \right]^2 + \left(\frac{150}{9} - 17.10 \right)^2 \right\}$$

$$= .0295$$

and

$$2\sqrt{\hat{V}(\hat{\mu}_{pps})} = .34$$

The estimate of the average number of defects per board, with a bound on the error of estimation, is then

$$1.71 \pm .34$$

The interval (1.37, 2.05) provides an approximate 95% confidence interval for the average number of defects per board.

4.7
SUMMARY

The objective of statistics is to make inferences about a population from information contained in a sample. Two factors affect the quantity of information in a given investigation. The first is the sample size. The larger the sample size, the more information we expect to obtain about the population. The second factor that affects the quantity of information is the amount of variation in the data. Variation can be controlled by the design of the sample survey, that is, the method by which observations are obtained.

In this chapter we discussed the simplest type of sample survey design, namely, simple random sampling. This design does not attempt to reduce the effect of data variation on the error of estimation. A simple random sample of size n occurs if each sample of n elements from the population has the same chance of being selected. Random number tables are quite useful in determining the elements that are to be included in a simple random sample.

In estimating a population mean μ and total τ, we use the sample mean $\bar{y}$ and sample total $N\bar{y}$, respectively. Both estimators are unbiased; that is, $E(\bar{y}) = \mu$ and $E(N\bar{y}) = \tau$. The estimated variance and the bound on the error of estimation are given for both estimators.

Sometime during the design of an actual survey, the experimenter must decide how much information is desired, that is, how large a bound on the error of estimation can be tolerated. Sample size requirements were presented for estimating μ and τ with a specified bound on the error of estimation.

The third parameter estimated was the population proportion p. The properties of $\hat{p}$ were presented and related to the properties of $\bar{y}$, the estimator of the population mean μ. Selecting the sample size to estimate p with a specified bound on the error of estimation was based on the same principle employed in selecting a sample size for estimating μ and τ.

If the population measurements vary considerably in size, and if some approximate measure of this size is available before sampling, then sampling with probabilities proportional to size may be advantageous. This method produces unbiased estimates of μ and τ, which may have much smaller variances than those produced through simple random sampling.

CASE STUDY REVISITED

THE IRS POLL

IN the opinion poll outlined at the beginning of this chapter, 54% of 2207 people surveyed said that they considered the IRS forms too complicated. An estimate of the true population proportion that consider the forms too complicated is found by using

$$\hat{p} \pm 2\sqrt{\frac{\hat{p}\hat{q}}{n-1}}$$

assuming the population size N to be very large. With $\hat{p} = .54$, the observed sample proportion, we have

$$.54 \pm 2\sqrt{\frac{(.54)(.46)}{2206}}$$

$$.54 \pm .02$$

or   .52   to   .56

as our estimate of the true population proportion. Thus we can be confident that most taxpayers do, indeed, find the IRS forms too complicated.

EXERCISES

4.1   List all possible simple random samples of size $n = 2$ that can be selected from the population $\{0, 1, 2, 3, 4\}$. Calculate σ^2 for the population and $V(\bar{y})$ for the sample mean $\bar{y}$. Thus, show by direct calculation that

$$V(\bar{y}) = \frac{N - n}{N - 1}\left(\frac{\sigma^2}{n}\right)$$

4.2   For the simple random samples generated in Exercise 1, calculate s^2 for each sample. Show numerically that

$$E(s^2) = \frac{N}{N - 1}\sigma^2$$

4.3   Suppose you were to estimate the number of weed clusters of a certain type in a field. What is the population, and what would you use for sampling units? How would you construct a frame? How would you select a simple random sample? If a sampling unit is an area, such as a square yard, does the size chosen for a sampling unit affect the accuracy of the results? What considerations would go into your choice of size of sampling unit?

4.4   The data set in the Appendix (Table 3) lists resident population figures per state from the 1980 census. Select a simple random sample of five states. Use the 1980 population figures for the sampled states to estimate the total United States population, and place a bound on the error of estimation. Does your interval answer include the total population figure given in the table? Do you think every possible 95% confidence interval based on samples of size 5 would include the true total? Why?

4.5   State park officials were interested in the proportion of campers who consider the campsite spacing adequate in a particular campground. They decided to take a simple random sample of $n = 30$ from the first $N = 300$ camping parties that visit the campground. Let $y_i = 0$ if the head of the ith party sampled does not think the campsite spacing is adequate and $y_i = 1$ if he does ($i = 1, 2, \ldots, 30$). Use the data in the accompanying table to estimate p, the proportion of campers who consider the campsite spacing adequate. Place a bound on the error of estimation.

Camper Sampled	Response, y_i
1	1
2	0
3	1
⋮	⋮
29	1
30	1

$$\sum_{i=1}^{30} y_i = 25$$

4.6 Use the data in Exercise 4.5 to determine the sample size required to estimate p with a bound on the error of estimation of magnitude $B = .05$.

4.7 A simple random sample of 100 water meters within a community is monitored to estimate the average daily water consumption per household over a specified dry spell. The sample mean and sample variance are found to be $\bar{y} = 12.5$ and $s^2 = 1252$. If we assume that there are $N = 10{,}000$ households within the community, estimate μ, the true average daily consumption, and place a bound on the error of estimation.

4.8 Using data in Exercise 4.7, estimate the total number of gallons of water, τ, used daily during the dry spell. Place a bound on the error of estimation.

4.9 Resources managers of forest game lands are concerned about the size of the deer and rabbit populations during the winter months in a particular forest. As an estimate of population size, they propose using the average number of pellet groups for rabbits and deer per 30-foot-square plots. From an aerial photograph the forest was divided into $N = 10{,}000$ 30-foot-square grids. A simple random sample of $n = 500$ plots was taken, and the number of pellet groups was observed for rabbits and for deer. The results of this study are summarized in the accompanying table. Estimate μ_1 and μ_2, the average number of pellet groups for deer and rabbits, respectively, per 30-square-foot plots. Place bounds on the errors of estimation.

Deer	Rabbits
Sample mean = 2.30	Sample mean = 4.52
Sample variance = .65	Sample variance = .97

4.10 A simple random sample of $n = 40$ college students was interviewed to determine the proportion of students in favor of converting from the semester to the quarter system. Twenty-five of the students answered affirmatively. Estimate the proportion of students on campus in favor of the change. (Assume $N = 2000$.) Place a bound on the error of estimation.

4.11 A dentist was interested in the effectiveness of a new toothpaste. A group of $N = 1000$ schoolchildren participated in a study. Prestudy records showed there was an average of 2.2 cavities every six months for the group. After three months on the study the dentist sampled $n = 10$ children to determine how they were progressing on the new toothpaste. Using the data in the accompanying table, estimate the mean number of cavities for the entire group, and place a bound on the error of estimation.

Child	Number of Cavities in the Three–Month Period
1	0
2	4
3	2
4	3
5	2
6	0
7	3
8	4
9	1
10	1

4.12 The Fish and Game Department of a particular state was concerned about the direction of its future hunting programs. To provide for a greater potential for future hunting, the department wanted to determine the proportion of hunters seeking any type of game bird. A simple random sample of $n = 1000$ of the $N = 99,000$ licensed hunters was obtained. Suppose 430 indicated they hunted game birds. Estimate p, the proportion of licensed hunters seeking game birds. Place a bound on the error of estimation.

4.13 Using the data in Exercise 4.12, determine the sample size the department must obtain to estimate the proportion of game bird hunters, given a bound on the error of estimation of magnitude $B = .02$.

4.14 A company auditor was interested in estimating the total number of travel vouchers that were incorrectly filed. In a simple random sample of $n = 50$ vouchers taken from a group of $N = 250$, 20 were filed incorrectly. Estimate the total number of vouchers from the $N = 250$ that have been filed incorrectly, and place a bound on the error of estimation. [*Hint*: If p is the population proportion of incorrect vouchers, then Np is the total number of incorrect vouchers. An estimator of Np is $N\hat{p}$, which has an estimated variance given by $N^2\hat{V}(\hat{p})$.]

4.15 A psychologist wishes to estimate the average reaction time to a stimulus among 200 patients in a hospital specializing in nervous disorders. A simple random sample of $n = 20$ patients was selected, and their reaction times were measured, with the following results:

$$\bar{y} = 2.1 \text{ seconds} \qquad s = .4 \text{ second}$$

Estimate the population mean μ, and place a bound on the error of estimation.

4.16 In Exercise 4.15, how large a sample should be taken in order to estimate μ with a bound of 1 second on the error of estimation? Use 1.0 second as an approximation of the population standard deviation.

4.17 The manager of a machine shop wishes to estimate the average time an operator needs to complete a simple task. The shop has 98 operators. Eight operators are selected at random and timed. The observed results are shown in the accompanying table. Estimate the average time for completion of the task among all operators, and place a bound on the error of estimation.

Time (in minutes)	
4.2	5.3
5.1	4.6
7.9	5.1
3.8	4.1

4.18 A sociological study conducted in a small town calls for the estimation of the proportion of households that contain at least one member over 65 years of age. The city has 621 households according to the most recent city directory. A simple random sample of $n = 60$ households was selected from the directory. At the completion of the fieldwork, out of the 60 households sampled, 11 contained at least one member over 65 years of age. Estimate the true population proportion p, and place a bound on the error of estimation.

4.19 In Exercise 4.18, how large a sample should be taken in order to estimate p with a bound of .08 on the error of estimation? Assume the true proportion p is approximately .2

4.20 An investigator is interested in estimating the total number of "count trees" (trees larger than a specified size) on a plantation of $N = 1500$ acres. This information is used to determine the total volume of lumber for trees on the plantation. A simple random sample of $n = 100$ 1-acre plots was selected, and each plot was examined for the number of count trees. The sample average for the $n = 100$ 1-acre plots was $\bar{y} = 25.2$ with a sample variance of $s^2 = 136$. Estimate the total number of count trees on the plantation. Place a bound on the error of estimation.

4.21 Using the results of the survey described in Exercise 4.20, determine the sample size required to estimate τ, the total number of trees on the plantation, with a bound on the error of estimation of magnitude $B = 1500$.

4.22 A large construction firm has 120 houses in various stages of completion. For estimation of the total dollar amount to be listed as inventory of construction in progress, a simple random sample of 12 of these houses is selected and accumulated costs determined on each. Assume the following costs were obtained for the 12 sample houses:

35,500	30,200	28,900
36,400	29,800	34,100
32,600	26,400	38,000
38,200	32,200	27,500

Estimate the total accumulated costs for the 120 houses, and place a bound on the error of estimation.

4.23 From the data of Table 3 in the Appendix, select a simple random sample of $n = 10$ states. Estimate the proportion of states with 1977 per capita income below $5500, and place a bound on the error of estimation.

4.24 Results of a public opinion poll reported in a newsmagazine (*Time*, January 2, 1984) showed that 51% of the respondents completely agreed with the following statement:

> The Soviets are just as afraid of nuclear war as we are, and therefore it is in our mutal interest to find ways to negotiate.

The article states that "the findings are based on a telephone survey of 1000 registered voters. . . . The potential sampling error is plus or minus 3%." How was the 3%

calculated, and what is its interpretation? Can we conclude that a majority of registered voters completely agree with the statement?

4.25 The Florida poll of February–March 1984 (*Gainesville Sun*, April 1, 1984) interviewed 871 adults from around the state. On one question 53% of the respondents favored strong support of Israel. Would you conclude that a majority of adults in Florida favor strong support of Israel?

4.26 The results of a Louis Harris poll state that 36% of Americans list football as their favorite sport. A note then states: "In a sample of this size (1091 adults) one can say with 95% certainty that the results are within plus or minus three percentage points of what they would be if the entire adult population had been polled" (*Gainesville Sun*, May 7, 1961). Do you agree?

4.27 The A. C. Nielsen Company has electronic monitors hooked up to about 1200 of the 80 million American homes. The data from the monitors provide estimates of the proportion of homes tuned to a particular TV program. Nielsen offers the following defense of this sample size (D. Cody, "Polls and Pollsters," *Sky*, October 1982, p. 116):

> Mix together 70,000 white beans and 30,000 red beans and then scoop out a sample of 1000. The mathematical odds are that the number of red beans will be between 270 and 330, or 27 to 33% of the sample, which translates to a "rating" of 30, plus or minus three, with a 20 to 1 assurance of statistical reliability. The basic statistical law wouldn't change even if the sampling came from 80 million beans rather than 100,000.

Interpret and justify this statement in terms of the results of this chapter.

4.28 An opinion poll on education questioned 1684 adults across the United States (*Gainesville Sun*, July 4, 1983). For results dealing with proportions of respondents favoring certain issues, the poll is reported to have a 6% margin of error. Do you agree?

4.29 An auditor detects that a certain firm is regularly overstating the dollar amounts of inventories because of delays in recording withdrawals. The auditor wants to estimate

Item Number	Audited Amount	Recorded Amount	Overstatement (Difference)
1	175	210	35
2	295	305	10
3	68	91	23
4	74	82	8
5	128	140	12
6	241	250	9
7	362	384	22
8	72	80	8
9	59	82	23
10	112	140	28
11	118	124	6
12	210	230	20
13	240	260	20
14	123	247	24
15	96	108	12

the total *overstated* amount on 1000 listed items by obtaining exact (audited) inventory amounts on a random sample of 15 items, and comparing these exact figures with the recorded amounts. The data for the sampled items are as shown in the table on page 72 (all data in dollars). Estimate the total overstated amount on the 1000 types of items, and place a bound on the error of estimation. (Ignore the fpc.)

4.30 An auditor randomly samples 20 accounts receivable from the 500 accounts of a certain firm. The auditor lists the amount of each account and checks to see whether the underlying documents are in compliance with stated procedures. The data are as follows (amounts in dollars, Y = yes, N = no):

Account	Amount	Compliance	Account	Amount	Compliance
1	278	Y	11	188	N
2	192	Y	12	212	N
3	310	Y	13	92	Y
4	94	N	14	56	Y
5	86	Y	15	142	Y
6	335	Y	16	37	Y
7	310	N	17	186	N
8	290	Y	18	221	Y
9	221	Y	19	219	N
10	168	Y	20	305	Y

Estimate the total accounts receivable for the 500 accounts of the firm, and place a bound on the error of estimation. Do you think that the *average* account receivable for the firm exceeds $250? Why?

4.31 Refer to Exercise 4.30. From the data given on the compliance checks, estimate the proportion of the firm's accounts that fail to comply with stated procedures. Place a bound on the error of estimation. Do you think the proportion of accounts that comply with stated procedures exceeds 80%? Why?

4.32 Refer to Exercise 4.30. Suppose the 20 accounts given there are now a population from which $n = 5$ accounts are to be sampled for further investigation. (The holder of the account will be contacted to verify the amounts.) Select a sample of 5 accounts with probabilities proportional to the amounts.

4.33 A state agriculture department wants to measure total yield of tomatoes for a sample of fields, with the goal of estimating total tomato yield for the state. Discuss the merits of simple random sampling as compared with sampling with probabilities proportional to size.

4.34 Refer to the United States population figures given in Table 3 of the Appendix. From the nine northeastern states, select a sample of four states with probabilities proportional to their total population sizes in 1970. Is this procedure an appropriate sampling scheme for estimating total unemployment in the Northeast? Is this procedure an appropriate sampling scheme for estimating acres of forest land in the Northeast?

4.35 For the $n = 4$ states selected in Exercise 4.34, record their 1980 population sizes from Table 3 in the Appendix. Use these data to estimate the total 1980 population in the northeastern states, and place a bound on the error of estimation. Is the total given in the table included in your interval estimate? Do you think this method of sampling is better than selecting a simple random sample of four states for purposes of estimating total population? Why?

4.36 The accompanying table shows personal consumption expenditures in the United States for a selection of goods and services (in billions of dollars). Select a sample of three categories with probabilities proportional to 1981 expenditures. Use the 1982 data for the sampled categories to estimate total expenditures across the nine categories for 1982. Place a bound on the error of estimation.

Category	1981	1982
Motor vehicles	101.6	109.9
Furniture and household equipment	93.3	93.5
Food	375.9	396.9
Clothing	115.3	119.0
Gasoline and oil	94.6	91.5
Fuel oil and coal	20.7	20.0
Housing	302.0	334.1
Housing operation	128.4	144.3
Transportation	65.5	68.4

Source: *The World Almanac & Book of Facts*, 1984 edition, copyright © Newspaper Enterprise Association, Inc., 1983, New York, NY 10166.

4.37 A study to assess the attitudes of accountants toward advertising their services involved sending questionnaires to 200 accountants selected from a list of 1400 names. A total of 82 usable questionnaires were returned. The data summary for one question is as follows:

Likelihood of advertising in the future (%)

	All Respondents (82)	Those Having Advertised in the Past (46)
Virtual certainty	22	35
Very likely	4	5
Somewhat likely	19	35
About 50–50	18	15
Somewhat unlikely	6	10
Very unlikely	12	0
Absolutely not	15	0
No response	4	0

Source: K. Traynor, "Accountant Advertising: Perceptions, Attitudes and Behaviors," *Journal of Advertising Research*, vol. 23, no. 6, 1984. © Copyright 1984, by the Advertising Research Foundation.

(a) Estimate the population proportion virtually certain to advertise.
(b) Estimate the population proportion having *at least* a 50–50 chance of advertising.
(c) Among those who advertised in the past, estimate the population proportion somewhat unlikely to advertise again.

(d) Among those who advertised in the past, estimate the population proportion having *at least* a 50–50 chance of advertising again.

Place bounds on the errors of estimation in all cases. Do parts (c) and (d) require further assumptions over those made for parts (a) and (b)?

4.38 A marketing research firm estimates the proportion of potential customers preferring a certain brand of lipstick by "randomly" selecting 100 women who come by their booth in a shopping mall. Of the 100 sampled, 65 women stated a preference for brand A.

(a) How would you estimate the true proportion of women preferring brand A, with a bound on the error of estimation?

(b) What is the target population in this study?

(c) Did the marketing research firm select a simple random sample?

(d) What additional problems do you see with this type of sampling?

4.39 A legal case is being formulated by a union of secretaries who claim that they are being paid unfairly low wages by their employer. The 64 secretaries in the firm have an average annual salary of $18,300, with a standard deviation of $400. The average salary for all secretaries in the city in which this firm is located is $20,100. Can you support the claim of the secretaries by statistical arguments? If so, carefully state these arguments and the assumptions underlying them.

4.40 The Equal Opportunity Employment Commission accuses a firm of violating minority-hiring standards, since of its 120 employees only 30 are nonwhite. In the labor market area for that firm, 36% of available employees are reported to be nonwhite. Can you support the accusation of the EOEC on statistical grounds? State your arguments for or against the accusation with careful attention to assumptions.

EXPERIENCES WITH REAL DATA

4.1 Table 4.7 lists some of the final statistics for the 1982–1983 season of the National Basketball Association. Use these data to complete the following exercises.

(a) Select a simple random sample of $n = 5$ teams from the $N = 23$ teams listed. Use the points scored for these 5 sampled teams to estimate the average number of points scored per team and the total number of points scored in the season for all teams. Place bounds on the errors of estimation in each case. Does your interval estimate of total points scored include the true total?

(b) On graph paper, plot the number of points scored versus the number of wins for each of the 23 teams. Does there seem to be a strong positive correlation between these two measurements? Suppose a sample of $n = 5$ teams is selected with probabilities proportional to number of wins, and this sample is used to estimate the total number of points scored in the league. Do you think the variance of this estimator will be larger or smaller than the variance of the estimator of total points scored used in part (a)? Why?

(c) Plot the number of wins versus the number of rebounds for each of the 23 teams. Does there appear to be a positive correlation between these two measurements?

(d) Select a sample of $n = 4$ teams with probabilities proportional to number of wins. Use the number of rebounds for these four teams to estimate the average number of rebounds per team in the league. Place a bound on the error of estimation.

TABLE 4.7 National Basketball Association 1982–1983 final statistics

	Wins	Losses	Points Scored	Points Allowed	Rebounds
Atlantic Division					
Philadelphia	65	17	9,191	8,562	3,920
Boston	56	26	9,191	8,752	3,805
New Jersey	49	33	8,672	8,445	3,693
New York	44	38	8,198	7,979	3,343
Washington	42	40	8,134	8,145	3,529
Central Division					
Milwaukee	51	31	8,740	8,379	3,572
Atlanta	43	39	8,335	8,413	3,572
Detroit	37	45	9,239	9,272	3,789
Chicago	28	54	9,102	9,403	3,794
Cleveland	23	59	7,964	8,574	3,587
Indiana	20	62	8,911	9,391	3,593
Midwest Division					
San Antonio	53	29	9,375	9,075	3,831
Denver	45	37	10,105	10,054	3,738
Kansas City	45	37	9,328	9,209	3,663
Dallas	38	44	9,243	9,277	3,677
Utah	30	52	8,938	9,282	3,643
Houston	14	68	8,145	9,096	3,466
Pacific Division					
Los Angeles	58	24	9,433	8,978	3,668
Phoenix	53	29	8,776	8,361	3,612
Seattle	48	34	9,019	8,756	3,721
Portland	46	36	8,808	8,633	3,560
Golden State	30	52	8,902	9,205	3,565
San Diego	25	57	8,903	9,299	3,502

Source: *The World Almanac & Book of Facts*, 1984 edition, copyright © Newspaper Enterprise Association, Inc., 1983, New York, NY 10166.

4.2 Identify a problem in your own area of interest for which you can actually draw a simple random sample to estimate a population mean, total, or proportion. Clearly define the population and the sampling units, and construct a frame. Select a simple random sample from the frame by using the random number table in the Appendix. Then collect the data and make the necessary calculations.

Some suggested projects are as follows.

Business: Estimate the average gross income for firms of a certain type in your area or the average amount spent for entertainment among college males.

Social sciences: Estimate the proportion of registered voters favoring some current political proposal, or estimate the average number of persons per household for a certain section of your city.

Physical sciences: Consider a laboratory experiment such as measuring the tensile strength of wire or the diameter of a machined rod. Take *n* independent observations on such an experiment and treat them as a simple random sample. Construct an interval estimate of the "population" mean. Here the population is merely conceptual (one could take many measurements of the phenomenon in question), and its mean represents the average strength of wire of this type or the average diameter of the rod.

Biological sciences: Estimate the average weight of animals fed on a certain diet for a specified time period, or estimate the average height of trees in a certain plot. As an example of working with totals instead of means, estimate the total number of insect colonies (of a certain type) infesting a plot. Be careful here in selecting the sampling units and constructing the frame.

If your real example involves a large data set, you may want to use a computer for calculation purposes. Most computing centers have standard programs that will compute sample means and variances. Four widely used packages of such programs are SPSS, SAS, BIOMED, and MINITAB (see the references in the Appendix).

CASE STUDY

CAN WE ESTIMATE THE TOTAL COST OF HEALTH CARE?

AN important problem of national concern involves the estimation of the cost of health care. These costs are studied by various agencies, in both government and private sectors, in order to establish government policies and to assess business decisions, such as rates for insurance policies.

A method of estimating hospital costs for one disease is considered in the article "Economic Impact of Kidney Stones in White Adult Males," by J. Shuster and R. L. Scheaffer (*Urology*, vol. 24, no. 4, 1984). In this work two regions of the United States, the Carolinas and the Rocky Mountain states, were singled out for special study. A sample of $n_1 = 363$ stone patients in the Carolinas had an average cost for first hospitalization of $1350; a sample of $n_2 = 258$ stone patients in the Rockies had an average cost for first hospitalization of $1150. Can we estimate the total annual hospitalization costs for this disease for both regions combined? The methods of Chapter 5 will show us how to do so if some additional information is available. The methods can then be used to find an estimate for the entire United States if sample information is available for other regions.

STRATIFIED RANDOM SAMPLING

5.1
INTRODUCTION

The purpose of sample survey design is to maximize the amount of information for a given cost. Simple random sampling, the basic sampling design, often provides good estimates of population quantities at low cost. In this chapter we define a second sampling procedure, stratified random sampling, which in many instances increases the quantity of informatioin for a given cost.

DEFINITION 5.1   A *stratified random sample* is one obtained by separating the population elements into nonoverlapping groups, called *strata*, and then selecting a simple random sample from each stratum.

Suppose a public opinion poll designed to estimate the proportion of voters who favor spending more tax revenue on an improved ambulance service is to be conducted in a certain county. The county contains two cities and a rural area. The population elements of interest for the poll are all men and women of voting age who reside in the county. A *stratified random sample* of adults residing in the county can be obtained by selecting a simple random sample of adults from each city and another simple random sample of adults from the rural area. That is, the two cities and the rural area represent three *strata* from which we obtain simple random samples.

In the county poll, why should we choose a stratified random sample rather than a simple random sample? First, keep in mind that our goal in designing surveys is to maximize the information obtained (or to minimize the bound on the error of estimation) for a fixed expenditure. Samples displaying small variability among the measurements will produce small bounds on the errors of estimation. Thus if all the adults in one city (say city A) tend to think alike on the ambulance service issue, we can obtain a very accurate estimate of the proportion in question with a relatively small sample. Similarly, if all the adults in the second city (city B) tend to think alike on this issue, though they may differ in opinion from those in city A, then we can again obtain an accurate estimate with a small sample. This situation may arise if city A has a hospital and hence has no great need for improved ambulance service, while city B does not have a hospital and hence has great need for an improved ambulance service. The opinions in the rural area may be more varied, but a smaller number of adults may reside here, and enough resources may be available for careful study of this area. When results of the stratified random sample are combined, the final estimate of the proportion of voters favoring more expenditures for an ambulance service may have a much smaller bound on the error of estimation than would an estimate from a simple random sample of comparable size.

Second, the cost of obtaining observations varies with the design of the survey. The cost of selecting the adults to be sampled, the cost of interviewer time and travel, and the cost of administering the overall sampling procedure may all be minimized by a carefully planned stratified random sample in compact, well-defined geographic areas. Such cost savings may allow the investigators to use a larger sample size than they could use for a simple random sample of the same total cost.

Third, estimates of a population parameter may be desired for certain subsets of the population. In the county poll each city commision may want to see an estimate of the proportion of voters favoring an expanded ambulance service for its own city. Stratified random sampling allows for separate estimates of population parameters within each stratum.

In summary, the principal reasons for using stratified random sampling rather than simple random sampling are as follows:

1. Stratification may produce a smaller bound on the error of estimation than would be produced by a simple random sample of the same size. This result is particularly true if measurements within strata are homogeneous.

2. The cost per observation in the survey may be reduced by stratification of the population elements into convenient groupings.

3. Estimates of population parameters may be desired for subgroups of the population. These subgroups should then be identifiable strata.

These three reasons for stratification should be kept in mind when one is deciding whether or not to stratify a population or deciding how to define strata. Sampling hospital patients on a certain diet to assess weight gain may be more efficient if the patients are stratified by sex, since men tend to weigh more than women. A poll of college students at a large university may be more conveniently administered and carried out if students are stratified into on-campus and off-campus residents. A quality control sampling plan in a manufacturing plant may be stratified by production lines because estimates of proportions of defective products may be required by the manager of each line.

Most major surveys have some degree of stratification incorporated into the design. As examples, we will look at three important groups of surveys conducted by the U.S. Bureau of Labor Statistics.

The consumer price index (CPI) is a measure of the average change in prices for a fixed collection of goods and services for urban consumers. The CPI is actually calculated from at least four different types of surveys: surveys of cities, surveys of urban families, surveys of outlets providing goods and services, and surveys of specific goods and services. In the design of most CPI surveys, 1166 sampling units (counties or groups of contiguous counties) are identified in the population and then grouped into 85 strata. Strata are chosen on the basis of geography, population size, percentage population increase from 1960 to 1970, major industry, percentage nonwhite, and percentage urban. The sampling units within a stratum are chosen to be as much alike as possible with regard to these characteristics.

The Current Population Survey (CPS) measures aspects of employment, unemployment, and persons not in the labor force. It stratifies 1931 sampling units across the United States into strata similar to those used in the CPI surveys, except rural sampling units are used and number of farms becomes an important quantity for stratification.

The Establishment Survey (ES) collects data on work hours and earnings for nonagricultural establishments in the United States. Establishments are stratified according to industry type and size, primarily for homogeneity of measurements but also for provision of estimates for various types of industries. For example, information is provided for such industrial categories as mining, construction, manufacturing, transportation, and finance, insurance, and real estate.

In this chapter stratification will always be used with simple random sampling in each stratum, as stated in Definition 5.1. However, stratification can be used with other types of sampling within strata. We will see some examples in later chapters.

5.2
How to Draw a Stratified Random Sample

The first step in the selection of a stratified random sample is to clearly specify the strata; then each sampling unit of the population is placed into its appropriate stratum. This step may be more difficult than it sounds. For example, suppose that you plan to stratify the sampling units, say households, into rural and urban units. What should be done with households in a town of 1000 inhabitants? Are these households rural or urban? They may be rural if the town is isolated in the country, or they may be urban if the town is adjacent to a large city. Hence to specify what is meant by *urban* and *rural* is essential so that each sampling unit clearly falls into only one stratum.

After the sampling units are divided into strata, we select a simple random sample from each stratum by using the techniques given in Chapter 4. We discuss the problem of choosing appropriate sample sizes for the strata later in this chapter. We must be certain that the samples selected from the strata are independent. That is, different random sampling schemes should be used within each stratum so that the observations chosen in one stratum do not depend upon those chosen in another.

Some additional notation is required for stratified random sampling. Let

L = number of strata

N_i = number of sampling units in stratum i

N = number of sampling units in the population

$\quad = N_1 + N_2 + \cdots + N_L$

The following example illustrates a situation in which stratified random sampling may be appropriate.

EXAMPLE 5.1

An advertising firm, interested in determining how much to emphasize television advertising in a certain county, decides to conduct a sample survey to estimate the average number of hours per week that households within the county watch television. The county contains two towns, town A and town B, and a rural area. Town A is built around a factory, and most households contain factory workers with school-aged children. Town B is an exclusive suburb of a city in a neighboring county and contains older residents with few children at home. There are 155 households in town A, 62 in town B, and 93 in the rural area. Discuss the merits of using stratified random sampling in this situation.

SOLUTION

The population of households falls into three natural groupings, two towns and a rural area, according to geographic location. Thus to use these divisions as three strata is quite natural simply for administrative convenience in selecting the samples and carrying out the fieldwork. In addition, each of the three groups of households should have similar behavior patterns among residents within the group. We expect to see relatively small variability in number of hours of television viewing among households within a group, and this is precisely the situation in which stratification produces a reduction in a bound on the error of estimation.

The advertising firm may wish to produce estimates on average television-viewing hours for each town separately. Stratified random sampling allows for these estimates.

For the stratified random sample, we have $N_1 = 155$, $N_2 = 62$, and $N_3 = 93$, with $N = 310$.

5.3
ESTIMATION OF A POPULATION MEAN AND TOTAL

How can we use the data from a stratified random sample to estimate the population mean? Let $\bar{y}_i$ denote the sample mean for the simple random sample selected from stratum i, n_i the sample size for stratum i, μ_i the population mean for stratum i, and τ_i the population total for stratum i. Then the population total τ is equal to $\tau_1 + \tau_2 + \cdots + \tau_L$. We have a simple random sample within each stratum. Therefore we know from Chapter 4 that $\bar{y}_i$ is an unbiased estimator of μ_i and $N_i\bar{y}_i$ is an unbiased estimator of the stratum total $\tau_i = N_i\mu_i$. It seems reasonable to form an estimator of τ, which is the sum of the τ_i's, by summing the estimators of the τ_i's. Similarly, since the population mean μ equals the population total τ divided by N, an unbiased estimator of μ is obtained by summing the estimators of the τ_i's over all strata and then dividing by N. We denote this estimator by $\bar{y}_{st}$, where the subscript st indicates that stratified random sampling is used.

Estimator of the population mean μ:

$$\bar{y}_{st} = \frac{1}{N}[N_1\bar{y}_1 + N_2\bar{y}_2 + \cdots + N_L\bar{y}_L] = \frac{1}{N}\sum_{i=1}^{L} N_i\bar{y}_i \qquad (5.1)$$

Estimated variance of $\bar{y}_{st}$:

$$\hat{V}(\bar{y}_{st}) = \frac{1}{N^2}[N_1^2\hat{V}(\bar{y}_1) + N_2^2\hat{V}(\bar{y}_2) + \cdots + N_L^2\hat{V}(\bar{y}_L)]$$

$$= \frac{1}{N^2}\left[N_1^2\left(\frac{N_1 - n_1}{N_1}\right)\left(\frac{s_1^2}{n_1}\right) + \cdots + N_L^2\left(\frac{N_L - n_L}{N_L}\right)\left(\frac{s_L^2}{n_L}\right)\right]$$

$$= \frac{1}{N^2}\sum_{i=1}^{L} N_i^2\left(\frac{N_i - n_i}{N_i}\right)\left(\frac{s_i^2}{n_i}\right) \qquad (5.2)$$

Bound on the error of estimation:

$$2\sqrt{\hat{V}(\bar{y}_{st})} = 2\sqrt{\frac{1}{N^2}\sum_{i=1}^{L} N_i^2\left(\frac{N_i - n_i}{N_i}\right)\left(\frac{s_i^2}{n_i}\right)} \qquad (5.3)$$

EXAMPLE 5.2

Suppose the survey planned in Example 5.1 is carried out. The advertising firm has enough time and money to interview $n = 40$ households and decides to select random samples of size $n_1 = 20$ from town A, $n_2 = 8$ from town B, and $n_3 = 12$ from the rural area. (We will discuss the choice of sample sizes later.) The simple random samples are selected and the interviews conducted. The results, with measurements of television-viewing time in hours per week, are shown on Tables 5.1 and 5.2.

Estimate the average television viewing time, in hours per week, for (a) all households in the county and (b) all households in town B. In both cases place a bound on the error of estimation.

The terms s_1^2, s_2^2, and s_3^2 in Table 5.2 are the sample variances for strata 1, 2, and 3, respectively; they are given by the formula

$$s_i^2 = \frac{\sum\limits_{j=1}^{n_i} (y_{ij} - \bar{y}_i)^2}{n_i - 1} = \frac{\sum\limits_{j=1}^{n_i} y_{ij}^2 - n_i \bar{y}_i^2}{n_i - 1}$$

for $i = 1, 2, 3$, where y_{ij} is the jth observation in stratum i. These variances estimate the corresponding true stratum variances σ_1^2, σ_2^2, and σ_3^2.

SOLUTION

(a) From Table 5.1 and Equation (5.1),

$$\bar{y}_{st} = \frac{1}{N}[N_1 \bar{y}_1 + N_2 \bar{y}_2 + N_3 \bar{y}_3]$$

$$= \frac{1}{310}[(155)(33.900) + (62)(25.125) + (93)(19.000)]$$

$$= 27.7$$

is the best estimate of the average number of hours per week that all households in the county spend watching television. Also,

$$\hat{V}(\bar{y}_{st}) = \frac{1}{N^2} \sum_{i=1}^{3} N_i^2 \left(\frac{N_i - n_i}{N_i}\right)\left(\frac{s_i^2}{n_i}\right)$$

$$= \frac{1}{(310)^2}\left[\frac{(155)^2(.871)(35.358)}{20} + \frac{(62)^2(.871)(232.411)}{8}\right.$$

$$\left. + \frac{(93)^2(.871)(87.636)}{12}\right]$$

$$= 1.97$$

The estimate of the population mean with an approximate two-standard-deviation bound on the error of estimation is given by

$$\bar{y}_{st} \pm 2\sqrt{\hat{V}(\bar{y}_{st})}, \qquad 27.675 \pm 2\sqrt{1.97}, \qquad 27.7 \pm 2.8$$

TABLE 5.1 Television viewing time, in hours per week

Stratum 1, Town A	Stratum 2, Town B	Stratum 3, Rural Area
35 28 26 41	27 4 49 10	8 15 21 7
43 29 32 37	15 41 25 30	14 30 20 11
36 25 29 31		12 32 34 24
39 38 40 45		
28 27 35 34		

TABLE 5.2 Calculations for Table 5.1

Stratum 1	Stratum 2	Stratum 3
$n_1 = 20$	$n_2 = 8$	$n_3 = 12$
$\bar{y}_1 = 33.900$	$\bar{y}_2 = 25.125$	$\bar{y}_3 = 19.000$
$s_1^2 = 35.358$	$s_2^2 = 232.411$	$s_3^2 = 87.636$
$N_1 = 155$	$N_2 = 62$	$N_3 = 93$

Thus we estimate the average number of hours per week that households in the county view television to be 27.7 hours. The error of estimation should be less than 2.8 hours with a probability approximately equal to .95.

(b) The $n_2 = 8$ observations from stratum 2 constitute a simple random sample; hence we can apply formulas from Chapter 4. The estimate of the average viewing time for town B with an approximate two-standard-deviation bound on the error of estimation is given by

$$\bar{y}_2 \pm 2\sqrt{\left(\frac{N_2 - n_2}{N_2}\right)\left(\frac{s_2^2}{n_2}\right)} \quad \text{or} \quad 25.1 \pm 2\sqrt{\left(\frac{62 - 8}{62}\right)\left(\frac{232.411}{8}\right)}$$

or

$$25.1 \pm 10.1$$

This estimate has a large bound on the error of estimation because s_2^2 is large and the sample size n_2 is small. Thus the estimate $\bar{y}_{st}$ of the population mean is quite good, but the estimate $\bar{y}_2$ of the mean of stratum 2 is poor. If an estimate is desired for a particular stratum, the sample from that stratum must be large enough to provide a reasonable bound on the error of estimation.

Procedures for the estimation of a population total τ follow directly from the procedures presented for estimating μ. Since τ is equal to $N\mu$, an unbiased estimator of τ is given by $N\bar{y}_{st}$.

Estimator of the population total τ:

$$N\bar{y}_{st} = N_1\bar{y}_1 + N_2\bar{y}_2 + \cdots + N_L\bar{y}_L = \sum_{i=1}^{L} N_i\bar{y}_i \qquad (5.4)$$

Estimated variance of $N\bar{y}_{st}$:

$$\hat{V}(N\bar{y}_{st}) = N^2 \hat{V}(\bar{y}_{st}) = \sum_{i=1}^{L} N_i^2 \left(\frac{N_i - n_i}{N_i}\right)\left(\frac{s_i^2}{n_i}\right) \tag{5.5}$$

Bound on the error of estimation:

$$2\sqrt{\hat{V}(N\bar{y}_{st})} = 2\sqrt{\sum_{i=1}^{L} N_i^2 \left(\frac{N_i - n_i}{N_i}\right)\left(\frac{s_i^2}{n_i}\right)} \tag{5.6}$$

EXAMPLE 5.3

Refer to Example 5.2 and estimate the total number of hours per week that households in the county view television. Place a bound on the error of estimation.

SOLUTION

For the data in Table 5.1,

$$N\bar{y}_{st} = 310(27.7) = 8587 \text{ hours}$$

The estimated variance of $N\bar{y}_{st}$ is given by

$$\hat{V}(N\bar{y}_{st}) = N^2 \hat{V}(\bar{y}_{st}) = (310)^2(1.97) = 189{,}278.560$$

The estimate of the population total with a bound on the error of estimation is given by

$$N\bar{y}_{st} \pm 2\sqrt{\hat{V}(N\bar{y}_{st})} \qquad \text{or} \qquad 8587 \pm 2\sqrt{189{,}278.560}$$

or $\qquad\qquad\qquad\qquad\qquad 8587 \pm 870$

Thus we estimate the total weekly viewing time for households in the county to be 8587 hours. The error of estimation should be less than 870 hours.

5.4
SELECTING THE SAMPLE SIZE FOR ESTIMATING POPULATION MEANS AND TOTALS

The amount of information in a sample depends on the sample size n, since $V(\bar{y}_{st})$ decreases as n increases. Let us examine a method of choosing the sample size to obtain a fixed amount of information for estimating a population parameter. Suppose we specify that the estimate $\bar{y}_{st}$ should lie within B units of the population mean, with probability approximately equal to .95. Symbolically, we want

$$2\sqrt{V(\bar{y}_{st})} = B$$

or

$$V(\bar{y}_{st}) = \frac{B^2}{4}$$

This equation contains the actual population variance of $\bar{y}_{st}$ rather than the estimated variance. For large N, the actual variance, $V(\bar{y}_{st})$, looks very similar to Equation (5.2), with $s_1^2, s_2^2, \ldots, s_L^2$ replaced by $\sigma_1^2, \sigma_2^2, \ldots, \sigma_L^2$.

Although we set $V(\bar{y}_{st})$ equal to $B^2/4$, we cannot solve for n unless we know something about the relationships among $n_1, n_2, \ldots, n_L$ and n. There are many ways of allocating a sample of size n among the various strata. In each case, however, the number of observations n_i allocated to the ith stratum is some fraction of the total sample size n. We denote this fraction by w_i. Hence we can write

$$n_i = nw_i \qquad i = 1, 2, \ldots, L \tag{5.7}$$

Using Equation (5.7), we can then set $V(\bar{y}_{st})$ equal to $B^2/4$ and solve for n.

Similarly, estimation of the population total τ with a bound of B units on the error of estimation leads to the equation

$$2\sqrt{V(N\bar{y}_{st})} = B$$

or, using Equation (5.5),

$$V(\bar{y}_{st}) = \frac{B^2}{4N^2}$$

Approximate sample size required to estimate μ or τ with a bound B on the error of estimation:

$$n = \frac{\sum\limits_{i=1}^{L} N_i^2 \sigma_i^2 / w_i}{N^2 D + \sum\limits_{i=1}^{L} N_i \sigma_i^2} \tag{5.8}$$

where w_i is the fraction of observations allocated to stratum i, σ_i^2 is the population variance for stratum i, and

$$D = \frac{B^2}{4} \qquad \text{when estimating } \mu$$

$$D = \frac{B^2}{4N^2} \qquad \text{when estimating } \tau$$

We must obtain approximations of the population variances $\sigma_1^2, \sigma_2^2, \ldots, \sigma_L^2$ before we can use Equation (5.8). One method of obtaining these approximations is to use the sample variances $s_1^2, s_2^2, \ldots, s_L^2$ from a previous experiment to estimate $\sigma_1^2, \sigma_2^2, \ldots, \sigma_L^2$. A second method requires knowledge of the range of the observations within each stratum. From Tchebysheff's theorem and the normal distribution the range should be roughly four to six standard deviations.

Methods of choosing the fractions $w_1, w_2, \ldots, w_L$ are given in Section 5.5.

EXAMPLE 5.4

A prior survey suggests that the stratum variances for Example 5.1 are approximately $\sigma_1^2 \approx 25$, $\sigma_2^2 \approx 225$, and $\sigma_3^2 \approx 100$. We wish to estimate the population mean by using $\bar{y}_{st}$. Choose the sample size to obtain a bound on the error of estimation equal to 2 hours if the allocation fractions are given by $w_1 = \frac{1}{3}$, $w_2 = \frac{1}{3}$, and $w_3 = \frac{1}{3}$. In other words, you are to take an equal number of observations from each stratum.

SOLUTION

A bound on the error of 2 hours means that

$$2\sqrt{V(\bar{y}_{st})} = 2 \qquad \text{or} \qquad V(\bar{y}_{st}) = 1$$

Therefore $D = 1$.

In Example 5.1, $N_1 = 155$, $N_2 = 62$, and $N_3 = 93$. Therefore

$$\sum_{i=1}^{3} \frac{N_i^2 \sigma_i^2}{w_i} = \frac{N_1^2 \sigma_1^2}{w_1} + \frac{N_2^2 \sigma_2^2}{w_2} + \frac{N_3^2 \sigma_3^2}{w_3}$$

$$= \frac{(155)^2(25)}{\left(\frac{1}{3}\right)} = \frac{(62)^2(225)}{\left(\frac{1}{3}\right)} + \frac{(93)^2(100)}{\left(\frac{1}{3}\right)}$$

$$= (24{,}025)(75) + (3844)(675) + (8649)(300)$$

$$= 6{,}991{,}275$$

$$\sum_{i=1}^{3} N_i \sigma_i^2 = N_1 \sigma_1^2 + N_2 \sigma_2^2 + N_3 \sigma_3^2$$

$$= (155)(25) + (62)(225) + (93)(100) = 27{,}125$$

$$N^2 D = (310)^2(1) = 96{,}100$$

From Equation (5.8) we then have

$$n = \frac{\sum\limits_{i=1}^{3} N_i^2 \sigma_i^2 / w_i}{N^2 D + \sum\limits_{i=1}^{3} N_i \sigma_i^2} = \frac{6{,}991{,}275}{96{,}100 + 27{,}125} = \frac{6{,}991{,}275}{123{,}225} = 56.7$$

Thus the experimenter should take $n = 57$ observations with

$$n_1 = n(w_1) = 57(\tfrac{1}{3}) = 19$$
$$n_2 = 19$$
$$n_3 = 19$$

EXAMPLE 5.5

As in Example 5.4, suppose the variances of Example 5.1 are approximated by $\sigma_1^2 \approx 25$, $\sigma_2^2 \approx 225$, and $\sigma_3^2 \approx 100$. We wish to estimate the population total τ with a bound of 400 hours on the error of estimation. Choose the appropriate sample size if an equal number of observations is to be taken from each stratum.

SOLUTION

The bound on the error of estimation is to be 400 hours and, therefore,

$$D = \frac{B^2}{4N^2} = \frac{(400)^2}{4N^2} = \frac{40,000}{N^2}$$

To calculate n from Equation (5.8), we need the following quantities:

$$\sum_{i=1}^{3} \frac{N_i^2 \sigma_i^2}{w_i} = 6,991,275 \qquad \text{(from Example 5.4)}$$

$$\sum_{i=1}^{3} N_i \sigma_i^2 = 27,125 \qquad \text{(from Example 5.4)}$$

$$N^2 D = N^2 \left(\frac{40,000}{N^2} \right) = 40,000$$

Using Equation (5.8) yields

$$n = \frac{\displaystyle\sum_{i=1}^{3} N_i^2 \sigma_i^2 / w_i}{N^2 D + \displaystyle\sum_{i=1}^{3} N_i \sigma_i^2} = \frac{6,991,275}{40,000 + 27,125} = 104.2 \text{ or } 105$$

Then $n_1 = n_2 = n_3 = 35$.

5.5
ALLOCATION OF THE SAMPLE

You recall that the objective of a sample survey design is to provide estimators with small variances at the lowest possible cost. After the sample size n is chosen, there are many ways to divide n into the individual stratum sample sizes, $n_1, n_2, \ldots, n_L$. Each division may result in a different variance for the sample mean. Hence our objective is to use an allocation that gives a specified amount of information at minimum cost.

In terms of our objective the best allocation scheme is affected by three factors. They are as follows:

1. The total number of elements in each stratum.

2. The variability of observations within each stratum.

3. The cost of obtaining an observation from each stratum.

The number of elements in each stratum affects the quantity of information in the sample. A sample of size 20 from a population of 200 elements should contain more information than a sample of 20 from 20,000 elements. Thus large sample sizes should be assigned to strata containing large numbers of elements.

Variability must be considered because a larger sample is needed to obtain a good estimate of a population parameter when the observations are less homogeneous.

If the cost of obtaining an observation varies from stratum to stratum, we will take small samples from strata with high costs. We will do so because our objective is to keep the cost of sampling at a minimum.

Approximate allocation that minimizes cost for a fixed value of $V(\bar{y}_{st})$ or minimizes $V(\bar{y}_{st})$ for a fixed cost:

$$n_i = n \left(\frac{N_i \sigma_i / \sqrt{c_i}}{N_1 \sigma_1 / \sqrt{c_1} + N_2 \sigma_2 / \sqrt{c_2} + \cdots + N_L \sigma_L / \sqrt{c_L}} \right) \qquad (5.9)$$

$$= n \left(\frac{N_i \sigma_i / \sqrt{c_i}}{\sum\limits_{k=1}^{L} N_k \sigma_k / \sqrt{c_k}} \right)$$

where N_i denotes the size of the ith stratum, σ_i^2 denotes the population variance for the ith stratum, and c_i denotes the cost of obtaining a single observation from the ith stratum.

One must approximate the variance of each stratum before sampling in order to use the allocation formula (5.9). The approximations can be obtained from earlier surveys or from knowledge of the range of the measurements within each stratum.

Substituting the n_i / n given by (5.9) for w_i in Equation (5.8) gives

$$n = \frac{\left(\sum\limits_{k=1}^{L} N_k \sigma_k / \sqrt{c_k} \right) \left(\sum\limits_{i=1}^{L} N_i \sigma_i \sqrt{c_i} \right)}{N^2 D + \sum\limits_{i=1}^{L} N_i \sigma_i^2} \qquad (5.10)$$

for optimal allocation with the variance of $\bar{y}_{st}$ fixed at D.

EXAMPLE 5.6

The advertising firm in Example 5.1 finds that obtaining an observation from a rural household costs more than obtaining a response in town A or B. The increase is due to costs of traveling from one rural household to another. The cost per observation in each town is estimated to be \$9.00 (that is, $c_1 = c_2 = 9$), and the costs per observation in the rural area to be \$16.00 (that is, $c_3 = 16$). The stratum standard deviations (approximated by the strata sample variances from a prior survey) are $\sigma_1 \approx 5$, $\sigma_2 \approx 15$, and $\sigma_3 \approx 10$. Find the overall sample size n and the stratum sample sizes, n_1, n_2, and n_3, that allow the firm to estimate, at minimum cost, the average television-viewing time with a bound on the error of estimation equal to 2 hours.

SOLUTION

We have

$$\sum_{k=1}^{3} \frac{N_k \sigma_k}{\sqrt{c_k}} = \frac{N_1 \sigma_1}{\sqrt{c_1}} + \frac{N_2 \sigma_2}{\sqrt{c_2}} + \frac{N_3 \sigma_3}{\sqrt{c_3}}$$

$$= \frac{155(5)}{\sqrt{9}} + \frac{62(15)}{\sqrt{9}} + \frac{93(10)}{\sqrt{16}} = 800.83$$

and

$$\sum_{i=1}^{3} N_i \sigma_i \sqrt{c_i} = N_1 \sigma_1 \sqrt{c_1} + N_2 \sigma_2 \sqrt{c_2} + N_3 \sigma_3 \sqrt{c_3}$$

$$= 155(5)\sqrt{9} + 62(15)\sqrt{9} + 93(10)\sqrt{16} = 8835$$

Thus

$$n = \frac{\left(\sum_{k=1}^{3} N_k \sigma_k / \sqrt{c_k} \right) \left(\sum_{i=1}^{3} N_i \sigma_i \sqrt{c_i} \right)}{N^2 D + \sum_{i=1}^{3} N_i \sigma_i^2}$$

$$= \frac{(800.83)(8835)}{(310)^2 (1) + 27,125} = 57.42 \text{ or } 58$$

Then

$$n_1 = n \left(\frac{N_1 \sigma_1 / \sqrt{c_1}}{\sum_{k=1}^{3} N_k \sigma_k / \sqrt{c_k}} \right) = n \left[\frac{155(5)/3}{800.83} \right] = .32n = 18.5 \text{ or } 18$$

Similarly,

$$n_2 = n \left[\frac{62(15)/3}{800.83} \right] = .39n = 22.6 \text{ or } 23$$

$$n_3 = n \left[\frac{93(10)/4}{800.83} \right] = .29n = 16.8 \text{ or } 17$$

Hence the experimenter should select 18 households at random from town

A, 23 from town B, and 17 from the rural area. He can then estimate the average number of hours spent watching television at minimum cost with a bound of 2 hours on the error of estimation.

In some stratified sampling problems the cost of obtaining an observation is the same for all strata. If the costs are unknown, we may be willing to assume that the costs per observation are equal. If $c_1 = c_2 = \cdots = c_L$, then the cost terms cancel in Equation (5.9) and

$$n_i = n \left(\frac{N_i \sigma_i}{\sum_{i=1}^{L} N_i \sigma_i} \right) \tag{5.11}$$

This method of selecting $n_1, n_2, \ldots, n_L$ is called *Neyman allocation*. Under Neyman allocation Equation (5.10) for the total sample size n becomes

$$n = \frac{\left(\sum_{i=1}^{L} N_i \sigma_i \right)^2}{N^2 D + \sum_{i=1}^{L} N_i \sigma_i^2} \tag{5.12}$$

EXAMPLE 5.7

The advertising firm of Example 5.1 decides to use telephone interviews rather than personal interviews because all households in the county have telephones, and this method reduces costs. The cost of obtaining an observation is then the same in all three strata. The stratum standard deviations are again approximated by $\sigma_1 \approx 5$, $\sigma_2 \approx 15$, *and* $\sigma_3 \approx 10$. The firm desires to estimate the population mean μ with a bound on the error of estimation equal to 2 hours. Find the appropriate sample size n and stratum sample sizes, n_1, n_2, and n_3.

SOLUTION

We will now use Equations (5.11) and (5.12), since the costs are the same in all strata. Therefore to find the allocation fractions, w_1, w_2, and w_3, we use Equation (5.11). Then

$$\sum_{i=1}^{3} N_i \sigma_i = N_1 \sigma_1 + N_2 \sigma_2 + N_3 \sigma_3$$

$$= (155)(5) + (62)(15) + (93)(10) = 2635$$

And from Equation (5.11)

$$n_1 = n \left(\frac{N_1 \sigma_1}{\sum_{i=1}^{3} N_i \sigma_i} \right) = n \left[\frac{(155)(5)}{2635} \right] = n(.30)$$

Similarly,
$$n_2 = n\left[\frac{(62)(15)}{2635}\right] = n(.35)$$

$$n_3 = n\left[\frac{(93)(10)}{2635}\right] = n(.35)$$

Thus $w_1 = .30$, $w_2 = .35$, and $w_3 = .35$.

Now let us use Equation (5.12) to find n. A bound of 2 hours on the error of estimation means that

$$2\sqrt{V(\bar{y}_{st})} = 2 \quad \text{or} \quad V(\bar{y}_{st}) = 1$$

Therefore,

$$D = \frac{B^2}{4} = 1 \quad \text{and} \quad N^2 D = (310)^2(1) = 96,100$$

Also,
$$\sum_{i=1}^{3} N_i \sigma_i^2 = 27,125$$

from Example 5.5, and Equation (5.12) gives

$$n = \frac{\left(\sum\limits_{i=1}^{3} N_i \sigma_i\right)^2}{N^2 D + \sum\limits_{i=1}^{3} N_i \sigma_i^2}$$

$$= \frac{(2635)^2}{96,100 + 27,125} = 56.34 \text{ or } 57$$

Then
$$n_1 = nw_1 = (57)(.30) = 17$$

$$n_2 = nw_2 = (57)(.35) = 20$$

$$n_3 = nw_3 = (57)(.35) = 20$$

The sample size n in Example 5.7 is nearly the same as in Example 5.6, but the allocation has changed. More observations are taken from the rural area because these observations no longer have a higher cost.

EXAMPLE 5.8

An experimenter wanted to estimate the average weight of 90 rats (50 male and 40 female) being fed a certain diet. The rats were separated by sex; hence to use stratified random sampling with two strata seemed appropriate. To approximate the variability within each stratum, the experimenter selected the smallest and largest rats in each stratum and weighed them. She found that the range was 10 grams for the males and 8 grams for the females. How large a sample should have been taken in order to estimate the population average with a bound of 1 gram on the error of estimation? Assume the cost of sampling was the same for both strata.

SOLUTION

Let us denote males as stratum 1 and females as stratum 2. To use Equation (5.11), we must first approximate σ_1 and σ_2. The standard deviation should be about one-fourth of the range, assuming that the weights have a bell-shaped distribution. Thus

$$\sigma_1 \approx \tfrac{10}{4} = 2.5 \quad \text{and} \quad \sigma_2 \approx \tfrac{8}{4} = 2.0$$

From Equation (5.11)

$$n_i = n \left(\frac{N_i\sigma_i}{\sum\limits_{i=1}^{2} N_i\sigma_i} \right)$$

where

$$\sum_{i=1}^{2} N_i\sigma_i = (50)(2.5) + (40)(2.0) = 125 + 80 = 205$$

Then

$$n_1 = n \left(\frac{N_1\sigma_1}{\sum\limits_{i=1}^{2} N_i\sigma_i} \right) = n\left(\frac{125}{205}\right) = .61n$$

and

$$n_2 = n\left(\frac{80}{205}\right) = .39n$$

Thus $w_1 = .61$ and $w_2 = .39$.

We must calculate the following quantities in order to find n:

$$\sum_{i=1}^{2} N_i\sigma_i^2 = (50)(2.5)^2 + (40)(2.0)^2 = 472.50$$

$$D = \frac{B^2}{4} = \frac{(1)^2}{4} = .25$$

Using Equation (5.12), we have

$$n = \frac{\left(\sum\limits_{i=1}^{2} N_i\sigma_i\right)^2}{N^2 D + \sum\limits_{i=1}^{2} N_i\sigma_i^2}$$

$$= \frac{(205)^2}{(90)^2(.25) + 472.50} = 16.83$$

The sample size n should have been 17 with

$$n_1 = nw_1 = (17)(.61) = 10$$

and

$$n_2 = nw_2 = (17)(.39) = 7$$

In addition to encountering equal costs, we sometimes encounter approximately equal variances, $\sigma_1^2, \sigma_2^2, \ldots, \sigma_L^2$. In that case the σ_i's cancel in Equation (5.11) and

$$n_i = n\left(\frac{N_i}{\sum\limits_{i=1}^{L} N_i}\right) = n\left(\frac{N_i}{N}\right)$$
(5.13)

This method of assigning sample sizes to the strata is called *proportional allocation* because sample sizes $n_1, n_2, \ldots, n_L$ are proportional to stratum sizes $N_1, N_2, \ldots, N_L$. Of course, proportional allocation can be, and often is, used when stratum variances and costs are not equal. One advantage to using this allocation is that the estimator $\bar{y}_{st}$ becomes simply the sample mean for the entire sample. This feature can be an important timesaving feature in some surveys.

Under proportional allocation Equation (5.8) for the value of n, which yields $V(\bar{y}_{st}) = D$, becomes

$$n = \frac{\sum\limits_{i=1}^{L} N_i \sigma_i^2}{ND + \dfrac{1}{N}\sum\limits_{i=1}^{L} N_i \sigma_i^2}$$
(5.14)

EXAMPLE 5.9

The advertising firm in Example 5.1 thinks that the approximate variances used in previous examples are in error and that the stratum variances are approximately equal. The common value of σ_i was approximated by 10 in a preliminary study. Telephone interviews are to be used, and hence costs will be equal in all strata. The firm desires to estimate the average number of hours per week that households in the county watch television, with a bound on the error of estimation equal to 2 hours. Find the sample size and stratum sample sizes necessary to achieve this accuracy.

SOLUTION

We have

$$\sum\limits_{i=1}^{3} N_i \sigma_i^2 = N_1 \sigma_1^2 + N_2 \sigma_2^2 + N_3 \sigma_3^2$$

$$= (155)(100) + (62)(100) + (93)(100)$$

$$= 310(100) = 31,000$$

Thus, since $D = 1$, Equation (5.14) gives

$$n = \frac{31,000}{310(1) + (1/310)(31,000)} = 75.6 \text{ or } 76$$

Therefore

$$n_1 = n\left(\frac{N_1}{\sum\limits_{i=1}^{3} N_i}\right) = n\left(\frac{N_1}{N}\right) = n\left(\frac{155}{310}\right) = n(.5) = 38$$

$$n_2 = n\left(\frac{N_2}{N}\right) = n\left(\frac{62}{310}\right) = n(.2) = 15$$

$$n_3 = n\left(\frac{N_3}{N}\right) = n\left(\frac{93}{310}\right) = n(.3) = 23$$

These results differ from those of Example 5.7 because here the variances are assumed to be equal in all strata and are approximated by a common value.

The amount of money to be spent on sampling is sometimes fixed before the experiment is started. Then the experimenter must find a sample size and allocation scheme that minimizes the variance of the estimator for a fixed expenditure.

EXAMPLE 5.10

In the television-viewing example, suppose the costs are as specified in Example 5.6. That is, $c_1 = c_2 = 9$ and $c_3 = 16$. Let the stratum variances be approximated by $\sigma_1 \approx 5$, $\sigma_2 \approx 15$, and $\sigma_3 \approx 10$. Given that the advertising firm has only \$500 to spend on sampling, choose the sample size and the allocation that minimize $V(\bar{y}_{st})$.

SOLUTION

The allocation scheme is still given by Equation (5.9). In Example 5.6 we found $w_1 = .32$, $w_2 = .39$, and $w_3 = .29$.

Since the total cost must equal \$500, we have

$$c_1 n_1 + c_2 n_2 + c_3 n_3 = 500$$

or

$$9n_1 + 9n_2 + 16n_3 = 500$$

Since $n_i = nw_i$, we can substitute as follows:

$$9nw_1 + 9nw_2 + 16nw_3 = 500$$

or

$$9n(.32) + 9n(.39) + 16n(.29) = 500$$

Solving for n, we obtain

$$11.03n = 500$$

$$n = \frac{500}{11.03} = 45.33$$

Therefore we must take $n = 45$ to ensure that the cost remains below \$500. The corresponding allocation is given by

$$n_1 = nw_1 = (45)(.32) = 14$$

$$n_2 = nw_2 = (45)(.39) = 18$$

$$n_3 = nw_3 = (45)(.39) = 13$$

We can make the following summary statement on stratified random sampling: In general, stratified random sampling with proportional allocation will produce an estimator with smaller variance than that produced by simple random sampling (with the same sample size) if there is considerable variability among the stratum means. If sampling costs are nearly equal from stratum to stratum, stratified random sampling with optimal allocation [Equation (5.8)] will yield estimators with smaller variance than will proportional allocation when there is variability among the stratum variances.

5.6
ESTIMATION OF A POPULATION PROPORTION

In our numerical examples we have been interested in estimating the average or the total number of hours per week spent watching television. In contrast, suppose that the advertising firm wants to estimate the proportion (fraction) of households that watches a particular show. The population is divided into strata, just as before, and a simple random sample is taken from each stratum. Interviews are then conducted to determine the proportion $\hat{p}_i$ of households in stratum i that view the show. This $\hat{p}_i$ is an unbiased estimator of p_i, the population proportion in stratum i (as described in Chapter 4). Reasoning as we did in Section 5.3, we conclude that $N_i\hat{p}_i$ is an unbiased estimator of the total number of households in stratum i that view this particular show. Hence $N_1\hat{p}_1 + N_2\hat{p}_2 + \cdots + N_L\hat{p}_L$ is a good estimator of the total number of viewing households in the population. Dividing this quantity by N, we obtain an unbiased estimator of the population proportion p of households viewing the show.

Estimator of the population proportion p:

$$\hat{p}_{st} = \frac{1}{N}(N_1\hat{p}_1 + N_2\hat{p}_2 + \cdots + N_L\hat{p}_L) = \frac{1}{N}\sum_{i=1}^{L} N_i\hat{p}_i \qquad (5.15)$$

Estimated variance of $\hat{p}_{st}$:

$$\hat{V}(\hat{p}_{st}) = \frac{1}{N^2} [N_1^2 \hat{V}(\hat{p}_1) + N_2^2 \hat{V}(\hat{p}_2) + \cdots + N_L^2 \hat{V}(\hat{p}_L)]$$

$$= \frac{1}{N^2} \sum_{i=1}^{L} N_i^2 \hat{V}(\hat{p}_i)$$

$$= \frac{1}{N^2} \sum_{i=1}^{L} N_i^2 \left(\frac{N_i - n_i}{N_i}\right)\left(\frac{\hat{p}_i \hat{q}_i}{n_i - 1}\right) \qquad (5.16)$$

Bound on the error of estimation:

$$2\sqrt{\hat{V}(\hat{p}_{st})} = 2\sqrt{\frac{1}{N^2} \sum_{i=1}^{L} N_i^2 \left(\frac{N_i - n_i}{N_i}\right)\left(\frac{\hat{p}_i \hat{q}_i}{n_i - 1}\right)} \qquad (5.17)$$

EXAMPLE 5.11

The advertising firm wanted to estimate the proportion of households in the county of Example 5.1 that view show X. The county is divided into three strata, town A, town B, and the rural area. The strata contain $N_1 = 155$, $N_2 = 62$, and $N_3 = 93$ households, respectively. A stratified random sample of $n = 40$ households is chosen with proportional allocation. In other words, a simple random sample is taken from each stratum; the sizes of the samples are $n_1 = 20$, $n_2 = 8$, and $n_3 = 12$. Interviews are conducted in the 40 sampled households; results are shown in Table 5.3. Estimate the proportion of households viewing show X, and place a bound on the error of estimation.

TABLE 5.3 Data for Example 5.11

Stratum	Sample Size	Number of Households Viewing Show X	$\hat{p}_i$
1	$n_1 = 20$	16	.80
2	$n_2 = 8$	2	.25
3	$n_3 = 12$	6	.50

SOLUTION

The estimate of the proportion of households viewing show X is given by $\hat{p}_{st}$. Using Equation (5.15), we calculate

$$\hat{p}_{st} = \frac{1}{310} [(155)(.80) + 62(.25) + 93(.50)] = .60$$

The variance of $\hat{p}_{st}$ can be estimated by using Equation (5.16). First, let us calculate the $\hat{V}(\hat{p}_i)$ terms. We have

$$\hat{V}(\hat{p}_1) = \left(\frac{N_1 - n_1}{N_1}\right)\left(\frac{\hat{p}_1\hat{q}_1}{n_1 - 1}\right) = \left(\frac{155 - 20}{155}\right)\left[\frac{(.8)(.2)}{19}\right]$$
$$= (.871)(.008) = .007$$

$$\hat{V}(\hat{p}_2) = \left(\frac{N_2 - n_2}{N_2}\right)\left(\frac{\hat{p}_2\hat{q}_2}{n_2 - 1}\right) = \left(\frac{62 - 8}{62}\right)\left[\frac{(.25)(.75)}{7}\right]$$
$$= (.871)(.027) = .024$$

$$\hat{V}(\hat{p}_3) = \left(\frac{N_3 - n_3}{N_3}\right)\left(\frac{\hat{p}_3\hat{q}_3}{n_3 - 1}\right) = \left(\frac{93 - 12}{93}\right)\left[\frac{(.5)(.5)}{11}\right]$$
$$= (.871)(.023) = .020$$

From Equation (5.16)

$$\hat{V}(\hat{p}_{st}) = \frac{1}{N^2}\sum_{i=1}^{3} N_i^2\hat{V}(\hat{p}_i)$$

$$= \frac{1}{(310)^2}[(155)^2(.007) + (62)^2(.024) + (93)^2(.020)]$$

$$= .0045$$

Then the estimate of proportion of households in the county that view show X, with a bound on the error of estimation, is given by

$$\hat{p}_{st} \pm 2\sqrt{\hat{V}(\hat{p}_{st})}, \qquad .60 \pm 2\sqrt{.0045}$$
$$.60 \pm 2(.07), \qquad .60 \pm .14$$

The bound on the error in Example 5.11 is quite large. We could reduce this bound and make the estimator more precise by increasing the sample size. The problem of choosing a sample size is considered in the next section.

5.7
SELECTING THE SAMPLE SIZE AND ALLOCATING THE SAMPLE TO ESTIMATE PROPORTIONS

To estimate a population proportion, we first indicate how much information we desire by specifying the size of the bound; the sample size is chosen accordingly.

The formula for the sample size n (for a given bound B on the error of estimation) is the same as Equation (5.8) except that σ_i^2 becomes p_iq_i.

Approximate sample size required to estimate p with a bound B on the error of estimation:

$$n = \frac{\sum\limits_{i=1}^{L} N_i^2 p_i q_i / w_i}{N^2 D + \sum\limits_{i=1}^{L} N_i p_i q_i} \tag{5.18}$$

where w_i is the fraction of observations allocated to stratum i, p_i is the population proportion for stratum i, and

$$D = \frac{B^2}{4}$$

The allocation formula that gives the variance of $\hat{p}_{st}$ equal to some fixed constant at minimum cost is the same as Equation (5.9) with σ_i replaced by $\sqrt{p_i q_i}$.

Approximate allocation that minimizes cost for a fixed value of $V(\hat{p}_{st})$ or minimizes $V(\hat{p}_{st})$ for a fixed cost:

$$n_i = n \frac{N_i \sqrt{p_i q_i / c_i}}{N_1 \sqrt{p_1 q_1 / c_1} + N_2 \sqrt{p_2 q_2 / c_2} + \cdots + N_L \sqrt{p_L q_L / c_L}}$$

$$= n \frac{N_i \sqrt{p_i q_i / c_i}}{\sum\limits_{k=1}^{L} N_k \sqrt{p_k q_k / c_k}} \tag{5.19}$$

where N_i denotes the size of the ith stratum, p_i denotes the population proportion for the ith stratum, and c_i denotes the cost of obtaining a single observation from the ith stratum.

EXAMPLE 5.12

The data of Table 5.2 were obtained from a sample conducted last year. The advertising firm now wants to conduct a new survey in the same county to estimate the proportion of households viewing show X. Although the fractions p_1, p_2, and p_3 that appear in Equations (5.18) and (5.19) are unknown, they can be approximated by the estimates from the earlier study, that is, $\hat{p}_1 = .80$, $\hat{p}_2 = .25$, and $\hat{p}_3 = .50$. The cost of obtaining an observation is \$9 for either town and \$16 for the rural area, that is, $c_1 = c_2 = 9$ and $c_3 = 16$. The number of households within the strata are $N_1 = 155$, $N_2 = 62$, and $N_3 = 93$. The firm wants to estimate the population proportion p with a bound on the error of estimation equal to .1. Find the sample size n and the strata sample sizes, n_1, n_2, and n_3, that will give the desired bound at minimum cost.

SOLUTION

We first use Equation (5.19) to find the allocation fractions w_i. Using $\hat{p}_i$ to approximate p_i, we have

$$\sum_{i=1}^{3} N_i \sqrt{\frac{\hat{p}_i \hat{q}_i}{c_i}} = N_1 \sqrt{\frac{\hat{p}_1 \hat{q}_1}{c_1}} + N_2 \sqrt{\frac{\hat{p}_2 \hat{q}_2}{c_2}} + N_3 \sqrt{\frac{\hat{p}_3 \hat{q}_3}{c_3}}$$

$$= 155 \sqrt{\frac{(.8)(.2)}{9}} + 62 \sqrt{\frac{(.25)(.75)}{9}} + 93 \sqrt{\frac{(.5)(.5)}{16}}$$

$$= \frac{62.000}{3} + \frac{26.846}{3} + \frac{46.500}{4}$$

$$= 20.667 + 8.949 + 11.625 = 41.241$$

and

$$n_1 = n \frac{N_1 \sqrt{\hat{p}_1 \hat{q}_1 / c_1}}{\sum\limits_{i=1}^{3} N_i \sqrt{\hat{p}_i \hat{q}_i / c_i}} = n \left(\frac{20.667}{41.241} \right) = n(.50)$$

Similarly,

$$n_2 = n \left(\frac{8.949}{41.241} \right) = n(.22)$$

$$n_3 = n \left(\frac{11.625}{41.241} \right) = n(.28)$$

Thus $w_1 = .50$, $w_2 = .22$, and $w_3 = .28$.

The next step is to use Equation (5.18) to find n. First, the following quantities must be calculated:

$$\sum_{i=1}^{3} \frac{N_i^2 \hat{p}_i \hat{q}_i}{w_i} = \frac{N_1^2 \hat{p}_1 \hat{q}_1}{w_1} + \frac{N_2^2 \hat{p}_2 \hat{q}_2}{w_2} + \frac{N_3^2 \hat{p}_3 \hat{q}_3}{w_3}$$

$$= \frac{(155)^2(.8)(.2)}{.50} + \frac{(62)^2(.25)(.75)}{.22} + \frac{(93)^2(.5)(.5)}{.28}$$

$$= 18,686.46$$

$$\sum_{i=1}^{3} N_i \hat{p}_i \hat{q}_i = N_1 \hat{p}_1 \hat{q}_1 + N_2 \hat{p}_2 \hat{q}_2 + N_3 \hat{p}_3 \hat{q}_3$$

$$= (155)(.8)(.2) + (62)(.25)(.75) + (93)(.5)(.5)$$

$$= 59.675$$

To find D, we let $2\sqrt{V(\hat{p}_{st})} = .1$ (the bound on the error of estimation). Then

$$V(\hat{p}_{st}) = \frac{(.1)^2}{4} = .0025 = D$$

and

$$N^2 D = (310)^2(.0025) = 240.25$$

Finally, from Equation (5.18) n is given approximately by

$$n = \frac{\sum_{i=1}^{3} N_i^2 \hat{p}_i \hat{q}_i / w_i}{N^2 D + \sum_{i=1}^{3} N_i \hat{p}_i \hat{q}_i} = \frac{18,686.46}{240.25 + 59.675} = 62.3 \text{ or } 63$$

Hence
$$n_1 = nw_1 = (63)(.50) = 31$$
$$n_2 = nw_2 = (63)(.22) = 14$$
$$n_3 = nw_3 = (63)(.28) = 18$$

If the cost of sampling does not vary from stratum to stratum, then the cost factors c_i cancel from Equation (5.19).

EXAMPLE 5.13

Suppose that in Example 5.12 telephone interviews are to be conducted, and hence the cost of sampling is the same in all strata. The fraction p_i will be approximated by $\hat{p}_i$, $i = 1, 2, 3$. We desire to estimate the population proportion p with a bound of .1 on the error of estimation. Find the appropriate sample size to achieve this bound at minimum cost.

SOLUTION

Equation (5.19) is used to find the fractions w_1, w_2, and w_3, but now all c_i terms can be replaced by 1. Hence

$$\sum_{i=1}^{3} N_i \sqrt{\hat{p}_i \hat{q}_i} = 155\sqrt{(.8)(.2)} + 62\sqrt{(.25)(.75)} + 93\sqrt{(.5)(.5)}$$

$$= 62.000 + 26.846 + 46.500 = 135.346$$

and
$$n_1 = n\left(\frac{N_1 \sqrt{\hat{p}_1 \hat{q}_1}}{\sum_{i=1}^{3} N_i \sqrt{\hat{p}_i \hat{q}_i}}\right) = n\left(\frac{62.000}{135.346}\right) = n(.46)$$

Similarly,
$$n_2 = n\left(\frac{26.846}{135.346}\right) = n(.20)$$

$$n_3 = n\left(\frac{46.500}{135.346}\right) = n(.34)$$

Thus $w_1 = .46$, $w_2 = .20$, and $w_3 = .34$.

Equation (5.18) or Equation (5.12) with $\sigma_i = \sqrt{p_i q_i}$ can be used to find n. Using (5.12), we have

$$\sum_{i=1}^{3} N_i \hat{p}_i \hat{q}_i = 59.675 \qquad \text{(from Example 5.12)}$$

$$N^2 D = 240.25 \qquad \text{(from Example 5.12)}$$

and

$$n = \frac{\left(\sum_{i=1}^{3} N_i \sqrt{\hat{p}_i \hat{q}_i} \right)^2}{N^2 D + \sum_{i=1}^{3} N_i \hat{p}_i \hat{q}_i}$$

$$= \frac{(135.346)^2}{240.25 + 59.675} = 61.08 \text{ or } 62$$

Hence we take a sample of 62 observations to estimate p with a bound on the error of magnitude $B = .1$. The corresponding allocation is given by

$$n_1 = nw_1 = 62(.46) = 29$$

$$n_2 = nw_2 = 62(.20) = 12$$

$$n_3 = nw_3 = 62(.34) = 21$$

These answers are close to those of Example 5.12. The changes in allocation result because costs do not vary in Example 5.13.

Recall that the allocation formula (5.9) assumes a very simple form when the variances as well as costs are equal for all strata. Equation (5.19) simplifies in the same way, provided all stratum proportions p_i are equal and all costs c_i are equal. Then Equation (5.19) becomes

$$n_i = n \left(\frac{N_i}{N} \right) \qquad i = 1, 2, \ldots, L \qquad (5.20)$$

As previously noted, this method for assignment of sample sizes to the strata is called *proportional allocation*.

EXAMPLE 5.14

In the television survey of Example 5.12 the advertising firm plans to use telephone interviews; therefore the cost of sampling will not vary from stratum to stratum. The stratum sizes are $N_1 = 155$, $N_2 = 62$, and $N_3 = 93$. The results of last year's survey (see Table 5.3) do not appear to hold for this year. The firm believes that the proportion of households viewing show X is close to .4 in each of the three strata. The firm desires to estimate the population proportion p with a bound of .1 on the error of estimation. Find the sample size n and the allocation that gives this bound at minimum cost.

SOLUTION

The allocation fractions are found by using Equation (5.19) with $p_1, \ldots, p_L$ and $c_1, \ldots, c_L$ replaced by 1. Thus

$$n_1 = n\left(\frac{N_1}{\sum_{i=1}^{3} N_i}\right) = n\left(\frac{N_1}{N}\right) = n\left(\frac{155}{310}\right) = n(.5)$$

$$n_2 = n\left(\frac{N_2}{N}\right) = n\left(\frac{62}{310}\right) = n(.2)$$

$$n_3 = n\left(\frac{N_3}{N}\right) = n\left(\frac{93}{310}\right) = n(.3)$$

or
$$w_1 = .5, \qquad w_2 = .2, \qquad w_3 = .3$$

The sample size n is found from Equation (5.18) using .4 as an approximation to p_1, p_2, and p_3, or it can be found by setting $\sigma_i^2 = p_i q_i$ in Equation (5.14). Using the latter approach, with $p_i = .4$, yields

$$\sum_{i=1}^{3} N_i p_i q_i = 155(.4)(.6) + 62(.4)(.6) + 93(.4)(.6)$$

$$= 74.4$$

$$ND = (310)(.0025) = .775$$

and
$$n = \frac{\sum_{i=1}^{3} N_i p_i q_i}{ND + \frac{1}{N}\sum_{i=1}^{3} N_i p_i q_i}$$

$$= \frac{74.4}{0.775 + (1/310)(74.4)} = 73.3 \text{ or } 74$$

Then
$$n_1 = nw_1 = (74)(.5) = 37$$

$$n_2 = nw_2 = (74)(.2) = 15$$

$$n_3 = nw_3 = (74)(.3) = 22$$

5.8
ADDITIONAL COMMENTS ON STRATIFIED SAMPLING

Stratified random sampling does not always produce an estimator with a smaller variance than that of the corresponding estimator in simple random sampling. The following example illustrates this point.

EXAMPLE 5.15

A wholesale food distributor in a large city wants to know whether demand is great enough to justify adding a new product to his stock. To aid in making his decision, he plans to add this product to a sample of the stores he services in order to estimate average monthly sales. He only services four large chains in the city. Hence for administrative convenience he decides to use stratified random sampling with each chain as a stratum. There are 24 stores in stratum 1, 36 in stratum 2, 30 in stratum 3, and 30 in stratum 4. Thus $N_1 = 24$, $N_2 = 36$, $N_3 = 30$, $N_4 = 30$, and $N = 120$. The distributor has enough time and money to obtain data on monthly sales in $n = 20$ stores. Because he has no prior information on the stratum variances, and because the cost of sampling is the same in each stratum, he decides to use proportional allocation, which gives

$$n_1 = n\left(\frac{N_1}{N}\right) = 20\left(\frac{24}{120}\right) = 4$$

Similarly,
$$n_2 = 20\left(\frac{36}{120}\right) = 6$$

$$n_3 = 20\left(\frac{30}{130}\right) = 5$$

$$n_4 = 5$$

The new product is introduced in four stores chosen at random from chain 1, six stores from chain 2, and five stores each from chains 3 and 4. The sales figures after a month show the results given in the accompanying table. Estimate the average sales for the month, and place a bound on the error of estimation.

Stratum 1	Stratum 2	Stratum 3	Stratum 4
94	91	108	92
90	99	96	110
102	93	100	94
110	105	93	91
	111	93	113
	101		
$\bar{y}_1 = 99$	$\bar{y}_2 = 100$	$\bar{y}_3 = 98$	$\bar{y}_4 = 100$
$s_1^2 = 78.67$	$s_2^2 = 55.60$	$s_3^2 = 39.50$	$s_4^2 = 112.50$

SOLUTION

From Equation (5.1)

$$\bar{y}_{st} = \frac{1}{N} \sum_{i=1}^{4} N_i \bar{y}_i = 99.3$$

Note that the estimate $\bar{y}_{st}$ of the population mean is the average of all sample observations when proportional allocation is used.

The estimated variance of $\bar{y}_{st}$, from Equation (5.2), is

$$\hat{V}(\bar{y}_{st}) = \frac{1}{N^2} \sum_{i=1}^{4} N_i^2 \left(\frac{N_i - n_i}{N_i}\right)\left(\frac{s_i^2}{n_i}\right)$$

where for this example

$$\left(\frac{N_i - n_i}{N_i}\right) = \frac{5}{6} \qquad i = 1, 2, 3$$

Then

$$\hat{V}(\bar{y}_{st}) = \frac{1}{(120)^2}\left(\frac{5}{6}\right)\left[(24)^2\left(\frac{78.67}{4}\right) + (36)^2\left(\frac{55.60}{6}\right)\right.$$

$$\left. + (30)^2\left(\frac{39.50}{5}\right) + (30)^2\left(\frac{112.50}{5}\right)\right]$$

$$= 2.93$$

and the estimate of average monthly sales with a bound on the error of estimation is

$$\bar{y}_{st} \pm 2\sqrt{\hat{V}(\bar{y}_{st})}, \qquad 99.3 \pm 2\sqrt{2.93}, \qquad 99.3 \pm 3.4$$

Suppose the distributor had decided to take a simple random sample of $n = 20$ stores and the same 20 stores as in Example 5.15 were selected. In other words, suppose the 20 stores constitute a simple random sample rather than a stratified random sample. Then the estimator of the population mean has the same value as that calculated in the example, that is,

$$\bar{y} = \bar{y}_{st} = 99.3$$

But the estimated variance becomes

$$\hat{V}(\bar{y}) = \left(\frac{N - n}{N}\right)\left(\frac{s^2}{n}\right) = \left(\frac{5}{6}\right)\left(\frac{59.8}{20}\right) = 2.49$$

We see that the estimated variance is *smaller* for simple random sampling. Thus we conclude simple random sampling may have been better than stratified random sampling for this problem. The experimenter did not consider the fact that sales vary greatly among stores within a chain when he stratified on chains. He could have obtained a smaller variance for his estimator by stratifying on amount of sales, that is, by putting stores with low monthly sales in one stratum, stores with high sales in another, and so forth.

In many sample survey problems more than one measurement is taken on each sampling unit in order to estimate more than one population parameter. This situation causes complications in selecting the appropriate sample size and allocation, as is illustrated in the following example.

EXAMPLE 5.16

A state forest service is conducting a study of the people who use state-operated camping facilities. The state has two camping areas, one located in the mountains and one located along the coast. The forest service wishes to estimate the average number of people per campsite and the proportion of campsites occupied by out-of-state campers during a particular weekend when all sites are expected to be used. The average number of people is to be estimated with a bound of 1 on the error of estimation, and the proportion of out-of-state users is to be estimated with a bound of .1. The two camping areas conveniently form two strata, the mountain location forming stratum 1 and the coastal location stratum 2. It is known that $N_1 = 120$ campsites and $N_2 = 80$ campsites. Find the sample size and allocation necessary to achieve both of the bounds.

SOLUTION

Assuming that the costs of sampling are the same in each stratum, we can achieve the smallest sample size by using Neyman allocation. However, this allocation depends on the stratum variances and gives different allocations for the two different types of measurements involved in the problem. Instead, we use proportional allocation because it is usually close to optimum and it gives the same allocation for any desired measurement. Thus

$$w_1 = \frac{N_1}{N} = \frac{120}{200} = .6$$

$$w_2 = \frac{N_2}{N} = \frac{80}{200} = .4$$

Now the sample size must be determined separately for each of the desired estimates. First, consider estimating the average number of persons per campsite. We must have an approximation of the stratum variances in order to use Equation (5.8) for the sample size. The forest service knows from experience that most sites contain from 1 to 9 persons. Therefore we can use the approximation

$$\sigma_i \approx \frac{9 - 1}{4} = 2 \qquad i = 1, 2$$

Hence

$$\sum_{i=1}^{2} \frac{N_i^2 \sigma_i^2}{w_i} = \frac{(120)^2(4)}{.6} + \frac{(80)^2(4)}{.4} = 160,000$$

$$\sum_{i=1}^{2} N_i \sigma_i^2 = (120)(4) + (80)(4) = 800$$

$$N^2 D = N^2 \left(\frac{B^2}{4}\right) = (200)^2 \left(\frac{1}{4}\right) = 10,000$$

From Equation (5.8)

$$n = \frac{\sum\limits_{i=1}^{2} N_i^2 \sigma_i^2/w_i}{N^2 D + \sum\limits_{i=1}^{2} N_i \sigma_i^2} = \frac{160,000}{10,000 + 800} = 14.8 \text{ or } 15$$

is the required sample size.

Now let us consider estimating the proportion of out-of-state users. No prior estimates of the stratum proportions p_i are available, so we let $p_1 = p_2 = .5$ to obtain a maximum sample size. We use Equation (5.18) to find n, and hence we must find

$$\sum_{i=1}^{2} \frac{N_i^2 p_i q_i}{w_i} = \frac{(120)^2(.5)(.5)}{.6} + \frac{(80)^2(.5)(.5)}{.4} = 10,000$$

$$N^2 D = N^2 \left(\frac{B^2}{4}\right) = (200)^2 \left(\frac{.01}{4}\right) = 100$$

$$\sum_{i=1}^{2} N_i p_i q_i = 120(.5)(.5) + 80(.5)(.5) = 50$$

From Equation (5.18)

$$n = \frac{\sum\limits_{i=1}^{2} N_i^2 p_i q_i/w_i}{N^2 D + \sum\limits_{i=1}^{2} N_i p_i q_i} = \frac{10,000}{100 + 50} = 67$$

Thus

$$n_1 = nw_1 = (67)(.6) = 40$$

$$n_2 = nw_2 = (67)(.4) = 27$$

are the sample sizes required in order to achieve both bounds. Note that these sample sizes give an estimate of the average number of persons per campsite with a much smaller bound than required.

5.9
AN OPTIMAL RULE FOR CHOOSING STRATA

If our only objective of stratification is to produce estimators with small variance, then the best criterion by which to define strata is the set of values that the response can take on. For example, suppose we wish to estimate the average income per household in a community. We could estimate this average quite accurately if we could put all low-income households in one stratum and all high-income households in another before actually sampling. Of course, this allocation is often impossible because detailed knowledge of incomes before sampling might make the statistical problem unnecessary in the first place. However, we sometimes have some relating frequency data on broad categories of the variable of interest or on some highly correlated variable. In these cases the "cumulative square root of the frequency method" works well for delineating strata. Rather than attempt to explain this method in theory, we will simply show how it works in practice.

EXAMPLE 5.17

An investigator wishes to estimate the average yearly sales for 56 firms, using a sample of $n = 15$ firms. Frequency data on these firms is available in the form of classification by $50,000 increments and appears in the accompanying table. How can we best allocate the firms to $L = 3$ strata?

Income (in thousands)	Frequency	$\sqrt{\text{Frequency}}$	Cumulative $\sqrt{\text{Frequency}}$
100–150	11	3.32	3.32
150–200	14	3.74	7.06
200–250	9	3.00	10.06
250–300	4	2.00	12.06
300–350	5	2.24	14.30
350–400	8	2.83	17.13
400–450	3	1.73	18.86
450–500	2	1.41	20.27
	56		

SOLUTION

Note that we have added two columns to the frequency data for the population, namely, the square root of the frequencies and the cumulative square root. The approximately optimal method for stratification is to mark off equal intervals on the cumulative square root scale. (*Note*: On this scale 7.06 is 3.32 + 3.74, and so on.) Thus $(20.27)/3 = 6.76$, and our stratum boundaries should be as close as possible to 6.76 and $2(6.76) = 13.52$. On the actual scale 7.06 is closest to 6.76 and 14.30 is closest to 13.52. Thus the following three strata result:

Stratum 1: Firms with scales from 100,000 to 200,000.

Stratum 2: Firms with scales from 200,001 to 350,000.

Stratum 3: Firms with scales from 350,001 to 500,000.

Assuming that firms in these strata can be identified before sampling, the sample of $n = 15$ can be allocated five to each stratum. (Equal stratum sample sizes are nearly optimal with this technique.)

5.10
STRATIFICATION AFTER SELECTION OF THE SAMPLE

Occasionally, sampling problems arise in which we would like to stratify on a key variable, but we cannot place the sampling units into their correct strata until after the sample is selected. For example, we may wish to stratify a public opinion poll by sex of the respondent. If the poll is conducted by sampling telephone numbers, then respondents cannot be placed into the male or female stratum until after they are contacted. Similarly, an auditor may want to stratify accounts according to whether they are wholesale or retail, but she may not have this information until after an account is actually pulled for the sample.

Suppose a simple random sample of n people is selected for a poll. The sample can be divided into n_1 males and n_2 females *after* the sample is interviewed. Then instead of using $\bar{y}$ to estimate μ, we can use $\bar{y}_{st}$ *provided* that N_i/N is known for both males and females. Note that in this situation n_1 and n_2 are *random* since they can change from sample to sample even though n is fixed. Thus this sample is not exactly a stratified random sample according to Definition 5.1. However, if N_i/N is known, and if $n_i \geq 20$ for each stratum, then this method of stratification after selection of the sample is nearly as accurate as stratified random sampling with proportional allocation.

Stratification after the selection of a sample is often appropriate when a simple random sample is not properly balanced according to major groupings of the population. Suppose, for example, that a simple random sample of $n = 100$ people is selected from a population that should be equally divided between men and women. The sample measurement of interest is the weight of the respondent, and the goal is to estimate the average weight of people in the population. The sample gives the following information:

Men	Women
$n_1 = 20$	$n_2 = 80$
$\bar{y}_1 = 180$ pounds	$\bar{y}_2 = 110$ pounds
	$\bar{y} = 124$

With men underrepresented in the sample, the estimate $\bar{y} = 124$ seems unduly low. We can adjust this estimate by calculating

$$\bar{y}_{st} = \left(\frac{N_1}{N}\right)\bar{y}_1 + \left(\frac{N_2}{N}\right)\bar{y}_2 = .5(180) + .5(110) = 145$$

This estimate seems to be more realistic, since men and women are now equally weighted. Note that N_i/N is known, to a good degree of approximation, even though neither N_1 nor N_2 is given.

EXAMPLE 5.18

A large firm knows that 40% of its accounts receivable are wholesale and 60% are retail. However, to identify individual accounts without pulling a file and looking at it is difficult. An auditor wishes to sample $n = 100$ of these accounts in order to estimate the average amount of accounts receivable for the firm. A simple random sample turns out to contain 70% wholesale accounts and 30% retail accounts. The data is separated into wholesale and retail accounts after sampling, with the following results (in dollars):

Wholesale	Retail
$n_1 = 70$	$n_2 = 30$
$\bar{y}_1 = 520$	$\bar{y}_2 = 280$
$s_1 = 210$	$s_2 = 90$

Estimate μ, the average amount of accounts receivable for the firm, and place a bound on the error of estimation.

SOLUTION

Since the observed proportion of wholesale accounts (.7) is far from the true population proportion (.4), stratifying *after* a simple random sample is selected seems appropriate. This procedure is justified since b_1 and b_2 both exceed 20.
Now

$$\bar{y}_{st} = \left(\frac{N_1}{N}\right)\bar{y}_1 + \left(\frac{N_2}{N}\right)\bar{y}_2 = (.4)(520) + (.6)(280) = 376$$

and ignoring the finite population correction, we have

$$\hat{V}(\bar{y}_{st}) = \left(\frac{N_1}{N}\right)^2 \hat{V}(\bar{y}_1) + \left(\frac{N_2}{N}\right)^2 \hat{V}(\bar{y}_2) = (.4)^2\left(\frac{s_1^2}{n_1}\right) + (.6)^2\left(\frac{s_2^2}{n_2}\right)$$

$$= (.16)\frac{(210)^2}{70} + (.36)\frac{(90)^2}{30} = 198$$

and

$$2\sqrt{\hat{V}(\bar{y}_{st})} = 2\sqrt{198} = 28$$

Hence

$$376 \pm 28$$

is our estimate of μ. We are quite confident that μ lies between $348 and $404.

Two notes of caution are in order. If N_i/N is not known, or if it cannot be closely approximated, this method of stratification should *not* be used since errors in the weights, N_i/N, can cause this stratified estimator to be very poor. Sometimes, this method of stratification is used to adjust for nonresponse. For example, if many nonrespondents to a simple random sample were males, then the sample proportion of males would be low, and an adjusted estimate could be produced by stratification after sampling. This method may still lead to serious bias in the result if the nonresponse biases the simple random sample. The point to remember is that the original sample must still be a simple random sample from the population.

5.11
SUMMARY

A stratified random sample is obtained by separating the population elements into groups, or strata, such that each element belongs to one and only one stratum, and then independently selecting a simple random sample from each stratum. This sample survey design has three major advantages over simple random sampling. First, the variance of the estimator of the population mean is usually reduced because the variance of observations within each stratum is usually smaller than the overall population variance. Second, the cost of collecting and analyzing the data is often reduced by the separation of a large population into smaller strata. Third, separate estimates can be obtained for individual strata without selecting another sample and, hence, without additional cost.

An unbiased estimator, $\bar{y}_{st}$, of the population mean is a weighted average of the sample means for the strata; it is given by Equation (5.1). An unbiased estimator of the variance of $\bar{y}_{st}$ is given by Equation (5.2); this estimator is used in placing bounds on the error of estimation. An unbiased estimator of the population total is also given, along with its estimated variance.

Before conducting a survey, experimenters should consider how large an error of estimation they will tolerate and then should select the sample size accordingly. The sample size n is given by Equation (5.8) for a fixed bound B on the error of estimation. The sample must then be allocated among the various strata. The allocation that gives a fixed amount of information at minimum cost is given by Equation (5.9); it is affected by the stratum sizes, the stratum variances, and the costs of obtaining observations.

The estimator $\hat{p}_{st}$ of a population proportion has the same form as $\bar{y}_{st}$ and is given by Equation (5.15). An unbiased estimator of $V(\hat{p}_{st})$ is given by Equation (5.16). The related allocation and sample size problems have the same solutions as before, except that σ^2 is replaced by $p_i q_i$.

CASE STUDY REVISITED _____

THE ESTIMATION OF HEALTH CARE COSTS

IN the problem of estimating total first hospitalization costs for kidney stone patients, the Carolinas and the Rockies were selected as strata because they have very different incident rates for the disease, and information was desired for each region separately. Also, this stratification into geographic regions simplified the sampling procedures.

The sample data is summarized as follows:

Carolinas	Rockies
$n_1 = 363$	$n_2 = 258$
$\bar{y}_1 = 1350$	$\bar{y}_2 = 1150$
$\dfrac{s_1^2}{n_1} = 3600$	$\dfrac{s_2^2}{n_2} = 3600$

To estimate the total annual cost for the regions, we must first find N_1 and N_2, the numbers of stone patients expected to be found in the respective regions in a typical year. We can approximate these figures if we can find the incident rates for the disease and if we know the total population of the regions.

A companion study showed the number of stone incidents in the Carolinas to be 454 out of 100,000 population and the number in the Rockies to be 263 out of 100,000. The population of the Carolinas is 8,993,000, and the population of the Rocky Mountain states is 7,351,000, according to the 1980 census. Thus

$$N_1 = 8,993,000\left(\frac{454}{100,000}\right) = 40,828$$

and

$$N_2 = 7,351,000\left(\frac{263}{100,000}\right) = 19,333$$

We can now estimate the total annual first hospitalization cost for stone patients in the two regions combined as

$$N_1\bar{y}_1 + N_2\bar{y}_2$$

or

$$(40,828)(1350) + (19,333)(1150) = 77,350,750$$

The bound on the error of estimation is (since population sizes are large compared with sample sizes)

$$2\sqrt{\frac{N_1^2(s_1^2)}{n_1} + \frac{N_2^2(s_2^2)}{n_2}} = 2\sqrt{(40,828)^2(3600) + (19,333)^2(3600)}$$
$$= 5,420,880$$

Thus we estimate the total annual cost for the two regions to be between, roughly, $72 million and $82 million.

This method can be used to estimate the total cost for the entire United States, but sample data would be required for the remaining geographic regions.

EXERCISES

5.1 A chain of department stores is interested in estimating the proportion of accounts receivable that are delinquent. The chain consists of four stores. So that the cost of sampling is reduced, stratified random sampling is used, with each store as a stratum. Since no information on population proportions is available before sampling, proportional allocation is used. From the accompanying table, estimate p, the proportion of delinquent accounts for the chain, and place a bound on the error of estimation.

	Stratum I	Stratum II	Stratum III	Stratum IV
Number of accounts receivable	$N_1 = 65$	$N_2 = 42$	$N_3 = 93$	$N_3 = 25$
Sample size	$n_1 = 14$	$n_2 = 9$	$n_3 = 21$	$n_4 = 6$
Sample number of delinquent accounts	4	2	8	1

5.2　A corporation desires to estimate the total number of man-hours lost, for a given month, because of accidents among all employees. Since laborers, technicians, and administrators have different accident rates, the researcher decides to use stratified random sampling, with each group forming a separate stratum. Data from previous years suggest the variances shown in the accompanying table for the number of man-hours lost per employee in the three groups, and current data give the stratum sizes. Determine the Neyman allocation for a sample of $n = 30$ employees.

I (Laborers)	II (Technicians)	III (Administrators)
$\sigma_1^2 = 36$	$\sigma_2^2 = 25$	$\sigma_3^2 = 9$
$N_1 = 132$	$N_2 = 92$	$N_3 = 27$

5.3　For Exercise 5.2, estimate the total number of man-hours lost during the given month and place a bound on the error of estimation. Use the data in the accompanying table, obtained from sampling 18 laborers, 10 technicians, and 2 administrators.

I (Laborers)			II (Technicians)		III (Administrators)
8	24	0	4	5	1
0	16	32	0	24	8
6	0	16	8	12	
7	4	4	3	2	
9	5	8	1	8	
18	2	0			

5.4　A zoning commission is formed to estimate the average appraised value of houses in a residential suburb of a city. Using the two voting districts in the suburb as strata is convenient because separate lists of dwellings are available for each district. From the data given in the acocmpanying table, estimate the average appraised value for all houses in the suburb, and place a bound on the error of estimation (note that proportional allocation was used).

Stratum I	Stratum II
$N_1 = 110$	$N_2 = 168$
$n_1 = 20$	$n_2 = 30$
$\sum\limits_{i=1}^{n_1} y_i = 240{,}000$	$\sum\limits_{i=1}^{n_2} y_i = 420{,}000$
$\sum\limits_{i=1}^{n_1} y_i^2 = 2{,}980{,}000{,}000$	$\sum\limits_{i=1}^{n_2} y_i^2 = 6{,}010{,}000{,}000$

5.5 A corporation wishes to obtain information on the effectiveness of a business machine. A number of division heads will be interviewed by telephone and asked to rate the equipment on a numerical scale. The divisions are located in North America, Europe, and Asia. Hence stratified sampling is used. The costs are larger for interviewing division heads located outside North America. The accompanying table gives the costs per interview, approximate variances of the ratings, and N_i's that have been established. The corporation wants to estimate the average rating with $V(\bar{y}_{st}) = .1$. Choose the sample size n that achieves this bound, and find the appropriate allocation.

Stratum I (North America)	Stratum II (Europe)	Stratum III (Asia)
$c_1 = \$9$	$c_2 = \$25$	$c_3 = \$36$
$\sigma_1^2 = 2.25$	$\sigma_2^2 = 3.24$	$\sigma_3^2 = 3.24$
$N_1 = 112$	$N_2 = 68$	$N_3 = 39$

5.6 A school desires to estimate the average score that may be obtained on a reading comprehension exam for students in the sixth grade. The school's students are grouped into three tracks, with the fast learners in track I and the slow learners in track III. The school decides to stratify on tracks since this method should reduce variability of test scores. The sixth grade contains 55 students in track I, 80 in track II, and 65 in track III. A stratified random sample of 50 students is proportionally allocated and yields simple random samples of $n_1 = 14$, $n_2 = 20$, and $n_3 = 16$ from tracks I, II, and III. The test is administered to the sample of students, with the

Track I		Track II		Track III	
80	92	85	82	42	32
68	85	48	75	36	31
72	87	53	73	65	29
85	91	65	78	43	19
90	81	49	69	53	14
62	79	72	81	61	31
61	83	53	59	42	30
		68	52	39	32
		71	61		
		59	42		

results as shown in the table. Estimate the average score for the sixth grade, and place a bound on the error of estimation.

5.7   Suppose the average test score for the class in Exercise 5.6 is to be estimated again at the end of the school year. The costs of sampling are equal in all strata, but the variances differ. Find the optimum (Neyman) allocation of a sample of size 50, using the data of Exercise 5.6 to approximate the variances.

5.8   Using the data of Exercise 5.6, find the sample size required to estimate the average score, with a bound of 4 points on the error of estimation. Use proportional allocation.

5.9   Repeat Exercise 5.8, by using Neyman allocation. Compare the result with the answer to Exercise 5.8.

5.10   A forester wants to estimate the total number of farm acres planted in trees for a state. Since the number of acres of trees varies considerably with the size of the farm, he decides to stratify on farm sizes. The 240 farms in the state are placed in one of four categories according to size. A stratified random sample of 40 farms, selected by using proportional allocation, yields the results shown in the accompanying table on number of acres planted in trees. Estimate the total number of acres of trees on farms in the state, and place a bound on the error of estimation.

Stratum I 0–200 Acres		Stratum II 200–400 Acres		Stratum III 400–600 Acres		Stratum IV Over 600 Acres	
$N_1 = 86$		$N_2 = 72$		$N_3 = 52$		$N_4 = 30$	
$n_1 = 14$		$n_2 = 12$		$n_3 = 9$		$n_4 = 5$	
97	67	125	155	142	256	167	655
42	125	67	96	310	440	220	540
25	92	256	47	495	510	780	
105	86	310	236	320	396		
27	43	220	352	196			
45	59	142	190				
53	21						

5.11   The study of Exercise 5.10 is to be made yearly, with the bound on the error of estimation of 5000 acres. Find an approximate sample size to achieve this bound if Neyman allocation is used. Use the data in Exercise 5.10.

5.12   A psychologist working with a group of mentally retarded adults desires to estimate their average reaction time to a certain stimulus. She feels that men and women probably will show a difference in reaction times, so she wants to stratify on sex. The group of 96 people contains 43 men. In previous studies of this type researchers have found that the times range from 5 to 20 seconds for men and from 3 to 14 seconds for women. The costs of sampling are the same for both strata. Using optimum allocation, find the approximate sample size necessary to estimate the average reaction time for the group to within 1 second.

5.13   A county government is interested in expanding the facilities of a day-care center for mentally retarded children. The expansion will increase the cost of enrolling a child in the center. A sample survey will be conducted to estimate the proportion of families with retarded children that will make use of the expanded facilities. The families are divided into those who use the existing facilities and those who do not. Some families live in the city in which the center is located, and some live in the surrounding suburban and rural areas. Thus stratified random sampling is used, with users in the city, users in the surrounding county, nonusers in the city, and nonusers in the county forming strata 1, 2, 3, and 4, respectively. Approximately 90% of the present users and 50% of the present nonusers will use the expanded facilities. The costs of obtaining an observation from a user is $4.00 and from a nonuser is $8.00. The difference in cost results because nonusers are difficult to locate.

Existing records give $N_1 = 97$, $N_2 = 43$, $N_3 = 145$, and $N_4 = 68$. Find the approximate sample size and allocation necessary to estimate the population proportion with a bound of .05 on the error of estimation.

5.14 The survey of Exercise 5.13 is conducted and yields the following proportion of families who will use the new facilities:

$$\hat{p}_1 = .87, \qquad \hat{p}_2 = .93, \qquad \hat{p}_3 = .60, \qquad \hat{p}_4 = .53$$

Estimate the population proportion p, and place a bound on the error of estimation. Was the desired bound achieved?

5.15 Suppose in Exercise 5.13 that the total cost of sampling is fixed at $400. Choose the sample size and allocation that minimizes the variance of the estimator $\hat{p}_{st}$ for this fixed cost.

5.16 Refer to the information on 56 business firms given in Example 5.17.
   (a) Suppose that the $n = 15$ observations are to comprise a stratified random sample with only two strata. Find the optimal dividing point between the strata. With $n_1 = 7$ and $n_2 = 8$, assume that the resulting sample measurements (in thousands of dollars) turn out to be 110, 142, 212, 227, 167, 130, 194 for stratum 1 and 387, 345, 465, 308, 280, 480, 355, 405 for stratum 2. Estimate μ by $\bar{y}_{st}$ and calculate the estimated variance of $\bar{y}_{st}$.
   (b) Now suppose the dividing point between the two strata is shifted to 300,000. Suppose the same 15 sample measurements are drawn in a stratified random sample with $n_1 = 8$ and $n_2 = 7$. Note that this sampling shifts the 280 value from stratum 2 to stratum 1. (This result would not be likely to happen in practice and is only used here for illustrative purposes.) Find $\bar{y}_{st}$ and calculate the estimated variance of $\bar{y}_{st}$. The numerical answer should indicate the superiority of the cumulative square root of frequencies method.

5.17 If no information is available on the variable of primary interest, say y, then optimal stratification can be approximated by looking at a variable, say x, that is highly correlated with y. Suppose an investigator wishes to estimate the average number

Number of Employees	Frequency
0–10	2
11–20	4
21–30	6
31–40	6
41–50	5
51–60	8
61–70	10
71–80	14
81–90	19
91–100	13
101–110	3
111–120	7

of days of sick leave granted by a certain group of firms in a given year. No information on sick leave is available, but data on the number of employees per firm can be found. Assume that for these firms total days of sick leave are highly correlated with number of employees. Use the frequency data in the accompanying table to optimally divide the 97 firms into $L = 4$ strata for which equal sample sizes can be used.

5.18 Refer to Exercise 4.30. The auditor now wants to subsample some accounts from the 20 for more detailed auditing. Separate the 20 accounts into two strata by applying the cumulative square root of frequencies method to the amounts given.

5.19 A standard quality control check on automobile batteries involves simply measuring their weight. One particular shipment from the manufacturer consisted of batteries produced in two different months, with the same number of batteires from each month. The investigator decides to stratify on months in the sampling inspection in order to observe month-to-month variation. Simple random samples of battery weights for the two months yielded the following measurements (in pounds):

Month A	Month B
61.5	64.5
63.5	63.8
63.5	63.5
64.0	66.5
63.8	63.5
64.5	64.0

Estimate the average weight of the batteries in the population (shipment), and place a bound on the error of estimation. Ignore the fpc. The manufacturing standard for this type of battery is 69 pounds. Do you think this shipment meets the standard on the average?

5.20 In Exercise 5.19, do you think stratifying on month is desirable, or would simple random sampling work just as well? Assume that taking a simple random sample is just as convenient as taking a stratified random sample.

5.21 A quality control inspector must estimate the proportion of defective microcomputer chips coming from two different assembly operations. She knows that among the chips in the lot to be inspected, 60% are from assembly operation A and 40% are from assembly operation B. In a random sample of 100 chips, 38 turn out to be from operation A and 62 from operation B. Among the sampled chips from operation A, 6 are defective. Among the sampled chips from operation B, 10 are defective.

(a) Considering only the simple random sample of 100 chips, estimate the proportion of defectives in the lot, and place a bound on the error of estimation.

(b) Stratifying the sample, after selection, into chips from operation A and B, estimate the proportion of defectives in the population, and place a bound on the error of estimation.

Ignore the fpc's in both cases. Which answer do you find more acceptable?

5.22 When does stratification produce large gains in precision over simple random sampling? (Assume costs of observations are constant under both designs.)

5.23 A market research analyst wants to estimate the proportion of people who favor his company's product over a similar product from a rival company. The test area for his research is the state of New York. He is also interested in separate estimates of this proportion for those between the ages of 18 and 25 and for those over age 25. Discuss possible designs for this survey.

5.24 A researcher wishes to estimate the average income of employees in a large firm. Records have the employees listed by seniority, and, generally speaking, salary increases with seniority. Discuss the relative merits of simple random sampling and stratified random sampling in this case. Which would you recommend, and how would you set up the sampling scheme?

5.25 In the use of $\bar{y}_{st}$ as an estimator of μ, finding an allocation and a sample size that minimizes the $V(\bar{y}_{st})$ for fixed cost c is sometimes advantageous. That is, the cost c allowed for the survey is fixed, and we want to find the best allocation of resources in terms of maximizing the information on μ. The optimum allocation in this case is still given by Equation (5.9). Show that the appropriate choice for n is

$$n = \frac{(c - c_0) \sum_{i=1}^{L} N_i \sigma_i / \sqrt{c_i}}{\sum_{i=1}^{L} N_i \sigma_i \sqrt{c_i}}$$

where c_0 is a fixed overhead cost for the survey.

EXPERIENCES WITH REAL DATA

5.1 Data from the 1980 census of the United States are given in Table 3 of the Appendix. Treating the four major divisions of the country (Northeast, North Central, South, and West) as strata, select a stratified random sample of states, and estimate the total population for 1980, with a bound on the error of estimation. In your design, choose a sample size and an allocation that you think are appropriate for producing a good estimate. What considerations go into your choice? Does the interval produced include the true total given in the table? Compare your answer with the answers of other students. Are all intervals the same length? Do all the intervals include the true population figure?

5.2 Using the same 1980 census data and the same strata as in Exercise 5.1, estimate the proportion of states having crude birth rates (annual births per 1000 population) in excess of the overall birth rate of 15.3 for 1978. Place a bound on the error of estimation. You may want to select a sample size and an allocation different from those used in Exercise 5.1.

5.3 Table 5.4 shows heights of tall buildings in selected United States cities. Using cities as strata, select a stratified random sample of buildings and estimate the average height for this population. Place a bound on the error of estimation. Compare your answer with the answers of others in the class.

5.4 Refer to Table 5.4. Using cities as strata, estimate the proportion of buildings on this list in excess of 500 feet tall. Place a bound on the error of estimation. Pay careful attention to the sample size and the allocation in order to efficiently obtain a small bound.

TABLE 5.4 Heights of tall buildings in selected United States cities (measurements in feet)

Atlanta, Ga.

Peachtree Center Plaza Hotel	723	Richard B. Russell Federal	383	Life of Georgia Tower	371
Georgia Pacific Tower	697	Building		Georgia Power Tower	349
Southern Bell Telephone	677	Atlanta Hilton Hotel	383	Peachtree Center South	332
First National Bank	556	Peachtree Center Harris	382	Gas Light Tower	331
Equitable Building	453	Building		Hyatt Regency Hotel	330
101 Marietta Tower	446	Southern Bell Telephone	380	100 Colony Square	328
Peachtree Summit No. 1	403	Trust Company Bank	377	Georgia Power Building	318
North Avenue Tower	403	Coastal States Insurance	377	Colony Square Hotel	310
Tower Place	401	Peachtree Center Cain Building	376		
National Bank of Georgia	390	Peachtree Center Building	374		

Chicago, Ill.

Sears Tower (world's tallest)	1454	1000 Lake Shore Plaza	590	Lincoln Tower	519
Standard Oil (Indiana)	1136	Apartments		Carbide & Carbon	503
John Hancock Center	1127	Marina City Apartments, 2		Walton Colonnade	500
Water Tower Place	859	buildings	588	LaSalle-Wacker	491
First National Bank	850	Mid Continental Plaza	580	American National Bank	479
Three First National Plaza	775	Pittsfield	557	Bankers	476
One Magnificent Mile	770	Kemper Insurance Building	555	Brunswick Building	475
Huron Apartments	723	Newberry Plaza	553	Continental Companies	475
IBM Building	695	One South Wacker Dr.	550	American Furniture Mart	474
Daley Center	662	Harbor Point	550	333 Wacker Dr.	472
Lake Point Tower	645	LaSalle National Bank	535	Sheraton Hotel	471
Board of Trade, including	605	One LaSalle St.	530	Playboy Building	468
81-foot statue		111 E. Chestnut St.	529	188 Randolph Tower	465
Prudential Building	601	River Plaza	524	Tribune Tower	462
Antenna tower, 311 feet,		Pure Oil	523	Chicago Marriott	460
makes total	912	United Insurance Building	522		

Dallas, Tex.

Main Centre	939	2001 Bryan St.	512	Southwestern Bell Toll Building	372
First International Building	710	San Jacinto Tower	456	Court House & Federal Office	362
LTV Center	686	Republic Bank Building,	452	Building	
Arco Tower	660	not including 150 foot		Mercantile Dallas Building	360
Thanksgiving Tower	645	ornamental tower		Sheraton Hotel	352
Two Dallas Centre	635	Wyndham Hotel	451	Plaza of The America's	344
First National Bank	625	One Main Place	445	(E. Tower)	
Republic Bank Tower	598	LTV Tower	434	Hyatt Hotel	343
First City Center	595	Mercantile National Bank	430	Elm Place	341
SW Bell Administration Tower	580	Building, not including		Main Tower	336
One Lincoln Plaza	579	115-foot weather beacon		Dallas Galleria Tower	333
Olympia York	562	Mobil Building	430	Plaza of the America's	332
Reunion Tower	560	Mart Hotel	400	(N. & S. Tower)	
Southland Life Tower	550	Fidelity Union Tower	400	Park Central No. 3	327
Diamond Shamrock	550	One Dallas Centre	386	Adolphus Tower	327

Detroit, Mich.

Detroit Plaza Hotel	720	David Stott	436	American Center	374
Penobscot Building	557	Michigan Consolidated Gas	430	Top of Troy Building	374
15000 Town Center Dr.	554	Company Building		Detroit Bank & Trust Building	370
Guardian	485	Fisher	420	Edison Plaza	365
Renaissance Center (4 buildings)	479	J. L. Hudson Building	397	Woodward Tower	358
Book Tower	472	McNamara Federal Office	393	Buhl	350
13000 Town Center Dr.	443	Building			
Cadillac Tower	437	Detroit Bank & Trust Building	374	*Continues*	

Continues

TABLE 5.4 *Continued*

Ford Building	346	1st Federal Savings & Loan	338	Commonwealth Building	325
Michigan Bell Telephone	340	Pontchartrain Motor Hotel	336	1300 Lafayette East	325

Houston, Tex.

Texas Commerce Tower	1002	Dresser Tower	550	City National Bank Building	395
Allied Bank Plaza	985	1415 Louisiana Tower	550	The Park Lane	390
Transco Tower	899	Pennzoil (2 buildings)	523	Five Post Oak Park	389
Republic Bank Center	780	Two Allen Center	521	Houston Natural Gas Building	386
Interfirst Plaza	744	Entex Building	518	Amoco Center	382
1600 Smith St.	729	Huntington	506	Bank of the Southwest	369
Gulf Tower	725	Tenneco Building	502	Lyric Center	365
One Shell Plaza	714	Conoco Tower	465	Warwick Towers	361
(not including 285-foot TV		One Allen Center	452	Sheraton-Lincoln Hotel	352
tower)		Summit Tower West	441	Allied Bank Tower	351
Four Allen Center	692	Coastal Tower	441	(4 Oaks Place)	
Capital National Bank Plaza	685	Four Leafs Towers (2 buildings)	439	West Tower (4 Oaks Place)	351
One Houston Center	678	Gulf Building	428	Two Shell Plaza	341
First City Tower	662	The Spires	426	American General Life	337
1100 Milam Building	651	Central Tower (4 Oaks Place)	420	Park West Tower One	337
Exxon Building	606	First City National Bank	410	Transco	333
The America Tower	577	Houston Lighting & Power	410	Four Seasons Hotel	330
Marathon Oil Tower	572	Neils Esperson Building	409	Allied Chemical Byilding	328
Two Houston Center	570	Hyatt Regency Houston	401		

Los Angeles, Calif.

First Interstate Bank	858	Union Bank Square	516	The Evian	390
Crocker Center, North	750	City Hall	454	Bonaventure Hotel	367
Security Pacific National Bank	735	Equitable Life Building	454	Beaudry Center	365
Atlantic Richfield Plaza	699	Transamerica Center	452	400 S. Hope St.	375
(2 buildings)		Mutual Benefit Life Insurance	435	California Federal Savings &	363
Wells Fargo Bank	625	Building		Loan Building	
Crocker-Citizen Plaza	620	Broadway Plaza	414	Century City Office Building	363
Century Plaza Towers	571	1900 Ave. of Stars	398	Bunker Hill Towers	349
(2 buildings)		1 Wilshire Building	395	International Industries Plaza	347

New York, N.Y.

World Trade Center (2 towers)	1350	Chemical Bank, N.Y. Trust	687	Waldorf-Astoria	625
Empire State	1250	Building		Burlington House	625
TV tower, 222 feet,		55 Water St.	687	Olympic Tower	620
makes total	1472	Chanin	680	10 E. 40th St.	620
Chrysler	1046	Gulf & Western Building	679	101 Park Ave.	618
American International Building	950	Marine Midland Building	677	New York Life	615
40 Wall Tower	927	McGraw Hill	674	Penney Building	609
Citicorp Center	914	Lincoln	673	IBM	603
RCA Building	850	1633 Broadway	670	780 3rd Ave.	600
1 Chase Manhattan Plaza	813	725 5th Ave.	664	560 Lexington Ave.	600
Pan Am Building	803	American Brands	648	Celanese Building	592
Woolworth	792	A. T. & T. Tower	648	U.S. Court House	590
1 Penn Plaza	764	General Electric	640	Federal Building	587
Exxon	750	Irving Trust	640	Time & Life	587
1 Liberty Plaza	743	345 Park Ave.	634	Cooper Bregstein Building	580
Citibank	741	Grace Plaza	630	1185 Avenue of Americas	580
One Astor Plaza	730	1 New York Plaza	630	Municipal	580
Union Carbide Building	707	Home Insurance Corporation	630	1 Madison Square Plaza	576
General Motors Building	705	Building		Westvaco Building	574
Metropolitan Life	700	N.Y. Telephone	630	Socony Mobil Building	572
500 5th Ave.	697	888 7th Ave.	628		
9 W. 57th St.	688	1 Hammarskjold Plaza	628		

Continues

TABLE 5.4 *Continued*

Sperry Rand Building	570	Transportation Building	546	North American Plywood	520
600 3rd Ave.	570	Equitable	545	Du Mont Building	520
Helmsley Building	565	1 Brooklyn Bridge Plaza	540	26 Broadway	520
1 Bankers Trust Plaza	565	Equitable Life	540	Newsweek Building	518
Palace Hotel	563	Ritz Tower	540	Sterling Drug Building	515
30 Broad St.	562	Bankers Trust	540	First National City Bank	515
Sherry-Netherland	560	1166 Avenue of Americas	540	Bank of New York	513
Continental Can	557	1700 Broadway	533	Navarre	513
Sperry & Hutchinson	555	Downtown Athletic Club	530	Williamsburgh Savings Bank,	512
Galleria	552	Nelson Towers	525	Brooklyn	
Interchem Building	552	767 3rd Ave.	525	ITT—American	512
151 E. 44th St.	550	Hotel Pierre	525	International	512
N.Y. Telephone	550	House of Seagram	525	1407 Broadway Realty Corp.	512
919 3rd Ave.	550	7 World Trade Center	525	United Nations	505
Burroughs Building	550	Random House	522		
Bankers Trust	547	3 Park Ave.	522		

Philadelphia, Pa.

City Hall Tower, including	548	Industrial Valley Bank Building	482	INA Annex	383
37-foot statue of Wm. Penn.		Philadelphia National Bank	475	Penn Mutual Life	375
1818 Market St.	500	Two Girard Plaza	450	The Drake	375
Provident Mutual Life	491	2000 Market St. Building	435	Medical Tower	364
Fidelity Mutual Life Insurance	490	One Reading Center	417	State Building	351
Building		Fidelity Bank Building	405	One Logan Square	350
Philadelphia Saving Fund	490	Lewis Tower	400	United Engineers	344
Society		1500 Locust St.	390	Land Title	344
Central Penn National Bank	490	Academy House	390	Packard	340
Centre Square (2 towers)	490/416	Philadelphia Electric Company	384	Inquirer Building	340

Pittsburgh, Pa.

U.S. Steel Building	841	Grant	485	Gateway Building No. 3	344
One Mellon Bank Center	725	Koppers	475	Centre City Tower	341
PPG Tower	623	Equibank Building	445	Federal Building	340
One Oxford Centre	615	Pittsburgh National Building	424	Bell Telephone	339
Gulf	582	Alcoa Building	410	Hilton Hotel	333
University of Pittsburgh	535	Liberty Tower	358	Frick	330
Mellon Bank Building	520	Westinghouse Building	355		
1 Oliver Plaza	511	Oliver	347		

San Francisco, Calif.

Transamerica Pyramid	853	Crocker National Bank	500	595 Market Building	410
Bank of America	778	Hilton Hotel	493	101 Montgomery St.	405
101 California St.	600	Pacific Gas & Electric	492	California State Automobile	399
5 Fremont Center	600	Union Bank	487	Assn.	
Embarcadero Center, No. 4	570	Pacific Insurance	476	Alcoa Building	398
Security Pacific Bank	569	Bechtel Building	475	St. Francis Hotel	395
One Market Plaza	565	333 Market Building	474	Shell Building	386
Wells Fargo Building	561	Hartford Building	465	Del Monte	378
Standard Oil	551	Mutual Benefit Life	438	Pacific 3–Apparel Mart	376
One Sansome–Citicorp	550	Russ Building	435	Meridien Hotel	374
Shaklee Building	537	Pacific Telephone Building	435	Union Square Hyatt House	355
Aetna Life	529	Pacific Gateway	416	Hotel	
First & Market Building	529	Embarcadero Center, No. 3	412		
Metropolitan Life	524	Embarcadero Center, No. 2	412		

Source: *The World Almanac & Book of Facts*, 1984 edition, copyright © Newspaper Enterprise Association, Inc., 1983, New York, NY 10166

5.5 Estimate the average retail price of a common grocery item (for example, coffee, bread, toothpaste, or sugar) in the city, or section of the city, in which you live. Set up three to five strata for the stores, giving some careful consideration to the best manner of stratification. Some suggestions are to stratify on type of store (large supermarket versus neighborhood convenience store), on geographic areas, or on a combination of the two. The latter method is important if you wish to compare estimates for small neighborhood stores in different sections of the city. Carefully construct a frame, looking over various possible sources for lists of stores that should be included in the population. Choose a sample size to achieve a fixed variance of the estimator at minimum cost. Produce estimates for each stratum as well as the entire population. Use a random number table in the actual selection of your samples.

CASE STUDY

HOW ACCURATE IS THE INVENTORY?

AN important task of an auditor is to assess the accuracy of a firm's reported inventory figures. This assessment is done by sampling items from the inventory list, determining appropriate dollar values for each sampled item, and then estimating the total inventory error. If the total error is denoted by τ_e, the total reported inventory amount (book amount) by τ_x and the total audited inventory amount by τ_y, then the auditor wants to estimate

$$\tau_e = \tau_x - \tau_y$$

Since τ_x is known (it is the firm's reported figure), the only problem is to estimate τ_y.

The techniques of Chapter 6 can be used to estimate total audited amounts of inventory and total inventory error. One such analysis is presented later in this chapter.

6

RATIO, REGRESSION, AND DIFFERENCE ESTIMATION

6.1
INTRODUCTION

Estimation of the population mean and total in preceding chapters was based on a sample of response measurements, $y_i, y_2, \ldots, y_n$, obtained by simple random sampling (Chapter 4) and stratified random sampling (Chapter 5). Sometimes other variables are closely related to the response y. By measuring y and one or more subsidiary variables, we can obtain additional information for estimating the population mean. You are probably familiar with the use of subsidiary variables to estimate the mean of a response y. It is basic to the concept of correlation and provides means for development of a prediction equation relating y and x by the method of least squares. This topic is ordinarily covered in introductory courses in statistics (Mendenhall, 1983, Chapter 10).

Chapters 4 and 5 presented simple estimators of population parameters utilizing the response measurements $y_1, y_2, \ldots, y_n$; however, primary emphasis was placed on the design of the sample survey (simple and stratified random sampling). In contrast, this chapter presents three new methods of estimation based on the use of a subsidiary variable x. The methods are called ratio, regression, and difference estimation. All three require the measurement of two variables, y and x, on each element of the sample. A variety of sampling designs can be employed in conjunction with ratio, regression, or difference estimation, but we will discuss mainly simple random sampling. The basic ideas of how these techniques carry over to stratified random sampling will, however, be illustrated for ratio estimation.

6.2
SURVEYS THAT REQUIRE THE USE OF RATIO ESTIMATORS

Estimating a population total sometimes requires the use of subsidiary variables. We illustrate the use of a *ratio estimator* for one of the situations. The wholesale price paid for oranges in large shipments is based on the sugar content of the load. The exact sugar content cannot be determined prior to the purchase and extraction of the juice from the entire load; however, it can be estimated. One method of estimating this quantity is to first estimate the mean sugar content per orange, μ_y, and then to multiply by the number of oranges N in the load. Thus we could randomly sample n oranges from the load to determine the sugar content y for each. The average of these sample measurements, $y_1, y_2, \ldots, y_n$, will estimate μ_y; $N\bar{y}$ will estimate the total sugar content for the load, τ_y. Unfortunately, this method is not feasible because it is too time-consuming and costly to determine N (that is, to count the total number of oranges in the load).

We can avoid the need to know N by noting the following two facts. First, the sugar content of an individual orange, y, is closely related to its weight x; second, the ratio of the total sugar content τ_y to the total weight of the truckload τ_x is equal to the ratio of the mean sugar content per orange, μ_y, to the mean weight μ_x. Thus

$$\frac{\mu_y}{\mu_x} = \frac{N\mu_y}{N\mu_x} = \frac{\tau_y}{\tau_x}$$

Solving for the total sugar content of the load, we have

$$\tau_y = \frac{\mu_y}{\mu_x}(\tau_x)$$

We can estimate μ_y and μ_x by using $\bar{y}$ and $\bar{x}$, the averages of the sugar contents and weights for the sample of n oranges. Also, we can measure τ_x, the total weight of the oranges on the truck. Then a *ratio estimate* of the total sugar content τ_y is

$$\hat{\tau}_y = \frac{\bar{y}}{\bar{x}}(\tau_x)$$

or, equivalently (multiplying numerator and denominator by n),

$$\hat{\tau}_y = \frac{n\bar{y}}{n\bar{x}}(\tau_x) = \frac{\sum\limits_{i=1}^{n} y_i}{\sum\limits_{i=1}^{n} x_i}(\tau_x)$$

In this case the number of elements in the population, N, is unknown, and therefore we cannot use the simple estimator $N\bar{y}$ of the population total τ_y (Section 4.3). Thus a ratio estimator or its equivalent is necessary to accomplish the estimation objective. However, if N is known, we have the choice of using the estimator $N\bar{y}$ or the ratio estimator to estimate τ_y. If y and x are highly correlated, that is, x contributes information for the prediction of y, the ratio estimator should be better than $N\bar{y}$, which depends solely on $\bar{y}$.

In addition to the population total τ_y, there are often other parameters of interest. We may want to estimate the population mean μ_y by using a ratio estimation procedure. For example, suppose we wish to estimate the average sugar content per orange in a large shipment. We could use the sample mean $\bar{y}$ to estimate μ_y. However, if x and y are correlated, a ratio estimator that uses information from the auxiliary variable x frequently provides a more precise estimator of μ_y.

The population ratio is another parameter that may be of interest to an investigator. For example, assume we want to estimate the ratio of total automobile sales for the first quarter of this year to the number of sales during the corresponding period of the previous year. Let τ_x be the total number of sales for the first quarter of last year, and let τ_y be the total number of sales for the same period this year. We are interested in estimating the ratio

$$R = \frac{\tau_y}{\tau_x}$$

The concept of ratio estimation is used in the analysis of data from many important and practical surveys used by government, business, and academic researchers. For instance, the consumer price index (CPI) is actually a ratio of costs of purchasing a fixed set of items of constant quality and quantity for two points in time. Currently, the CPI compares today's prices with those of 1967. The CPI is based, in part, on data collected every month or every other month from 24,000 establishments (stores, hospitals, filling stations, and so on) selected from 85 urban areas around the country. The CPI is used mainly as a measure of inflation (see Chapter 1).

The Current Population Survey adjusts unemployment figures for age, sex, and race by a ratio estimation technique. For example, the ratio of number of unemployed blacks to number of blacks in the work force for a sample area can be expanded to a measure of number of unemployed blacks in a larger area by simply multiplying that sample ratio by the number of blacks in the work force of the larger area.

The Nielsen Retail Index can provide ratios of average sales prices for two competing brands of a product or for a single product at two points in time. The SAMI can provide total stock volume ratios for two competing brands.

Forecasting often employs a ratio estimation technique. For example, the ratio of total first-period sales for the current year to a similar total for last year can be multiplied by last year's total sales to estimate this year's total sales. Similar methods are used to forecast population growth.

In academic research, sociologists are interested in measures like the ratio of total monthly food budget to total monthly income per family, or the ratio of number of children to total number of people residing in a housing unit. Medical researchers can measure the relative potency of a new drug by looking at the ratio of the average amount of new drug required to evoke a certain response to the average amount of a standard drug required for the same response.

As you can see, the possible applications of ratio estimation are endless. However, we will now shift our emphasis to the construction of estimators for μ_y, τ_y, and R; and we will provide numerical examples of each. Whenever appropriate, comparisons will be made to the estimators of these parameters presented in previous chapters.

6.3
RATIO ESTIMATION USING SIMPLE RANDOM SAMPLING

Let us assume that a simple random sample of size n is to be drawn from a finite population containing N elements. How, then, do we estimate a population mean μ_y, a total τ_y, or a ratio R, utilizing sample information on y and a subsidiary variable x?

Estimator of the population ratio R:

$$r = \frac{\sum\limits_{i=1}^{n} y_i}{\sum\limits_{i=1}^{n} x_i} \tag{6.1}$$

Estimated variance of r:

$$\hat{V}(r) = \hat{V}\left(\frac{\sum\limits_{i=1}^{n} y_i}{\sum\limits_{i=1}^{n} x_i}\right) = \left(\frac{N-n}{nN}\right)\left(\frac{1}{\mu_x^2}\right)\frac{\sum\limits_{i=1}^{n}(y_i - rx_i)^2}{n-1} \tag{6.2}$$

Bound on the error of estimation:

$$2\sqrt{\hat{V}(r)} = 2\sqrt{\left(\frac{N-n}{nN}\right)\left(\frac{1}{\mu_x^2}\right)\frac{\sum\limits_{i=1}^{n}(y_i - rx_i)^2}{n-1}} \tag{6.3}$$

[If the population mean for x, μ_x, is unknown, we use $\bar{x}^2$ to approximate μ_x^2 in Equations (6.2) and (6.3).]

EXAMPLE 6.1

In a survey to examine trends in real estate, an investigator is interested in the relative change over a two-year period in the assessed value of homes in a particular community. A simple random sample of $n = 20$ homes is selected from the $N = 1000$ homes in the community. From tax records the investigator obtains the assessed value for this year (y) and the corresponding value for two years ago (x) for each of the $n = 20$ homes included in the sample. He wishes to estimate R, the relative change in assessed value for the $N = 1000$ homes, using information contained in the sample.

TABLE 6.1 Data and calculation for the real estate valuation survey (figures in $10,000 units)

Home	Assessed Value Two Years Ago, x_i	Current Value, y_i	x_i^2	y_i^2	$x_i y_i$
1	6.7	7.1	44.89	50.41	47.57
2	8.2	8.4	67.24	70.56	68.88
3	7.9	8.2	62.41	67.24	74.78
4	6.4	6.9	40.96	47.61	44.16
5	8.3	8.4	68.89	70.56	69.72
6	7.2	7.9	51.84	62.41	56.88
7	6.0	6.5	36.00	42.25	39.00
8	7.4	7.6	54.76	57.76	56.24
9	8.1	8.9	65.61	79.21	72.09
10	9.3	9.9	86.49	98.01	92.07
11	8.2	9.1	67.24	82.81	74.62
12	6.8	7.3	46.24	53.29	49.64
13	7.4	7.8	54.76	60.84	57.72
14	7.5	8.3	56.25	68.89	62.25
15	8.3	8.9	68.89	79.21	73.87
16	9.1	9.6	82.81	92.16	87.36
17	8.6	8.7	73.96	75.69	74.82
18	7.9	8.8	62.41	77.44	69.52
19	6.3	7.0	39.69	49.00	44.10
20	8.9	9.4	79.21	88.36	83.66
	154.5	164.7	1210.55	1373.71	1288.95

The data for the real estate survey are presented in Table 6.1. We have added the x_i^2, y_i^2, and $x_i y_i$ columns, which are useful in the calculation of $\hat{V}(r)$.

Using the data in Table 6.1, estimate R, the relative change in real estate valuation over the given two-year period. Place a bound on the error of estimation.

SOLUTION

The estimate of R using the sample data is given by

$$r = \frac{\sum\limits_{i=1}^{20} y_i}{\sum\limits_{i=1}^{20} x_i} = \frac{\text{total current valuation of the 20 homes}}{\text{total valuation of the 20 homes 2 years ago}}$$

Using Table 6.1,

$$r = \frac{164.7}{154.5} = 1.07$$

Hence we estimate that real estate valuation has increased approximately 20% over a two-year period in the area studied.

The bound on the error of estimation is found by using Equation (6.3). A shortcut method for calculating $\sum_{i=1}^{n} (y_i - rx_i)^2$ is given by

$$\sum\limits_{i=1}^{n} (y_i - rx_i)^2 = \sum\limits_{i=1}^{n} y_i^2 + r^2 \sum\limits_{i=1}^{n} x_i^2 - 2r \sum\limits_{i=1}^{n} x_i y_i \qquad (6.4)$$

These quantities can be obtained from Table 6.1:

$$\sum\limits_{i=1}^{20} (y_i - rx_i)^2 = 1373.71 + (1.07)^2(1210.55) - 2(1.07)(1288.95)$$

$$= 1.3157$$

Using Equation (6.3) yields

$$2\sqrt{\hat{V}(r)} = 2\sqrt{\left(\frac{N-n}{nN}\right)\left(\frac{1}{\bar{x}^2}\right)\frac{\sum\limits_{i=1}^{n} (y_i - rx_i)^2}{n-1}}$$

$$= 2\sqrt{\frac{1000-20}{20(1000)}\left[\frac{1}{(7.725)^2}\right]\left(\frac{1.3157}{19}\right)} = .02$$

Thus we estimate the ratio of current real estate valuation to that of two years ago to be $r = 1.07$, and we are quite confident that the error of estimation is less than .02. That is, the true ratio R for the population should be between 1.05 and 1.09. Note that the bound on the error of estimation is quite small. Hence r should be a fairly accurate estimate of R.

The large-sample confidence intervals based on normal distribution theory, as introduced in Chapter 2, apply in the ratio estimation case as well. Thus, for example, an approximate 90% confidence interval for the ratio R is of the form

$$r \pm 1.645\sqrt{\hat{V}(r)}$$

The ratio technique for estimating a population total τ_y was applied in estimating the total sugar content of a truckload of oranges. The simple estimator $N\bar{y}$ is not applicable because we do not know N, the total number of oranges in the truck. The following ratio estimation procedure can be applied in estimating τ_y whether or not N is known.

Ratio estimator of the population total $\hat{\tau}_y$:

$$\hat{\tau}_y = \frac{\sum\limits_{i=1}^{n} y_i}{\sum\limits_{i=1}^{n} x_i}(\tau_x) = r\tau_x \tag{6.5}$$

Estimated variance of $\hat{\tau}_y$:

$$\hat{V}(\hat{\tau}_y) = (\tau_x)^2 \hat{V}(r) = \tau_x^2 \left(\frac{N-n}{nN}\right)\left(\frac{1}{\mu_x^2}\right)\frac{\sum\limits_{i=1}^{n}(y_i - rx_i)^2}{n-1} \tag{6.6}$$

where μ_x and τ_x are the population mean and total, respectively, for the random variable x.

Bound on the error of estimation:

$$2\sqrt{\hat{V}(\hat{\tau}_y)} = 2\sqrt{\tau_x^2 \left(\frac{N-n}{nN}\right)\left(\frac{1}{\mu_x^2}\right)\frac{\sum\limits_{i=1}^{n}(y_i - rx_i)^2}{n-1}} \tag{6.7}$$

Note that although we do not need to know N or μ_x, we must know τ_x in order to estimate τ_y by use of the ratio estimation procedure.

EXAMPLE 6.2

In a study to estimate the total sugar content of a truckload of oranges, a random sample of $n = 10$ oranges was juiced and weighed (see Table 6.2). The total weight of all the oranges, obtained by first weighing the truck loaded and then unloaded, was found to be 1800 pounds. Estimate τ_y, the total sugar content for the oranges, and place a bound on the error of estimation.

SOLUTION

The sugar content of an orange is usually recorded in degrees brix, which is a measure of the number of pounds of solids (mostly sugar) per 100 pounds of juice. For our calculations we will use the actual pounds per

TABLE 6.2 Data for Example 6.2

Orange	Sugar Content (in pounds)	Weight of Orange (in pounds)
1	.021	.40
2	.030	.48
3	.025	.43
4	.022	.42
5	.033	.50
6	.027	.46
7	.019	.39
8	.021	.41
9	.023	.42
10	.025	.44
	$\sum\limits_{i=1}^{10} y_i = .246$	$\sum\limits_{i=1}^{10} x_i = 4.35$

orange. An estimate of τ_y can be obtained by using Equation (6.5):

$$\hat{\tau}_y = r\tau_x = \frac{\sum\limits_{i=1}^{10} y_i}{\sum\limits_{i=1}^{10} x_i} (\tau_x) = \frac{.246}{4.35}(1800) = 101.79 \text{ pounds}$$

A bound on the error of estimation can be found if we use a modified version of Equation (6.7). Because N is unknown in this example, we assume that the finite population correction, $(N - n)/N$, is near unity. This assumption is reasonable because we expect at least $N = 4000$ oranges even in a small truckload. The sample mean $\bar{x}$ must be used in place of μ_x in Equation (6.7), because μ_x is unknown. With these adjustments Equation (6.7) becomes

$$2\sqrt{\hat{V}(\hat{\tau}_y)} = 2\sqrt{\tau_x^2 \left(\frac{1}{n}\right)\left(\frac{1}{\bar{x}^2}\right) \frac{\sum\limits_{i=1}^{n} (y_i - rx_i)^2}{n - 1}}$$

Use Equation (6.4), for computation ease:

$$\sum\limits_{i=1}^{10} (y_i - rx_i)^2 = \sum\limits_{i=1}^{10} y_i^2 + r^2 \sum\limits_{i=1}^{10} x_i^2 - 2r \sum\limits_{i=1}^{10} x_i y_i$$

where

$$r = \frac{\sum\limits_{i=1}^{10} y_i}{\sum\limits_{i=1}^{10} x_i} = \frac{.246}{4.35} = .0566$$

From the data,

$$\sum_{i=1}^{10} y_i^2 = (.021)^2 + (.030)^2 + \cdots + (.025)^2 = .006224$$

$$\sum_{i=1}^{10} x_i^2 = (.40)^2 + (.48)^2 + \cdots + (.44)^2 = 1.9035$$

$$\sum_{i=1}^{10} y_i x_i = (.021)(.40) + (.030)(.48) + \cdots + (.025)(.44) = .10839$$

$$\bar{x} = \frac{4.35}{10} = .435$$

Substituting into Equation (6.4) gives

$$\sum_{i=1}^{10} (y_i - rx_i)^2 = \sum_{i=1}^{10} y_i^2 + r^2 \sum_{i=1}^{10} x_i^2 - 2r \sum_{i=1}^{10} x_i y_i$$

$$= .006224 + (.0566)^2(1.9035) - 2(.0566)(.10839)$$

$$= .000052285$$

Then the bound on the error of estimation is

$$2\sqrt{\hat{V}(\hat{\tau}_y)} = 2\sqrt{\tau_x^2 \left(\frac{1}{n}\right)\left(\frac{1}{\bar{x}^2}\right) \frac{\sum\limits_{i=1}^{n} (y_i - rx_i)^2}{n - 1}}$$

$$= 2\sqrt{(1800)^2\left(\frac{1}{10}\right)\left[\frac{1}{(.435)^2}\right]\left(\frac{.000052285}{9}\right)} = 6.3$$

To summarize, the ratio estimate of the total sugar content of the truckload of oranges is $\hat{\tau}_y = 101.79$ pounds, with a bound on the error of estimation of 6.3. We are confident that the total sugar content τ_y lies in the interval

$$101.79 \pm 6.3$$

that is, the interval 95.49 to 108.09 pounds.

You will recall that the population size N is frequently known. Consequently, the investigator must decide under what conditions use of the ratio estimator $\hat{\tau}_y = r\tau_x$ is better than use of the corresponding estimator $N\bar{y}$, where both estimators are based on simple random sampling (see Section 6.5). Generally, $r\tau_x$ possesses a smaller variance than $N\bar{y}$ when there is a strong positive correlation between x and y (where ρ, the correlation coefficient between x and y, is greater than $\frac{1}{2}$). Intuitively, this statement makes sense because in ratio estimation we are using the additional information provided by the subsidiary variable x.

If an investigator is interested in a population mean rather than a population total, the corresponding ratio estimation procedure is shown in Equations (6.8), (6.9), and (6.10).

Ratio estimator of a population mean μ_y:

$$\hat{\mu}_y = \frac{\sum\limits_{i=1}^{n} y_i}{\sum\limits_{i=1}^{n} x_i}(\mu_x) = r\mu_x \tag{6.8}$$

Estimated variance of $\hat{\mu}_y$:

$$\hat{V}(\hat{\mu}_y) = \mu_x^2 \hat{V}(r) = \mu_x^2 \left(\frac{N-n}{nN}\right)\left(\frac{1}{\mu_x^2}\right)\frac{\sum\limits_{i=1}^{n}(y_i - rx_i)^2}{n-1} \tag{6.9}$$

Bound on the error of estimation:

$$2\sqrt{\hat{V}(\hat{\mu}_y)} = 2\sqrt{\left(\frac{N-n}{nN}\right)\frac{\sum\limits_{i=1}^{n}(y_i - rx_i)^2}{n-1}} \tag{6.10}$$

Note that we do not need to know τ_x or N to estimate μ_y when using the ratio procedure; however, we must know μ_x.

EXAMPLE 6.3

A company wishes to estimate the average amount of money μ_y paid to employees for medical expenses during the first three months of the current calendar year. Average quarterly reports are available in the fiscal reports of the previous year. A random sample of 100 employee records is taken from the population of 1000 employees. The sample results are summarized below. Use the data to estimate μ_y and to place a bound on the error of estimation.

$$n = 100, \qquad N = 1000$$

Total for the current quarter:

$$\sum_{i=1}^{100} y_i = 1750$$

Total for the corresponding quarter of the previous year:

$$\sum_{i=1}^{100} x_i = 1200$$

Population total τ_x for the corresponding quarter of the previous year

$$\tau_x = 12,500$$

$$\sum_{i=1}^{100} y_i^2 = 31,650, \qquad \sum_{i=1}^{100} x_i^2 = 15,620, \qquad \sum_{i=1}^{100} y_i x_i = 22,059.35$$

SOLUTION

The estimate of μ_y is

$$\hat{\mu}_y = r\mu_x$$

where

$$\mu_x = \frac{T_x}{N} = \frac{12{,}500}{1000} = 12.5$$

Then

$$\hat{\mu}_y = \frac{\sum\limits_{i=1}^{100} y_i}{\sum\limits_{i=1}^{100} x_i}(\mu_x) = \frac{1750}{1200}(12.5) = 18.23$$

The bound on the error of estimation can be found by using Equation (6.10); however, we must first calculate

$$\sum_{i=1}^{100} (y_i - rx_i)^2 = \sum_{i=1}^{100} y_i^2 + r^2 \sum_{i=1}^{100} x_i^2 - 2r \sum_{i=1}^{100} y_i x_i$$

$$= 31{,}650 + (1.4583)^2(15{,}620) - (2.9166)(22{,}059.35)$$

$$= 441.68$$

Substituting into Equation (6.10) gives the bound on the error of estimation:

$$2\sqrt{\hat{V}(\hat{\mu}_y)} = 2\sqrt{\left(\frac{N-n}{nN}\right)\frac{\sum\limits_{i=1}^{n}(y_i - rx_i)^2}{n-1}}$$

$$= 2\sqrt{\frac{1000-100}{100(1000)}\left(\frac{441.68}{99}\right)} = .42$$

Thus we estimate the average amount of money paid to employees for medical expenses to be $18.23. We are very confident that the error for estimating μ_y is less than $.42.

To remember the formulas for ratio estimation of a population mean, total, or ratio, we make the following associations. The sample ratio r is given by the formula

$$r = \frac{\sum\limits_{i=1}^{n} y_i}{\sum\limits_{i=1}^{n} x_i} \tag{6.11}$$

The estimators of R, τ_y, and μ_y are then

$$\hat{R} = r \tag{6.12}$$

$$\hat{\tau}_y = r\tau_x \tag{6.13}$$

$$\hat{\mu}_y = r\mu_x \tag{6.14}$$

Thus we need know only the formula for r and its relationship to $\hat{\mu}_y$ and $\hat{\tau}_y$.

Approximate variances can be obtained if you remember the basic formula,

$$\hat{V}(r) = \left(\frac{N-n}{nN}\right)\left(\frac{1}{\mu_x^2}\right)\frac{\sum_{i=1}^{n}(y_i - rx_i)^2}{n-1} \tag{6.15}$$

Thus $$\hat{V}(\hat{\tau}_y) = \tau_x^2 \hat{V}(r) \tag{6.16}$$

$$\hat{V}(\hat{\mu}_y) = \mu_x^2 \hat{V}(r) \tag{6.17}$$

6.4
SELECTING THE SAMPLE SIZE

We stated previously that the amount of information contained in the sample depends on the variation in the data (which is frequently controlled by the sample survey design) and the number of observations n included in the sample. Once the sampling procedure (design) has been chosen, the investigator must determine the number of elements to be drawn. We will consider the sample size required to estimate a population parameter R, μ_y, or τ_y to within B units for simple random sampling using ratio estimators.

Note that the procedure for choosing the sample size n is identical to that presented in Section 4.4. The number of observations required to estimate R, a population ratio, with a bound on the error of estimation of magnitude B is determined by setting two standard deviations of the ratio estimator r equal to B and solving this expression for n. That is, we must solve

$$2\sqrt{V(r)} = B \tag{6.18}$$

for n. Although we have not discussed the form of $V(r)$, you recall that $\hat{V}(r)$, the estimated variance of r, is given by the formula

$$\hat{V}(r) = \left(\frac{N-n}{nN}\right)\left(\frac{1}{\mu_x^2}\right)\frac{\sum_{i=1}^{n}(y_i - rx_i)^2}{n-1} \tag{6.19}$$

We can rewrite Equation (6.19) as

$$\hat{V}(r) = \left(\frac{N-n}{nN}\right)\left(\frac{1}{\mu_x^2}\right)s^2 \tag{6.20}$$

In this instance we define

$$s^2 = \frac{\sum\limits_{i=1}^{n} (y_i - rx_i)^2}{n - 1}$$

An approximate population variance, $V(r)$, can be obtained from $\hat{V}(r)$ by replacing s^2 with the corresponding population variance σ^2. Thus the number of observations required to estimate R with a bound B on the error of estimation is determined by solving the following equation for n:

$$2\sqrt{V(r)} = 2\sqrt{\left(\frac{N - n}{nN}\right)\left(\frac{1}{\mu_x^2}\right)\sigma^2} = B \qquad (6.21)$$

Sample size required to estimate R with a bound on the error of estimation B:

$$n = \frac{N\sigma^2}{ND + \sigma^2} \qquad (6.22)$$

where

$$D = \frac{B^2 \mu_x^2}{4}$$

In a practical situation we are faced with a problem in determing the appropriate sample size because we do not know σ^2. If no past information is available to calculate s^2 as an estimate of σ^2, we take a preliminary sample of size n' and compute

$$\hat{\sigma}^2 = \frac{\sum\limits_{i=1}^{n'} (y_i - rx_i)^2}{n' - 1}$$

Then we substitute this quantity for σ^2 in Equation (6.22), and we find an *approximate* sample size. If μ_x is also unknown, it can be replaced by the sample mean $\bar{x}$, calculated from the n' preliminary observations.

EXAMPLE 6.4

A manufacturing company wishes to estimate the ratio of change from last year to this year in the number of man-hours lost due to sickness. A preliminary study of $n' = 10$ employee records is made, and the results are given in the accompanying table. The company records show that the total number of man-hours lost because of sickness for the previous year was $T_x = 16,300$. Use the data to determine the sample size required to estimate R, the rate of change for the company, with a bound on the error of estimation of magnitude $B = .01$. Assume the company has 1000 employees ($N = 1000$).

Employee	Man-Hours Lost in Previous Year, x	Man-Hours Lost in Current Year, y
1	12	13
2	24	25
3	15	15
4	30	32
5	32	36
6	26	24
7	10	12
8	15	16
9	0	2
10	14	12
	178	187

SOLUTION

First, we calculate an estimate of σ^2 by using the data from the preliminary study. Thus

$$\hat{\sigma}^2 = \frac{\sum\limits_{i=1}^{10} (y_i - rx_i)^2}{9}$$

where
$$\sum_{i=1}^{10} (y_i - rx_i)^2 = \sum_{i=1}^{10} y_i^2 + r^2 \sum_{i=1}^{10} x_i^2 - 2r \sum_{i=1}^{10} x_i y_i$$

Next, from the given data we determine

$$\sum_{i=1}^{10} y_i^2 = (13)^2 + (25)^2 + \cdots + (12)^2 = 4463$$

$$\sum_{i=1}^{10} x_i^2 = (12)^2 + (24)^2 + \cdots + (14)^2 = 4066$$

$$\sum_{i=1}^{10} x_i y_i = (12)(13) + (24)(25) + \cdots + (14)(12) = 4245$$

$$r = \frac{\sum\limits_{i=1}^{10} y_i}{\sum\limits_{i=1}^{10} x_i} = \frac{187}{178} = 1.05$$

Hence

$$\sum_{i=1}^{10} (y_i - rx_i)^2 = \sum_{i=1}^{10} y_i^2 + r^2 \sum_{i=1}^{10} x_i^2 - 2r \sum_{i=1}^{10} x_i y_i$$
$$= 4463 + (1.05)^2(4066) - 2(1.05)(4245) = 31.265$$

and
$$\hat{\sigma}^2 = \frac{\sum\limits_{i=1}^{10} (y_i - rx_i)^2}{9} = \frac{31.265}{9} = 3.474$$

The required sample size can now be found by using Equation (6.22). Note that

$$\mu_x = \frac{T_x}{N} = \frac{16,300}{1000} = 16.3$$

and
$$D = \frac{B^2\mu_x^2}{4} = \frac{(.01)^2(16.3)^2}{4} = .006642$$

Thus
$$n = \frac{N\hat{\sigma}^2}{ND + \hat{\sigma}^2} = \frac{1000(3.474)}{1000(.006642) + 3.474} = 343.416$$

Therefore we should sample approximately 344 employee records to estimate R, the rate of change in man-hours lost due to sickness, with a bound on the error of estimation of .01 hour.

Similarly, we can determine the number of observations n needed to estimate a population mean μ_y, with a bound on the error of estimation of magnitude B. The required sample size is found by solving the following equation for N:

$$2\sqrt{V(\hat{\mu}_y)} = B \qquad (6.23)$$

Stated differently,

$$2\mu_x\sqrt{V(r)} = B \qquad \text{[from Equation (6.17)]}$$

The solution is shown in Equation (6.24).

Sample size required to estimate μ_y with a bound on the error of estimation B:

$$n = \frac{N\sigma^2}{ND + \sigma^2} \qquad (6.24)$$

where
$$D = \frac{B^2}{4}$$

Note that we need not know the value of μ_x to determine n in Equation (6.24); however, we do need an estimate of σ^2, either from prior information if it is available or from information obtained in a preliminary study.

EXAMPLE 6.5

An investigator wishes to estimate the average number of trees μ_y per acre on an $N = 1000$-acre plantation. She plans to sample n 1-acre plots and count the number of trees y on each plot. She also has aerial photographs of the plantation from which she can estimate the number of trees x on each plot for the entire plantation. Hence she knows μ_x. Therefore to use a ratio estimator of μ_y seems appropriate. Determine the sample size needed to estimate μ_y with a bound on the error of estimation of magnitude $B = 1.0$.

SOLUTION

Assuming no prior information is available, we must conduct a preliminary study to estimate σ^2. Since an investigator can readily examine ten 1-acre plots in a day to determine the total number of trees y per plot, conducting a preliminary study of $n' = 10$ plots is convenient. The results of such a study are given in the accompanying table, with the corresponding aerial estimates x.

Plot	Aerial Estimate, x	Actual Number, y
1	23	25
2	14	15
3	20	22
4	25	24
5	12	13
6	18	18
7	30	35
8	27	30
9	8	10
10	31	29
	208	221

An estimate of σ^2 is given by

$$\hat{\sigma}^2 = \frac{\sum\limits_{i=1}^{10} (y_i - rx_i)^2}{9}$$

Use Equation (6.4):

$$\sum_{i=1}^{10} (y_i - rx_i)^2 = \sum_{i=1}^{10} y_i^2 + r^2 \sum_{i=1}^{10} x_i^2 - 2r \sum_{i=1}^{10} x_i y_i$$

From the preliminary study,

$$\sum_{i=1}^{10} y_i^2 = (25)^2 + (15)^2 + \cdots + (29)^2 = 5469$$

$$\sum_{i=1}^{10} x_i^2 = (23)^2 + (14)^2 + \cdots + (31)^2 = 4872$$

$$\sum_{i=1}^{10} x_i y_i = (23)(25) + (14)(15) + \cdots + (31)(29) = 5144$$

$$r = \frac{\sum_{i=1}^{10} y_i}{\sum_{i=1}^{10} x_i} = \frac{221}{208} = 1.06$$

Thus

$$\sum_{i=1}^{10} (y_i - rx_i)^2 = \sum_{i=1}^{10} y_i^2 + r^2 \sum_{i=1}^{10} x_i^2 - 2r \sum_{i=1}^{10} x_i y_i$$

$$= 5469 + (1.06)^2(4872) - 2(1.06)(5144) = 37.8992$$

$$\hat{\sigma}^2 = \frac{\sum_{i=1}^{10} (y_i - rx_i)^2}{9} = \frac{37.8992}{9} = 4.21$$

We now determine n from Equation (6.24), where $D = B^2/4 = \frac{1}{4}$:

$$n = \frac{N\sigma^2}{ND + \sigma^2} = \frac{100(4.21)}{1000(.25) + 4.21} = 16.56$$

To summarize, we need to examine approximately 17 plots to estimate μ_y, the average number of trees per 1-acre plot, with a bound on the error of estimation of $B = 1.0$. We only need 7 additional observations since we have 10 from the preliminary study.

The sample size required to estimate τ_y with a bound on the error of estimation of magnitude B can be found by solving the following expression for n:

$$2\sqrt{V(\hat{\tau}_y)} = B \tag{6.25}$$

or, equivalently,

$$2\tau_x \sqrt{V(r)} = B \qquad \text{[from Equation (6.16)]}$$

Sample size required to estimate τ_y with a bound on the error of estimation B:

$$n = \frac{N\sigma^2}{ND + \sigma^2} \tag{6.26}$$

where

$$D = \frac{B^2}{4N^2}$$

EXAMPLE 6.6

An auditor wishes to compare the actual dollar value of an inventory of a hospital, τ_y, with the recorded inventory, τ_x. The recorded inventory τ_x can be summarized from computer-stored hospital records. The actual inventory τ_y could be determined by examining and counting all hospital supplies, but this process would be very time-consuming and costly. Hence the auditor plans to estimate τ_y from a sample of n different items randomly selected from the hospital's supplies.

Records in the computer list $N = 2100$ different item types and the number of each particular item in the hospital inventory. With these data a total value for each item, x, can be obtained by multiplying the total number of each recorded item by the unit value per item. The total dollar value of the inventory obtained from the computer is given by

$$\tau_x = \text{sum of the dollar values for the } N = 2100 \text{ items} = \sum_{i=1}^{2100} x_i$$

In this instance τ_x was found to be \$950,000. Determine the sample size (number of items) needed to estimate τ_y with a bound on the error of estimation of magnitude $B = \$500$.

SOLUTION

Because there is no prior information available, a preliminary study must be conducted to estimate σ^2. Two people can determine the actual dollar value y for each of 15 items in a day. For this example we will use the data from a single day's inventory ($n' = 15$) as a preliminary study to obtain a rough

Item	Dollar Value from Computer, x	Actual Dollar Value, y
1	15.0	14.0
2	9.5	9.0
3	14.2	12.5
4	20.5	22.0
5	6.7	6.3
6	9.8	8.4
7	25.7	28.5
8	12.6	10.0
9	15.1	14.4
10	30.9	28.2
11	7.3	15.5
12	28.6	26.3
13	14.7	13.1
14	20.5	19.5
15	10.9	9.8
	242.0	237.5

estimate of σ^2 and, consequently, a rough approximation of the required sample size n. Actually, the investigator would probably take a preliminary study of two or three days' inventory to provide a good approximation to σ^2 and hence n; however, to simplify computations, we will consider a preliminary study of $n' = 15$ items. These data are summarized in the accompanying table along with the corresponding computer figures (entries in hundreds of dollars).

To determine an estimate of σ^2, we must calculate

$$\sum_{i=1}^{15} (y_i - rx_i)^2 = \sum_{i=1}^{15} y_i^2 + r^2 \sum_{i=1}^{15} x_i^2 - 2r \sum_{i=1}^{15} x_i y_i$$

Using the data from the preliminary study, we obtain

$$\sum_{i=1}^{15} y_i^2 = (14.0)^2 + (9.0)^2 + \cdots + (9.8)^2 = 4522.19$$

$$\sum_{i=1}^{15} x_i^2 = (15.0)^2 + (9.5)^2 + \cdots + (10.9)^2 = 4706.54$$

$$\sum_{i=1}^{15} x_i y_i = (15.0)(14.0) + (9.5)(9.0) + \cdots + (10.9)(9.8) = 4560.27$$

$$r = \frac{\sum_{i=1}^{15} y_i}{\sum_{i=1}^{15} x_i} = \frac{237.5}{242} = .9814 \approx .98$$

Thus

$$\sum_{i=1}^{15} (y_i - rx_i)^2 = \sum_{i=1}^{15} y_i^2 + r^2 \sum_{i=1}^{15} x_i^2 - 2r \sum_{i=1}^{15} x_i y_i$$

$$= 4522.19 + (.98)^2(4706.54) - 2(.98)(4560.27)$$

$$= 104.2218$$

$$\hat{\sigma}^2 = \frac{\sum_{i=1}^{15} (y_i - rx_i)^2}{14} = \frac{104.2218}{14} = 7.4444$$

The required sample size now can be found by using Equation (6.26). We have

$$D = \frac{B^2}{4N^2} = \frac{(500)^2}{4(2100)^2} = .01417$$

and hence

$$n = \frac{N\sigma^2}{ND + \sigma^2} = \frac{2100(7.4444)}{2100(.01417) + 7.4444} = 420.2326$$

Thus the auditor must sample approximately 421 items to estimate τ_y, the actual dollar value of the inventory, to within $B = \$500$.

6.5
WHEN TO USE RATIO ESTIMATION

Use of the ratio estimator is most effective when the relationship between the response y and a subsidiary variable x is linear through the origin and the variance of y is proportional to x. The following example illustrates this point. An automobile tire distributor wishes to estimate the average cash receipts for his 1570 stores ($N = 1570$) during a particular sales period. From a simple random sample of $n = 50$ stores, the corresponding cash receipts y_i ($i = 1, 2, \ldots, 50$) are observed. One possible estimator of μ_y, the average cash receipts for the company, is $\bar{y}$, the sample mean.

In addition to obtaining cash receipts y_i, suppose the distributor can obtain x_i ($i = 1, 2, \ldots, 50$), the number of customers who made purchases in store i during the sales period. To determine the relationship between y and x, he can plot the sales and customer data for the $n = 50$ sampled stores.

If the plot is similar to the one presented in Figure 6.1, we can assume that the cash receipts y are linearly related to the number of customers purchasing goods, x. In fact, we could depict this relationship with a straight line passing through the intersection of the x and y axes, and hence we can say it is linear through the origin. In addition, you will note from Figure 6.1 that the "scatter" of y values widens as x increases. Hence we can say that the variance of y is proportional to x. Under these conditions the ratio estimator of μ_y, the average amount of cash receipts per store, should have a smaller variance and, hence, be more precise than $\bar{y}$.

FIGURE 6.1 Plot of cash receipts versus number of customers

Sometimes, a plot of y versus x does not clearly indicate that ratio estimation should be used. The strength of the correlation ρ between y and x is another good indicator of the effectiveness of the ratio estimator. For $\rho > \frac{1}{2}$, the ratio estimator should provide a more precise estimate of μ_y or τ_y than would $\bar{y}$ or $N\bar{y}$.

Unlike the estimation procedures discussed previously, ratio estimation usually leads to biased estimators. Thus we must consider the magnitude of the bias to decide which estimation procedure to use. Although there are no exact formulas to determine the bias of these estimators, it can be shown that the absolute value of the bias is less than or equal to the product of the standard deviation of the sample mean of the subsidiary variable x and the standard deviation of the ratio estimator, all divided by μ_x. That is,

$$|E(\hat{\theta}) - \theta| \le \frac{\sigma_{\bar{x}}\sigma_{\hat{\theta}}}{\mu_x} \tag{6.27}$$

where $\hat{\theta}$ can be the ratio estimator r, $\hat{\mu}_y$, or $\hat{\tau}_y$, and θ is the corresponding parameter estimated. If estimates of $\sigma_{\bar{x}}$, $\sigma_{\hat{\theta}}$, and μ_x are known from prior experimentation, we can estimate the maximum bias for a given physical situation by using Equation (6.27).

Generally, for a large sample size ($n > 30$) and for $(\sigma_{\bar{x}}/\mu_x) \le .10$, the bias is negligible. Note also that ratio estimators are unbiased when the relationship between y and x is linear through the origin.

Finally, we must consider the cost of obtaining information on the subsidiary variable x. If the physical situation suggests the use of ratio estimation, the experimenter must decide whether the increased precision of the ratio estimator justifies the additional cost.

6.6
RATIO ESTIMATION IN STRATIFIED RANDOM SAMPLING

For the same reasons indicated in Chapter 5, stratifying the population before using a ratio estimator is sometimes advantageous. We will assume that we can take a large enough sample of both x's and y's in each stratum for the variance approximations to work fairly well.

There are two different methods for constructing estimators of a ratio in stratified sampling. One is to estimate the ratio of μ_y to μ_x within each stratum and then form a weighted average of these *separate* estimates as a single estimate of the population ratio. The result of this procedure is called a *separate ratio estimator*.

The other method involves first estimating μ_y by the usual $\bar{y}_{st}$ and similarly estimating μ_x by $\bar{x}_{st}$. Then $\bar{y}_{st}/\bar{x}_{st}$ can be used as an estimator of μ_y/μ_x. This estimator is called a *combined ratio estimator*.

We will not introduce a general (and cumbersome) notation for these estimators but will illustrate their use by a numerical example. The derivation in the Appendix shows that the variance of a sum of random variables is the sum of the variances, if the variables are independent. This fact will allow us to use a sum of terms similar to those in Equation (6.9) for the variance of either the combined or the separate ratio estimator. The next two examples illustrate the techniques used.

EXAMPLE 6.7

Refer to Example 6.4. Treat the 10 observations given there on man–hours lost due to sickness as a simple random sample from company A. Thus $n_A = 10$, $\bar{y}_A = 18.7$, $\bar{x}_A = 17.8$, $r_A = 1.05$, $N_A = 1000$, and $\tau_{xA} = 16{,}300$.

A simple random sample of $n_B = 10$ measurements was taken from company B within the same industry. (Assume companies A and B together form the population of workers of interest in this problem.) The data are given in the accompanying table. It is known that $N_B = 1500$ employees and $\tau_{xB} = 12{,}800$. Find the separate ratio estimate of μ_y and its estimated variance.

SOLUTION

The ratio estimator of μ_{yA} is $(\bar{y}_A/\bar{x}_A)(\mu_{xA})$ [see Equation (6.8)], and its estimated variance is given by Equation (6.9). The corresponding estimator of μ_{yB} is $(\bar{y}_B/\bar{x}_B)(\mu_{xB})$, with a similar estimated variance.

To obtain an estimator of μ_y, the population mean of the y's, we need to average the estimators together choosing weights proportional to the stratum sizes, as in Chapter 5. Thus $\hat{\mu}_{yRS}$, given by

$$\hat{\mu}_{yRS} = \left(\frac{N_A}{N}\right)\left(\frac{\bar{y}_A}{\bar{x}_A}\right)(\mu_{xA}) + \left(\frac{N_B}{N}\right)\left(\frac{\bar{y}_B}{\bar{x}_B}\right)(\mu_{xB})$$

will be the estimator of μ_y, with estimated variance

$$\hat{V}(\hat{\mu}_{yRS}) = \left(\frac{N_A}{N}\right)^2\left(\frac{N_A - n_A}{N_A n_A}\right)\frac{\displaystyle\sum_{i=1}^{n_A}(y_i - r_A x_i)^2}{n_A - 1}$$

$$+ \left(\frac{N_B}{N}\right)^2\left(\frac{N_B - n_B}{N_B n_B}\right)\frac{\displaystyle\sum_{i=1}^{n_B}(y_i - r_B x_i)^2}{n_B - 1}$$

The observed value of $\hat{\mu}_{yRS}$ from the data is

$$\left(\frac{1000}{2500}\right)\left(\frac{18.7}{17.8}\right)(16.3) + \left(\frac{1500}{2500}\right)\left(\frac{4.6}{7.8}\right)(8.53) = 9.87$$

Since we already have

$$\sum_{i=1}^{n_A}(y_i - r_A x_i)^2 = 31.26$$

and by similar calculations for company B,

$$\sum_{i=1}^{n_B}(y_i - r_B x_i)^2 = 87.45$$

Employee	Man-Hours Lost in Previous Year, x_B	Man-Hours Lost in Current Year, y_B
1	10	8
2	8	0
3	0	4
4	14	6
5	12	10
6	6	0
7	4	2
8	0	4
9	8	4
10	16	8
	78	46

we can substitute into $\hat{V}(\hat{\mu}_{yRS})$ to obtain

$$\hat{V}(\hat{\mu}_{yRS}) = .40$$

EXAMPLE 6.8

Refer to the data of Example 6.7 and find a combined ratio estimate of μ_y.

SOLUTION

Here we use $\bar{y}_{st}$ to estimate μ_y, $\bar{x}_{st}$ to estimate μ_x, and

$$\hat{\mu}_{yRC} = \frac{\bar{y}_{st}}{\bar{x}_{st}}(\mu_x)$$

as the combined ratio estimator of μ_y. If we denote $(\bar{y}_{st}/\bar{x}_{st})$ by r_C, the estimated variance of $\hat{\mu}_{yRC}$ is

$$\hat{V}(\hat{\mu}_{yRC}) = \left(\frac{N_A}{N}\right)^2 \left(\frac{N_A - n_A}{N_A n_A}\right) \left(\frac{1}{n_A - 1}\right) \sum_{i=1}^{n_A} [(y_i - \bar{y}_A) - r_C(x_i - \bar{x}_A)]^2$$

$$+ \left(\frac{N_B}{N}\right)^2 \left(\frac{N_B - n_B}{N_B n_B}\right) \left(\frac{1}{n_B - 1}\right) \sum_{i=1}^{n_B} [(y_i - \bar{y}_B) - r_C(x_i - \bar{x}_B)]^2$$

For the data given,

$$\bar{y}_{st} = (.4)(18.7) + (.6)(4.6) = 10.24$$

$$\bar{x}_{st} = (.4)(17.8) + (.6)(7.8) = 11.80$$

$$\mu_x = \frac{16,300 + 12,800}{2500} = 11.64$$

Hence the observed value of $\hat{\mu}_{yRC}$ is

$$\frac{10.24}{11.80}(11.64) = 10.13$$

Also,

$$\sum_{i=1}^{n_A} [(y_i - \bar{y}_A) - r_C(x_i - \bar{x}_A)]^2 = 51.56$$

$$\sum_{i=1}^{n_B} [(y_i - \bar{y}_B) - r_C(x_i - \bar{x}_B)]^2 = 144.21$$

and upon substitution into $\hat{V}(\hat{\mu}_{yRC})$, we have

$$\hat{V}(\hat{\mu}_{yRC}) = .66$$

On comparing Examples 6.7 and 6.8, we see that the combined ratio estimator gives the larger estimated variance. This result is generally the case, and so we should employ the separate ratio estimator most of the time. However, the separate ratio estimator may have a larger bias since each stratum ratio estimate contributes to that bias. In summary, if the stratum sample sizes are large enough (say 20 or so) so that the separate ratios do not have large biases and so that the variance approximations work adequately, then use the separate ratio estimator. If stratum sample sizes are very small, or if the within-stratum ratios are all approximately equal, then the combined ratio estimator may perform better.

Of course, an estimator of the population total can be found by multiplying either of the estimators above by the population size N, and the variances can be adjusted accordingly. Thus we might use the notation

$$\hat{\tau}_{yRS} = N\hat{\mu}_{yRS}$$

6.7
REGRESSION ESTIMATION

We saw in Section 6.5 that the ratio estimator is most appropriate when the relationship between y and x is linear through the origin. If there is evidence of a linear relationship between the observed y's and x's, but not necessarily one that would pass through the origin, the this extra information provided by the auxiliary variable x may be taken into account through a regression estimator of the mean μ_y. One must still have knowledge of μ_x before the estimator can be employed, as it was in the case of ratio estimation of μ_y.

The underlying line that shows the basic relationship between the y's and x's is simetimes referred to as the *regression line* of y upon x. Thus the subscript L in the ensuing formulas is used to denote *linear regression*.

The estimator given next assumes the x's to be fixed in advance and the y's to be random variables. We can think of the x value as something that has already been observed, like last year's first-quarter earnings, and the y response as a random variable yet to be observed, such as the current quarterly earnings of a company for which x is already known. The probabilistic properties of the estimator then depend only on y for a given set of x's.

Regression estimator of a population mean μ_y:

$$\hat{\mu}_{yL} = \bar{y} + b(\mu_x - \bar{x}) \tag{6.28}$$

where

$$b = \frac{\sum\limits_{i=1}^{n} (y_i - \bar{y})(x_i - \bar{x})}{\sum\limits_{i=1}^{n} (x_i - \bar{x})^2}$$

Estimated variance of $\hat{\mu}_{yL}$:

$$\hat{V}(\hat{\mu}_{yL}) = \left(\frac{N-n}{Nn}\right)\left(\frac{1}{n-2}\right)\left[\sum_{i=1}^{n} (y_i - \bar{y})^2 - b^2 \sum_{i=1}^{n} (x_i - \bar{x})^2\right] \tag{6.29}$$

Bound on the error of estimation:

$$2\sqrt{\hat{V}(\hat{\mu}_{yL})} = 2\sqrt{\left(\frac{N-n}{Nn}\right)\left(\frac{1}{n-2}\right)\left[\sum_{i=1}^{n} (y_i - \bar{y})^2 - b^2 \sum_{i=1}^{n} (x_i - \bar{x})^2\right]} \tag{6.30}$$

When calculating b from observed pairs $(y_1, x_1), \ldots, (y_n, x_n)$, we may use the fact that

$$\frac{\sum\limits_{i=1}^{n} (y_i - \bar{y})(x_i - \bar{x})}{\sum\limits_{i=1}^{n} (x_i - \bar{x})^2} = \frac{\sum\limits_{i=1}^{n} y_i x_i - n\bar{x}\bar{y}}{\sum\limits_{i=1}^{n} x_i^2 - n\bar{x}^2}$$

EXAMPLE 6.9

A mathematics achievement test was given to 486 students prior to their entering a certain college. From these students a simple random sample of $n = 10$ students was selected and their progress in calculus observed. Final calculus grades were then reported, as given in the accompanying table. It is known that $\mu_x = 52$ for all 486 students taking the achievement test. Estimate μ_y for this population, and place a bound on the error of estimation.

Student	Achievement Test Score, x	Final Calculus Grade, y
1	39	65
2	43	78
3	21	52
4	64	82
5	57	92
6	47	89
7	28	73
8	75	98
9	34	56
10	52	75

SOLUTION

Calculations yield $\bar{y} = 76$, $\bar{x} = 46$,

$$b = \frac{\sum\limits_{i=1}^{n} y_i x_i - n\bar{x}\bar{y}}{\sum\limits_{i=1}^{n} x_i^2 - n\bar{x}^2} = \frac{36{,}854 - 10(46)(76)}{23{,}634 - 10(46)^2} = .766$$

$$\sum_{i=1}^{n} (y_i - \bar{y})^2 = \sum_{i=1}^{n} y_i^2 - n\bar{y}^2 = 2056$$

$$\sum_{i=1}^{n} (x_i - \bar{x})^2 = \sum_{i=1}^{n} x_i^2 - n\bar{x}^2 = 2474$$

The observed value of $\hat{\mu}_{yL}$ is then

$$\bar{y} + b(\mu_x - \bar{x}) = 76 + (.766)(52 - 46) = 80$$

Also,

$$\hat{V}(\hat{\mu}_{yL}) = \frac{N - n}{Nn}\left(\frac{1}{n-2}\right)\left[\sum_{i=1}^{n} (y_i - \bar{y})^2 - b^2 \sum_{i=1}^{n} (x_i - \bar{x})^2\right]$$

$$= \frac{486 - 10}{486(10)}\left(\frac{1}{8}\right)[2056 - (.766)^2(2474)] = 7.397$$

and the bound on the error of estimation is

$$2\sqrt{\hat{V}(\hat{\mu}_{yL})} = 5.4$$

Notice that the regression estimator of μ_y inflates the value of $\bar{y}$ since $\bar{x}$ turns out to be less than μ_x and b is positive.

A close examination of the data on sugar content and weight of oranges given in Example 6.2 might suggest that a regression estimator is more appropriate than a ratio estimator. (A plot of the points will show that the regression line does not appear to go through the origin.) However, the regression estimator of a total is of the form $N\hat{\mu}_{yL}$, specifically requiring knowledge of N. Since the ratio estimator also works well in this case, determining the number of oranges in the truckload may not be worth the extra cost and time.

In other cases N may be known or easily found. Thus one should carefully consider the choice between ratio and regression estimators when estimating population means or totals.

6.8
DIFFERENCE ESTIMATION

The difference method of estimating a population mean or total is similar to the regression method in that it adjusts the $\bar{y}$ value up or down by an amount depending on the difference $(\mu_x - \bar{x})$. However, the regression coefficient b is not computed. In effect, b is set equal to unity.

The difference method is, then, easier to employ than the regression method and frequency works just as well. It is commonly employed in auditing procedures, and we will consider such an example in this section.

The following formulas hold provided that simple random sampling was employed.

Difference estimator of a population μ_y:

$$\hat{\mu}_{yD} = \bar{y} + (\mu_x - \bar{x}) = \mu_x + \bar{d} \qquad (6.31)$$

where
$$\bar{d} = \bar{y} - \bar{x}$$

Estimated variance of $\hat{\mu}_{yD}$:

$$\hat{V}(\hat{\mu}_{yD}) = \left(\frac{N-n}{Nn}\right)\frac{\sum_{i=1}^{n}(d_i - \bar{d})^2}{n-1} \qquad (6.32)$$

where
$$d_i = y_i - x_i$$

Bound on the error of estimation:

$$2\sqrt{\hat{V}(\hat{\mu}_{yD})} = 2\sqrt{\left(\frac{N-n}{Nn}\right)\frac{\sum_{i=1}^{n}(d_i - \bar{d})^2}{n-1}} \qquad (6.33)$$

EXAMPLE 6.10

Auditors are often interested in comparing the audited value of items with the book value. Generally, book values are known for every item in the population, and audit values are obtained for a sample of these items. The book values can then be used to obtain a good estimate of the total or average audit value for the population.

Suppose a population contains 180 inventory items with a stated book value of \$13,320. Let x_i denote the book value and y_i the audit value of the ith item. A simple random sample of $n = 10$ items yields the results shown in the accompanying table. Estimate the mean audit value of μ_y by the difference method and estimate the variance of $\hat{\mu}_{yD}$.

Sample	Audit Value, y_i	Book Value, x_i	d_i
1	9	10	−1
2	14	12	+2
3	7	8	−1
4	29	26	+3
5	45	47	−2
6	109	112	−3
7	40	36	+4
8	238	240	−2
9	60	59	+1
10	170	167	+3

SOLUTION

Since $\bar{y} = 72.1$, $\bar{x} = 71.7$, and $\mu_x = 74.0$,

$$\hat{\mu}_{yD} = \mu_x + \bar{d} = 74.0 + (72.1 - 71.7) = 74.4$$

Also,

$$\left(\frac{1}{n-1}\right) \sum_{i=1}^{n} (d_i - \bar{d})^2 = \left(\frac{1}{n-1}\right) \sum_{i=1}^{n} d_i^2 - n\bar{d}^2$$

$$= \frac{58 - 10(.4)^2}{9} = 6.27$$

Thus

$$\hat{V}(\hat{\mu}_{yD}) = \left(\frac{N-n}{Nn}\right) \frac{\sum_{i=1}^{n} (d_i - \bar{d})^2}{n-1} = \left[\frac{180 - 10}{(180)10}\right](6.27) = .59$$

The type of problems that difference estimators are designed to solve can also be solved by regression and ratio estimators. We will first compare the calculations for the three estimators, and then we will talk about how to choose one over another for certain situations.

EXAMPLE 6.11

Refer to the problem of Example 6.10. Estimate μ_y by using a regression estimator and ratio estimator. Calculate an estimate of the variance in each case.

SOLUTION

Starting with the regression estimator, we have

$$b = \frac{\sum\limits_{i=1}^{n} y_i x_i - n\bar{x}\bar{y}}{\sum\limits_{i=1}^{n} x_i^2 - n\bar{x}^2} = \frac{105,881 - 10(71.7)(72.1)}{106,003 - 10(71.7)^2} = .99$$

Thus

$$\hat{\mu}_{yL} = \bar{y} + b(\mu_x - \bar{x}) = 72.1 + .99(74.0 - 71.7) = 74.38$$

Using Equation (6.29) and following the computations through yields

$$\hat{V}(\hat{\mu}_{yL}) = 2.24$$

For the ratio estimation of μ_y, Equation (6.8) yields

$$\hat{\mu}_y = \frac{\sum\limits_{i=1}^{n} y_i}{\sum\limits_{i=1}^{n} x_i}(\mu_x) = \frac{721}{717}(74) = 74.41$$

Following Equation (6.9),

$$\hat{V}(\hat{\mu}_y) = .66$$

Note that all three estimates of μ_y are very close together, but in this case the difference estimator has the smallest estimated variance, with the regression estimator having by far the largest.

How do you choose the best estimator in a given situation? The difference estimator will work well when the plot of y versus x shows the

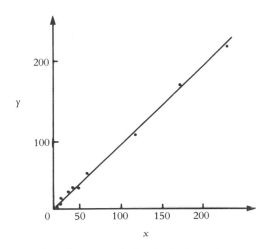

FIGURE 6.2 Plot of y versus x for Example 6.10

points falling along a straight line with unit slope. Checking such a plot for the data of Example 6.10 (see Figure 6.2) reveals that the data do indeed lie close to a straight line with a slope of unity. Thus the difference estimator is the best of the three for this case.

We have already seen in Section 6.5 that the ratio estimator is very good in cases for which the dispersion of points becomes greater as the x and y values increase. (See Figure 6.1.) In the terms of the auditing example the ratio method is preferable if the differences between audit and book values are proportional to the book values.

What, then, can be said about the regression estimator? If the plot of y versus x falls along a straight line with slope far different from unity, then the regression estimator may pay big dividends. To dramatize the point in an overly simplified example, suppose that the y value is always two times the x value, as in the following five points:

y_i	x_i	d_i
2	1	1
4	2	2
6	3	3
8	4	4
10	5	5

Then the regression estimator will yield $\hat{V}(\hat{\mu}_{yL}) = 0$ [following Equation (6.29) with $b = 2$]. The difference estimator, in contrast, will have an estimated variance based on

$$\sum_{i=1}^{n} (d_i - \bar{d})^2 = 10$$

and certainly far greater than zero.

Of course, an estimated variance of zero will almost never occur in practice; but if the plot of y versus x is linear with a slope different from unity, calculating the regression estimate and its variance may be worth the extra effort.

A caution is in order here, as well as in other places where more than one estimation method can be employed. The method to be used should be chosen on the basis of the theoretical considerations of the problem and perhaps some preliminary sampling. The experimenter should not collect data and then look for an estimator that gives small variance computations.

6.9
SUMMARY

This chapter has briefly presented ratio estimation of a population mean, total, and ratio for simple random sampling. By measuring a variable y and a subsidiary variable x on each element in the sample, we obtain additional information for estimating the population parameter of interest. When a strong positive correlation exists between the variables x and y, the ratio estimation procedure usually provides more precise estimators of μ_y and τ_y than do the standard techniques presented in Chapter 4.

Sample size requirements were presented for estimating μ_y, τ_y, and R with a bound on the error of estimation equal to B. In each case one must obtain an estimate of σ^2 from prior information or from a preliminary study to approximate the required sample size.

Regression estimation is another technique for incorporating information on a subsidiary variable. This method is usually better than ratio estimation if the relationship between the y's and the x's is a straight line not through the origin.

Although these methods can be employed with any sampling design, we have concentrated on simple random sampling, while mentioning stratified random sampling for the ratio case.

The method of difference estimation is similar in principle to regression estimation. It works well when the plot of y versus x reveals points lying uniformly close to a straight line with unit slope.

CASE STUDY REVISITED

THE ESTIMATION OF INVENTORY ERROR

AN auditor can use the method of ratio estimation to obtain an estimate of total audited amount, which can then be used to estimate the total inventory error, as outlined at the beginning of this chapter.

An electronics firm reports that their inventory of $N = 100$ computer terminals amounts to $150,000. An auditor decides to estimate the total error on this item by sampling five terminals and determining their actual value. The sample data and pertinent calculations are as follows (in units of $1000):

Book Amount, x_i	Audit Amount, y_i	rx_i	$y_i - rx_i$	$(y_i - rx_i)^2$
1.3	1.1	1.17	−.07	.0049
1.2	1.3	1.08	.22	.0484
1.5	1.4	1.35	.05	.0025
1.7	1.5	1.53	−.03	.0009
1.3	1.0	1.17	−.17	.0289
7.0	6.3			.0856

$$r = \frac{\sum\limits_{i=1}^{n} y_i}{\sum\limits_{i=1}^{n} x_i} = \frac{7.0}{6.3} = .9$$

$$\bar{x} = \frac{7.0}{5} = 1.4$$

The estimate of τ_y, the total audited amount, is

$$\hat{\tau}_y = r\tau_x = (.9)(150) = 135$$

(in \$1000 units), and the bound on the error is

$$2\sqrt{(\tau_x)^2\left(\frac{N-n}{nN}\right)\left(\frac{1}{\bar{x}}\right)^2\left(\frac{1}{n-1}\right)\sum_{i=1}^{n}(y_i - rx_i)^2} = 2\sqrt{(150)^2\left(\frac{95}{500}\right)\left(\frac{1}{1.4}\right)^2\left(\frac{1}{4}\right)(.0856)}$$

$$= 14$$

Thus we estimate the total audited amount to be between $135 - 14 = 121$ and $135 + 14 = 149$. The total error, $\tau_e = \tau_x - \tau_y$, is then estimated to be between

$$150 - 149 = 1 \quad \text{and} \quad 150 - 121 = 29$$

Note that it is quite likely that the total error is positive.

EXERCISES

6.1   A forester is interest in estimating the total volume of trees in a timber sale. He records the volume for each tree in a simple random sample. In addition, he measures the basal area for each tree marked for sale. He then uses a ratio estimator of total volume.

   The forester decides to take a simple random sample of $n = 12$ from the $N = 250$ trees marked for sale. Let x denote basal area and y the cubic-foot volume for a tree. The total basal area for all 250 trees, τ_x, is 75 square feet. Use the data in the accompanying table to estimate τ_y, the total cubic-foot volume for those trees marked for sale, and place a bound on the error of estimation.

Tree Sampled	Cubic-Foot Basal Area, x	Volume, y
1	.3	6
2	.5	9
3	.4	7
4	.9	19
5	.7	15
6	.2	5
7	.6	12
8	.5	9
9	.8	20
10	.4	9
11	.8	18
12	.6	13

6.2   Use the y data in Exercise 6.1 to compute an estimate of τ_y, using $N\bar{y}$. Place a bound on the error of estimation. Compare your results with those obtained in Exercise 6.1. Why is the estimate $N\bar{y}$, which does not use any basal-area data, much larger than the ratio estimate? (Look at μ_x and $\bar{x}$. Can you speculate about the reason for this discrepancy?)

6.3   A consumer survey was conducted to determine the ratio of the money spent on food to the total income per year for households in a small community. A simple random sample of 14 households was selected from 150 in the community. Sample data are given in the accompanying table. Estimate R, the population ratio, and place a bound on the error of estimation.

Household	Total Income, x	Amount Spent on Food, y
1	25,100	3800
2	32,200	5100
3	29,600	4200
4	35,000	6200
5	34,400	5800
6	26,500	4100
7	28,700	3900
8	28,200	3600
9	34,600	3800
10	32,700	4100
11	31,500	4500
12	30,600	5100
13	27,700	4200
14	28,500	4000

6.4   A corporation is interested in estimating the total earnings from sales of color television sets at the end of a three-month period. The total earnings figures are available for all districts within the corporation for the corresponding three-month period of the previous year. A simple random sample of 13 district offices is selected from the 123 offices within the corporation. Using a ratio estimator, estimate τ_y and place a bound on the error of estimation. Use the data in the accompanying table, and take $\tau_x = 128,200$.

Office	Three–Month Data from Previous Year, x_i	Three–Month Data from Current Year, y_i
1	550	610
2	720	780
3	1500	1600
4	1020	1030
5	620	600
6	980	1050
7	928	977
8	1200	1440
9	1350	1570
10	1750	2210
11	670	980
12	729	865
13	1530	1710

6.5   Use the data in Exercise 6.4 to estimate the mean earnings for offices within the corporation. Place a bound on the error of estimation.

6.6   An investigator has a colony of $N = 763$ rats that have been subjected to a standard drug. The average length of time to thread a maze correctly under influence of the standard drug was found to be $\mu_x = 17.2$ seconds. The investigator now would like to subject a random sample of 11 rats to a new drug. Estimate the average time required to thread the maze while under the influence of the new drug. (See the data in the accompanying table.) Place a bound on the error of estimation. (*Hint*: To employ a ratio estimator for μ_y is reasonable if we assume that the rats will react to the new drug in much the same way as they reacted to the standard drug.)

Rat	Standard Drug, x_1	New Drug, y_i
1	14.3	15.2
2	15.7	16.1
3	17.8	18.1
4	17.5	17.6
5	13.2	14.5
6	18.8	19.4
7	17.6	17.5
8	14.3	14.1
9	14.9	15.2
10	17.9	18.1
11	19.2	19.5

6.7   A group of 100 rabbits is being used in a nutrition study. A prestudy weight is recorded for each rabbit. The average of these weights is 3.1 pounds. After two months the experimenter wants to obtain a rough approximation of the average weight of the rabbits. She selects $n = 10$ rabbits at random and weighs them. The original weights and current weights are presented in the accompanying table. Estimate the average current weight, and place a bound on the error of estimation.

Rabbit	Original Weight	Current Weight
1	3.2	4.1
2	3.0	4.0
3	2.9	4.1
4	2.8	3.9
5	2.8	3.7
6	3.1	4.1
7	3.0	4.2
8	3.2	4.1
9	2.9	3.9
10	2.8	3.8

6.8   A social worker wants to estimate the ratio of the average number of rooms per apartment to the average number of people per apartment in an urban ghetto area. He selects a simple random sample of 25 apartments from the 275 in the ghetto area. Let x_i denote the number of people in apartment i, and let y_i denote the number of rooms in apartment i. From a count of the number of rooms and number of people in each apartment, the following data are obtained:

$$\bar{x} = 9.2, \qquad \bar{y} = 2.6,$$

$$\sum_{i=1}^{25} x_i^2 = 2240, \qquad \sum_{i=1}^{25} x_i y_i = 522, \qquad \sum_{i=1}^{25} y_i^2 = 169.0$$

Estimate the ratio of average number of rooms to average number of people for this area, and place a bound on the error of estimation.

6.9   A forest resource manager is interested in estimating the number of dead fir trees in a 300-acre area of heavy infestation. Using an aerial photo, he divides the area into 200 one-and-a-half-acre plots. Let x denote the photo count of dead firs and y the actual ground count for a simple random sample of $n = 10$ plots. The total number of dead fir trees obtained from the photo count is $\tau_x = 4200$. Use the sample data in the accompanying table to estimate τ_y, the total number of dead firs in the 300-acre area. Place a bound on the error of estimation.

Plot Sampled	Photo Count, x_i	Ground Count, y_i
1	12	18
2	30	42
3	24	24
4	24	36
5	18	24
6	30	36
7	12	14
8	6	10
9	36	48
10	42	54

6.10 Members of a teachers' association are concerned about the salary increases given to high school teachers in a particular school system. A simple random sample of $n = 15$ teachers is selected from an alphabetical listing of all high school teachers in the system. All 15 teachers are interviewed to determine their salaries for this year and the previous year (see the accompanying table). Use these data to estimate R, the rate of change, for $N = 750$ high school teachers in the community school system. Place a bound on the error of estimation.

Teacher	Past Year's Salary	Present Year's Salary
1	15,400	16,500
2	16,700	17,600
3	17,792	18,920
4	19,956	21,400
5	16,355	17,020
6	15,108	16,308
7	17,891	19,100
8	15,216	16,320
9	15,416	16,420
10	15,397	16,600
11	18,152	19,560
12	16,436	17,750
13	19,192	20,800
14	17,006	18,300
15	17,311	18,920

6.11   An experimenter was investigating a new food additive for cattle. Midway through
the two-month study, she was interested in estimating the average weight for the
entire herd of $N = 500$ steers. A simple random sample of $n = 12$ steers was selected
from the herd and weighed. These data and prestudy weights are presented in the
accompanying table for all cattle sampled. Assume μ_x, the prestudy average, was
880 pounds. Estimate μ_y, the average weight for the herd, and place a bound on
the error of estimation.

Steer	Prestudy Weight (in pounds)	Present Weight (in pounds)
1	815	897
2	919	992
3	690	752
4	984	1093
5	200	768
6	260	828
7	1323	1428
8	1067	1152
9	789	875
10	573	642
11	834	909
12	1049	1122

6.12   An advertising firm is concerned about the effect of a new regional promotional
campaign on the total dollar sales for a particular product. A simple random sample
of $n = 20$ stores is drawn from the $N = 452$ regional stores in which the product
is sold. Quarterly sales data are obtained for the current three-month period and
the three-month period prior to the new campaign. Use these data (see the accompanying table) to estimate τ_y, the total sales for the current period, and place a bound
on the error of estimation. Assume $\tau_x = 216,256$.

Store	Precampaign Sales	Present Sales	Store	Precampaign Sales	Present Sales
1	208	239	11	599	626
2	400	428	12	510	538
3	440	472	13	828	888
4	259	276	14	473	510
5	351	363	15	924	998
6	880	942	16	110	171
7	273	294	17	829	889
8	487	514	18	257	265
9	183	195	19	388	419
10	863	897	20	244	257

6.13   Use the data of Exercise 6.12 to determine the sample size required to estimate τ_y,
with a bound on the error of estimation equal to \$3800.

6.14   Refer to Exercises 6.4 and 6.5. By using a regression estimator, estimate the mean
earnings μ_y, and place a bound on the error of estimation. Compare your answer

with that of Exercise 6.5. Are there any advantages to using the regression estimator here?

6.15 Show how to adjust Equations (6.28) and (6.29) for estimating a total τ_y rather than a mean μ_y.

6.16 Refer to Exercise 6.9. Estimate τ_y by using a regression estimator, and place a bound on the error of estimation. Do you think the regression estimator is better than the ratio estimator for this problem?

6.17 Traders on the futures market are interested in relative prices of certain commodities rather than specific price levels. These relative prices can be presented in terms of a ratio. One such important ratio in agriculture is the cattle/hog ratio. From 64 trading days in the first quarter of 1977, the cattle and hog prices were sampled on 18 days, with the results as shown in the accompanying table. Estimate the true value of (μ_y/μ_x) for this period, and place a bound on the error of estimation.

Cattle, y_i	Hogs, x_i	Cattle, y_i	Hogs, x_i
42.40	47.80	39.65	49.40
41.40	48.60	38.45	44.30
39.60	48.20	37.80	43.90
39.45	46.75	37.20	42.70
37.00	46.50	37.60	43.25
37.80	45.40	37.50	44.55
38.55	47.30	36.90	45.10
38.60	48.20	37.30	45.00
38.80	49.40	38.60	45.25

6.18 Under what conditions should one employ a ratio estimator of τ_y rather than an estimator of the form $N\bar{y}$?

6.19 Discuss the relative merits of ratio, regression, and difference estimation.

6.20 The number of persons below the poverty level (in thousands) for all races and for blacks alone is given in the accompanying table for a random sample of $n = 6$ states. Estimate the ratio of number of blacks below the poverty level to total number of people below the poverty level for all states combined. Place a bound on the error of estimation.

State	All Races	Black
Arkansas	417	149
Georgia	869	472
Illinois	1284	545
Massachusetts	547	57
New Jersey	699	407
Oklahoma	391	59

Source: U. S. Bureau of the Census, *Statistical Abstract of the United States: 1982–1983* (103d edition). Washington, D.C., 1984. Data on page 424.

6.21 A traditional audit expresses retail sales as opening inventory plus store purchases minus closing inventory. Thus such an audit will look at these three items for a retail store over a period of time (say, six weeks) in order to report total sales. Such data combined from several stores and collected for a variety of competing brands allow one to estimate market shares (percentage of the total market held by a certain brand).

Faster methods of estimating market shares are the weekend selldown and store purchase audit methods. The first eliminates the store purchases, since purchases are minimal on a weekend, but uses a shorter time frame and is subject to distortion by weekend specials. The second uses only purchase information to compute market share and involves no audit of inventories.

Data on market shares calculated by the three methods, traditional (T), weekend (W), and purchases (P), are given in the accompanying table for one brand of beer. Observations were taken in six different time periods within a year.

T	W	P
15	16	12
18	17	14
16	17	20
14	16	11
13	12	8
16	18	15

(a) Estimate the ratio of the average market share calculated by the weekend method to that calculated by the tradiditional method. Place a bound on the error of estimation.

(b) Estimate the ratio of the average market share calculated by the purchase method to that calculated by the traditional method. Place a bound on the error of estimation.

(c) Which of the less costly methods (W or P) compares more favorably with the traditional method?

6.22 From the data given in the accompanying table on expenditures from six different areas of the United States health care field, estimate the ratio of health care expenditures in 1982 to those for 1981, and place a bound on the error of estimation. What are the shortcomings of this estimate of the true ratio of health care expenditures in the United States?

Area	1981	1982
Hospital care	118.0	135.5
Physician's services	54.8	61.8
Dentist's services	17.3	19.5
Nursing home care	24.2	27.3
Drugs	21.3	22.4
Eyeglasses and appliances	5.7	5.7

Source: *The World Almanac & Book of Facts*, 1984 edition, copyright © Newspaper Enterprise Association, Inc., 1983, New York, NY 10166.

6.23 National income for 1981 is to be estimated from a sample of $n = 10$ industries that report their 1981 income earlier than the remaining 35. (There are 45 industries used to determine the total national income.) Income data for 1980 is available for all 45 industries and totals 2174.2 (in billions). The data are given in the accompanying table.

Industry	1980	1981
Textile mill products	13.6	14.5
Chemicals and allied	37.7	42.7
Lumber and wood	15.2	15.1
Electrical and electronic equipment	48.4	53.6
Motor vehicles and equipment	19.6	25.4
Trucking and warehousing	33.5	35.9
Banking	44.4	48.5
Real estate	198.3	221.2
Health services	99.2	114.0
Educational services	15.4	17.0

Source: U. S. Bureau of the Census, *Statistical Abstract of the United States: 1982–83* (103d edition). Washington, D.C., 1984. Data on page 444.

(a) Find a ratio estimator of the 1981 total income, and place a bound on the error of estimation.

(b) Find a regression estimator of the 1981 total income, and place a bound on the error of estimation.

(c) Find a difference estimator of the 1981 total income, and place a bound on the error of estimation.

(d) Which of the three methods, (a), (b), or (c), is most appropriate in this case? Why?

6.24 The sales manager of a firm wants to measure the relationship between monthly sales and monthly advertising costs. What parameter would you suggest he estimate? Why? What data should he collect?

6.25 A certain manufacturing firm produces a product that is packaged under two brand names, for marketing purposes. These two brands serve as strata for estimating potential sales volume for the next quarter. A simple random sample of customers for each brand is contacted and asked to provide a potential sales figure y (in number of units) for the coming quarter. Last year's true sales figure, for the same quarter, is available for each of the sampled customers and is denoted by x. The data are given in the accompanying table. The sample for brand I was taken from a list of 120 customers for whom the total sales in the same quarter of last year was 24,500 units. The brand II sample came from 180 customers with a total quarterly sales last year of 21,200 units. Find a ratio estimate of the total potential sales for next quarter. Estimate the variance of your estimator.

Brand I		Brand II	
x_i	y_i	x_i	y_i
204	210	137	150
143	160	189	200
82	75	119	125
256	280	63	60
275	300	103	110
198	190	107	100
		159	180
		63	75
		87	90

Experiences with Real Data

6.1   Table 6.3 shows normal temperature (T) and amount of precipitation (P) for weather stations around the United States. Use the January and March precipitation data for a sample of n stations to determine the following estimates.

(a)   Estimate the ratio of the average March precipitation to the average January precipitation.

(b)   Estimate the average March precipitation for all stations, making use of the January and March data. Choose one of the three possible estimators from this chapter, and give reasons for your choice.

(c)   Select an appropriate sample size and place a bound on the error of estimation in both (a) and (b).

TABLE 6.3 Monthly normal temperature and precipitation

	Jan.		Feb.		March		April		May	
Station	T	P	T	P	T	P	T	P	T	P
Albany, N.Y.	22	2.2	24	2.1	33	2.6	47	2.7	58	3.3
Albuquerque, N.M.	35	0.3	40	0.4	46	0.5	56	0.5	65	0.5
Anchorage, Alaska	12	0.8	18	0.8	24	0.6	35	0.6	46	0.6
Asheville, N.C.	38	3.4	39	3.6	46	4.7	56	3.5	64	3.3
Atlanta, Ga.	42	4.3	45	4.4	51	5.8	61	4.6	69	3.7
Baltimore, Md.	33	2.9	35	2.8	43	3.7	54	3.1	64	3.6
Barrow, Alaska	−15	0.2	−19	0.2	−15	0.2	−1	0.2	19	0.2
Birmingham, Ala.	44	4.8	47	5.3	53	6.2	63	4.6	71	3.6
Bismarck, N.D.	8	0.5	14	0.4	25	0.7	43	1.4	54	2.2
Boise, Idaho	29	1.5	36	1.2	41	1.0	49	1.1	57	1.3
Boston, Mass.	29	3.7	30	3.5	38	4.0	49	3.5	59	3.5
Buffalo, N.Y.	24	2.9	24	2.6	32	2.9	45	3.2	55	3.0
Burlington, Vt.	17	1.7	19	1.7	29	1.9	43	2.6	55	3.0
Caribou, Maine	11	2.0	13	2.1	24	2.2	37	2.4	50	3.0
Charleston, S.C.	49	2.9	51	3.3	56	4.8	65	3.0	72	3.8
Chicago, Ill.	24	1.9	27	1.6	37	2.7	50	3.8	60	3.4
Cincinnati, Ohio	32	3.4	34	3.0	43	4.1	55	3.9	64	4.0
Cleveland, Ohio	27	2.6	28	2.2	36	3.1	48	3.5	58	3.5
Columbus, Ohio	28	2.9	30	2.3	39	3.4	51	3.7	61	4.1
Dallas–Ft. Worth, Tex.	45	1.8	49	2.4	55	2.5	65	4.3	73	4.5
Denver, Col.	30	0.6	33	0.7	37	1.2	48	1.9	57	2.6
Des Moines, Iowa	19	1.1	24	1.1	34	2.3	50	2.9	61	4.2
Detroit, Mich.	26	1.9	27	1.8	35	2.3	48	3.1	58	3.4
Dodge City, Kans.	31	0.5	35	0.6	41	1.1	54	1.7	64	3.1
Duluth, Minn.	9	1.2	12	0.9	24	1.8	39	2.6	49	3.4
Eureka, Calif.	47	7.4	48	5.2	48	4.8	50	3.0	53	2.1
Fairbanks, Alaska	−12	0.6	−3	0.5	10	0.5	29	0.3	47	0.7
Fresno, Calif.	45	1.8	50	1.7	54	1.6	60	1.2	67	0.3
Galveston, Tex.	54	3.0	56	2.7	61	2.6	69	2.6	76	3.2

Continues

TABLE 6.3 *Continued*

	Jan.		Feb.		March		April		May	
Station	T	P	T	P	T	P	T	P	T	P
Grand Junction, Colo.	27	0.6	34	0.6	41	0.8	52	0.8	62	0.6
Grand Rapids, Mich.	23	1.9	25	1.5	33	2.5	47	3.4	57	3.2
Hartford, Conn.	25	3.3	27	3.2	36	3.8	48	3.8	58	3.5
Helena, Mont.	18	0.6	25	0.4	31	0.7	43	0.9	52	1.8
Honolulu, Hawaii	72	4.4	72	2.5	73	3.2	75	1.4	77	1.0
Houston, Tex.	52	3.6	55	3.5	61	2.7	69	3.5	76	5.1
Huron, S.D.	13	0.4	18	0.8	29	1.1	46	2.0	57	2.8
Indianapolis, Ind.	28	2.9	31	2.4	40	3.8	52	3.9	62	4.1
Jackson, Miss.	47	4.5	50	4.6	56	5.6	66	4.7	73	4.4
Jacksonville, Fla.	55	2.8	56	3.6	61	3.6	68	3.1	74	3.2
Juneau, Alaska	24	3.9	28	3.4	32	3.6	39	3.0	47	3.3
Kansas City, Mo.	27	1.3	32	1.3	41	2.6	54	3.5	64	4.3
Knoxville, Tenn.	41	4.7	43	4.7	50	4.9	60	3.6	68	3.3
Lander, Wyo.	20	0.5	26	0.7	31	1.2	43	2.4	53	2.6
Little Rock, Ark.	40	4.2	43	4.4	50	4.9	62	5.3	70	5.3
Los Angeles, Calif.	57	3.0	58	2.8	59	2.2	62	1.3	65	0.1
Louisville, Ky.	33	3.5	36	3.5	44	5.1	56	4.1	65	4.2
Marquette, Mich.	18	1.5	20	1.5	27	1.9	40	2.6	50	2.9
Memphis, Tenn.	41	4.9	44	4.7	51	5.1	63	5.4	71	4.4
Miami, Fla.	67	2.2	68	2.0	71	2.1	75	3.6	78	6.1
Milwaukee, Wis.	19	1.6	23	1.1	31	2.2	45	2.8	54	2.9
Minneapolis, Minn.	12	0.7	17	0.8	28	1.7	45	2.0	57	3.4
Mobile, Ala.	51	4.7	54	4.8	59	7.1	68	5.6	75	4.5
Moline, Ill.	22	1.7	26	1.3	36	2.6	51	3.8	61	3.9
Nashville, Tenn.	38	4.8	41	4.4	49	5.0	60	4.1	69	4.1
Newark, N.J.	31	2.9	33	3.0	41	3.9	52	3.4	62	3.6
New Orleans, La.	53	4.5	56	4.8	61	5.5	69	4.2	75	4.2
New York, N.Y.	32	2.7	33	2.9	41	3.7	52	3.3	62	3.5
Nome, Alaska	6	0.9	5	0.8	7	0.8	19	0.7	35	0.7
Norfolk, Va.	41	3.4	41	3.3	48	3.4	58	2.7	67	3.3
Okla. City, Okla.	37	1.1	41	1.3	48	2.1	60	3.5	68	5.2
Omaha, Nebr.	23	0.8	28	1.0	37	1.6	52	3.0	63	4.1
Parkersburg, W. Va.	33	3.1	35	2.8	43	3.8	55	3.5	64	3.6
Philadelphia, Pa.	32	2.8	34	2.6	42	3.7	53	3.3	63	3.4
Phoenix, Ariz.	51	0.7	55	0.6	60	0.8	68	0.3	76	0.1
Pittsburgh, Pa.	28	2.8	29	2.4	38	3.6	50	3.4	60	3.6
Portland, Maine	22	3.4	23	3.5	32	3.6	43	3.3	53	3.3
Portland, Oreg.	38	5.9	43	4.1	46	3.6	51	2.2	57	2.1
Providence, R.I.	28	3.5	29	3.5	37	4.0	47	3.7	57	3.5
Raleigh, N.C.	41	3.2	42	3.3	49	3.4	60	3.1	67	3.3
Rapid City, S.D.	22	0.5	26	0.6	31	1.0	45	2.1	55	2.8
Reno, Nev.	32	1.2	37	0.9	40	0.7	47	0.5	55	0.7
Richmond, Va.	38	2.9	39	3.0	47	3.4	58	2.8	67	3.4

Continues

TABLE 6.3 *Continued*

Station	Jan.		Feb.		March		April		May	
	T	P	T	P	T	P	T	P	T	P
St. Louis, Mo.	31	1.9	35	2.1	43	3.0	57	3.9	66	3.9
Salt Lake City, Utah	28	1.3	33	1.2	40	1.6	49	2.1	58	1.5
San Antonio, Tex.	51	1.7	55	2.1	61	1.5	70	2.5	76	3.1
San Diego, Calif.	55	1.9	57	1.5	58	1.6	61	0.8	63	0.2
San Francisco, Calif.	48	4.4	51	3.0	53	2.5	55	1.6	58	0.4
San Juan, P.R.	75	3.7	75	2.5	76	2.0	78	3.4	79	6.5
Sault Ste. Marie, Mich.	14	1.9	15	1.5	24	1.7	38	2.2	49	3.0
Savannah, Ga.	50	2.9	52	2.9	58	4.4	66	2.9	73	4.2
Seattle, Wash.	38	5.8	42	4.2	44	3.6	49	2.5	55	1.7
Spokane, Wash.	25	2.5	32	1.7	38	1.5	46	1.1	55	1.5
Springfield, Mo.	33	1.7	37	2.2	44	3.0	57	4.3	65	4.9
Syracuse, N.Y.	24	2.7	25	2.8	33	3.0	47	3.1	57	3.0
Tampa, Fla.	60	2.3	62	2.9	66	3.9	72	2.1	77	2.4
Trenton, N.J.	32	2.8	33	2.7	41	3.8	52	3.2	62	3.4
Washington, D.C.	36	2.6	37	2.5	45	3.3	56	2.9	66	3.7
Wilmington, Del.	32	2.9	34	2.8	42	3.7	52	3.2	62	3.4

Source: *The World Almanac & Book of Facts*, 1984 edition, copyright © Newspaper Enterprise Association, Inc., 1983, New York, NY 10166.

6.2   Data from the 1980 United States census is given in Table 3 of the Appendix. Using the four regions of the country as strata, form a stratified ratio estimate of the 1980 population total making use of the 1970 data for states and for the United States as a whole. Choose an appropriate sample size and allocation. Place a bound on the error of estimation.

6.3   An interesting project is to estimate what proportion of the money spent for entertainment by students in your community goes to a specific type of entertainment, such as movie theaters. You can obtain this estimate by listing a simple random sample of *n* students, calling them on the telephone (or interviewing them personally), and recording the total amount spent on entertainment (x_i) as well as the amount spent on movies (y_i). Then estimate the ratio (μ_y/μ_x), and place a bound on the error.

    Think about sample size before you begin the study. Also, concentrating on students in one locality, such as an apartment building or group of fraternity houses, rather than students at large may be most convenient. Nonresponse is always a problem when dealing with human populations, so try to think of ways to minimize it.

6.4   Ratio estimation is often a convenient method of estimating properties of physical objects that are difficult to measure directly. Gather a box of rocks or other irregular shaped objects. You wish to estimate the total volume of the rocks. Volume of irregularly shaped objects is somewhat difficult to measure, but volume is related to weight, which is quite easy to measure. Thus volume can be estimated by using the ratio of volume to weight.

Select a sample of n rocks. Measure the weight and the volume for each rock in the sample. (You may want to use water displacement as a method of measuring volume.) Then obtain the total weight of all rocks in the box. Use these data to construct an estimate, with a bound on the error, of the total volume of the rocks.

CASE STUDY

IS QUALITY BEING MAINTAINED BY THE MANUFACTURER?

CONSUMERS and manufacturers are both concerned about the quality of items purchased or produced. Consumers want assurances that they are buying a product that will perform according to specifications, and manufacturers want evidence that their products are meeting certain standards. Such evidence is most often provided by quality control sampling plans within the manufacturing operation. These sampling plans often involve selecting items for inspection from a continuously moving production line. The selection process may require the sampling of every 100th manufactured item, one item every hour, or some similar systematic plan. The observed quality characteristic for each item may be a measurement, such as weight or time to failure, or simply a classification into the categories of "conforming" and "nonconforming." The average value of the observed characteristic is then compared with a standard value to see whether quality is being maintained.

In the manufacture of certain hydraulic equipment, one important component is a bronze casting. The main quality characteristic of the casting is the percentage of copper it contains. In the manufacturing process one casting is selected after each half hour of production, and the data on percentage copper is accumulated over an 8-hour day. Thus 16 measurements are obtained. For one day's production the percentages of copper averaged 87, with a variance of 18. The standard for the bronze was 90% copper. Is the standard being met? Techniques presented in this chapter will help us answer the question.

7

SYSTEMATIC
SAMPLING

7.1
INTRODUCTION

As we have seen in Chapters 4 and 5, both simple random sampling and stratified random sampling require very detailed work in the sample selection process. Sampling units on an adequate frame must be numbered (or otherwise identified) so that a randomization device, such as a random number table, can be used to select specific units for the sample. A sample survey design that is widely used primarily because it simplifies the sample selection process is called *systematic sampling*.

The basic idea of systematic sampling is as follows: Suppose a sample of n names is to be selected from a long list. A simple way to make this selection is to choose an appropriate interval and to select names at equal intervals along the list. Thus every tenth name might be selected, for example. If the starting point for this regular selection process is random, the result is a systematic sample.

DEFINITION 7.1   A sample obtained by randomly selecting one element from the first k elements in the frame and every kth element thereafter is called a *1-in-k systematic sample*.

As in previous chapters, we present methods for estimating a population mean, total, and proportion. We will also discuss appropriate bounds on the error of estimation and sample size requirements.

Systematic sampling provides a useful alternative to simple random sampling for the following reasons:

1. Systematic sampling is easier to perform in the field and hence is less subject to selection errors by field-workers than are either simple random samples or stratified random samples, especially if a good frame is not available.

2. Systematic sampling can provide greater information per unit cost than simple random sampling can provide.

In general, systematic sampling involves random selection of one element from the first k elements and then selection of every kth element thereafter. This procedure is easier to perform and usually less subject to interviewer error than is simple random sampling. For example, using simple random sampling to select a sample of $n = 50$ shoppers on a city street corner would be difficult. The interviewer could not determine which shoppers to include in the sample, because the population size N would not be known until all shoppers had passed the corner. In contrast, the interviewer could take a systematic sample (say 1 in 20 shoppers) until the required sample size was obtained. This procedure would be an easy one for even an inexperienced interviewer.

In addition to being easier to perform and less subject to interviewer error, systematic sampling *frequently* provides more information per unit cost than does simple random sampling. A systematic sample is generally spread more uniformly over the entire population and thus may provide more information about the population than an equivalent amount of data contained in a simple random sample. Consider the following illustration: We wish to select a 1-in-5 systematic sample of travel vouchers from a stack of $N = 1000$ (that is, sample $n = 200$ vouchers) to determine the proportion of vouchers filed incorrectly. A voucher is drawn at random from the first 5 vouchers (for example, no. 3), and every fifth voucher thereafter is included in the sample. (See the accompanying table.)

Suppose that most of the first 500 vouchers have been correctly filed, but because of a change in clerks, the second 500 have all been incorrectly filed. Simple random sampling could accidentally select a large number (perhaps all) of the 200 vouchers from either the first or the second 500 vouchers and hence yield a very poor estimate of p. In contrast, systematic sampling would select an equal number of vouchers from each of the two groups and would give a very accurate estimate of the fraction of vouchers incorrectly filed.

Additional examples are discussed in Section 7.3 to illustrate how to choose between systematic and simple random sampling in a given situation. Note, however, that the accuracy of estimates from systematic sampling depend upon the *order* of the sampling units in the frame. If the incorrect vouchers had been randomly dispensed among all vouchers, then the advantage of systematic sampling would have been lost.

Systematic sampling is very commonly used in a wide variety of contexts. The United States census directs only a minimal number of questions to every resident, but it gathers much more information from a systematic

Voucher	Voucher Sampled
1	
2	
3	3
4	
5	
6	
7	
8	8
9	
10	
⋮	⋮
996	
997	
998	998
999	
1000	

sample of residents. In the 1980 census 14 items were on the short form distributed to all residents. Another 42 items were on the long form that was distributed to, approximately, a 1-in-5 systematic sample of residents.

The Gallup poll begins its sampling process by listing 200,000 election districts in the United States and then systematically selecting 300 for a follow-up study of households. The households, or dwellings, within a sampled district may again be selected systematically—by choosing the second dwelling in every other block when moving east to west, for example.

Industrial quality control sampling plans are most often systematic in structure. An inspection plan for manufactured items moving along an assembly line may call for inspecting every 50th item. An inspection of cartons of products stored in a warehouse may suggest sampling the second carton from the left in the third row down from the top in every 5th stack. In the inspection of work done at fixed stations, the inspection plan may call for walking up and down the rows of workstations and inspecting the machinery at every 10th station. The time of day is often important in assessing quality of worker performance, and so an inspection plan may call for sampling the output of a workstation at systematically selected times throughout the day.

Auditors are frequently confronted with the problem of sampling a list of accounts to check compliance with accounting procedures or to verify dollar amounts. The most natural way to sample these lists is to choose accounts systematically.

Market researchers and opinion pollsters who sample people on the move very often employ a systematic design. Every 20th customer at a checkout counter may be asked his or her opinion on the taste, color, or texture of a food product. Every 10th person boarding a bus may be asked to fill out a questionnaire on bus service. Every 100th car entering an amusement park may be stopped and the driver questioned on various advertising policies of the park or on ticket prices. All of these samples are systematic samples.

Crop yield estimates often result from systematic samples of fields and small plots within fields. Similarly, foresters may systematically sample field plots to estimate the proportion of diseased trees or may systematically sample trees themselves to study growth patterns.

Thus systematic sampling is a popular design. We will now investigate the construction of these designs and the properties of resulting estimators of means, totals, and proportions.

7.2
HOW TO DRAW A SYSTEMATIC SAMPLE

Although simple random sampling and systematic sampling both provide useful alternatives to one another, the methods of selecting the sample data are different. A simple random sample from a population is selected by using a table of random numbers, as noted in Section 4.3. In contrast, various methods are possible in systematic sampling. The investigator can select a 1-in-3, a 1-in-5, or, in general, a 1-in-k systematic sample. For example, a medical investigator is interested in obtaining information about the average number of times 15,000 specialists prescribed a certain drug in the previous year ($N = 15,000$). To obtain a simple random sample of $n = 1600$ specialists, we would use the methods of Section 4.3 and refer to a table of random numbers; however, this procedure would require a great deal of work. Alternatively, we could select one name (specialist) at random from the first $k = 9$ names appearing on the list and then select every ninth name thereafter until a sample of size 1600 is selected. This sample is called a *1-in-9 systematic sample*.

Perhaps you wonder how k is chosen in a given situation. If the population size N is known, we can determine an approximate sample size n for the survey (see Section 7.5) and then choose k to achieve that sample size. There are $N = 15,000$ specialists in the population for the medical survey. Suppose the required sample size is $n = 100$. We must then choose k to be 150 or less. For $k = 150$, we will obtain exactly $n = 100$ observations, while for $k < 150$, the sample size will be greater than 100.

In general, for a systematic sample of n elements from a population of size N, k must be less than or equal to N/n (that is, $k \leq N/n$). Note in the preceding example that $k \leq 15,000/100$; that is, $k \leq 150$.

We cannot accurately choose k when the population size is unknown. We can determine an approximate sample size n, but we must guess the value of k needed to achieve a sample of size n. If too large a value of k is chosen, the required sample size n will not be obtained by using a 1-in-k systematic sample from the population. This result presents no problem if the experimenter can return to the population and conduct another 1-in-k systematic sample until the required sample size is obtained. However, in some situations obtaining a second systematic sample is impossible. For example, conducting another 1-in-20 systematic sample of shoppers is impossible if the required sample of $n = 50$ shoppers is not obtained at the time they pass the corner.

7.3
ESTIMATION OF A POPULATION MEAN AND TOTAL

As we have repeatedly stressed, the objective of most sample surveys is to estimate one or more population parameters. We can estimate a population mean μ from a systematic sample by using the sample mean $\bar{y}$. This outcome is shown in Equation (7.1).

Estimator of the population mean μ:

$$\hat{\mu} = \bar{y}_{sy} = \frac{\sum\limits_{i=1}^{n} y_i}{n} \tag{7.1}$$

where the subscript sy signifies that systematic sampling was used.

Estimated variance of $\bar{y}_{sy}$:

$$\hat{V}(\bar{y}_{sy}) = \frac{s^2}{n}\left(\frac{N-n}{N}\right) \tag{7.2}$$

Bound on the error of estimation:

$$2\sqrt{\hat{V}(\bar{y}_{sy})} = 2\sqrt{\frac{s^2}{n}\left(\frac{N-n}{N}\right)} \tag{7.3}$$

If N is unknown, we eliminate the fpc, $(N-n)/N$, in Equations (7.2) and (7.3).

You will recognize that the estimated variance of $\bar{y}_{sy}$ given in Equation (7.2) is identical to the estimated variance of $\bar{y}$ obtained by using simple random sampling (Section 4.3). This result does not imply that the corresponding population variances are equal. The variance of $\bar{y}$ is given by

$$V(\bar{y}) = \frac{\sigma^2}{n}\left(\frac{N-n}{N-1}\right) \tag{7.4}$$

Similarly, the variance of $\bar{y}_{sy}$ is given by

$$V(\bar{y}_{sy}) = \frac{\sigma^2}{n}[1 + (n - 1)\rho] \tag{7.5}$$

where ρ is a measure of the correlation between pairs of elements within the same systematic sample. If ρ is close to one, then the elements within the sample are all quite similar with respect to the characteristic being measured, and systematic sampling will yield a higher variance of the sample mean than will simple random sampling. If ρ is negative, then systematic sampling may be better than simple random sampling. The correlation may be negative if elements within the systematic sample tend to be extremely different. (Note that ρ cannot be so large negatively that the variance expression becomes negative.) For ρ close to zero and N fairly large, systematic sampling is roughly equivalent to simple random sampling.

An unbiased estimate of $V(\bar{y}_{sy})$ cannot be obtained by using the data from only one systematic sample. This statement does not imply that we can never obtain an estimate of $V(\bar{y}_{sy})$. When systematic sampling is equivalent to simple random sampling, we can take $V(\bar{y}_{sy})$ to be approximately equal to the estimated variance of $\bar{y}$ based on simple random sampling.

For which populations does this relationship occur? To answer this question, we must consider the following three types of populations:

1. Random population,
2. Ordered population,
3. Periodic population.

DEFINITION 7.2   A population is *random* if the elements of the population are in random order.

Elements of a systematic sample drawn from a random population are expected to be heterogeneous with ρ approximately equal to zero. Thus when N is large, the variance of $\bar{y}_{sy}$ is approximately equal to the variance of $\bar{y}$ based on simple random sampling. Systematic sampling in this case is equivalent to simple random sampling. For example, an investigator wishes to determine the average number of prescriptions written by certain doctors during the previous year. If the frame consists of a current alphabetical listing of doctors, the assumption that the names on the list are unrelated to the number of prescriptions written for a particular drug is reasonable. Hence we consider the population random. A systematic sample will be equivalent to a simple random sample in this case.

DEFINITION 7.3   A population is *ordered* if the elements within the population are ordered in magnitude according to some scheme.

In a survey to estimate the effectiveness of instruction in a large introductory course, students are asked to evaluate their instructor according to a numerical scale. A sample is then drawn from a list of evaluations that are arranged in ascending numerical order. The population of measurements from which the sample is drawn is considered an *ordered* population.

A systematic sample drawn from an ordered population is generally heterogeneous with $\rho \leq 0$. It can be shown, using Equations (7.4) and (7.5), that when N is large and $\rho \leq 0$,

$$V(\bar{y}_{sy}) \leq V(\bar{y})$$

Thus a systematic sample from an ordered population provides more information per unit cost than does a simple random sample, because the variance of $\bar{y}_{sy}$ is less than the corresponding variance of $\bar{y}$.

Since we cannot obtain an estimate of $V(\bar{y}_{sy})$ from the sample data, a conservative estimate (one that is larger than we would expect) of $V(\bar{y}_{sy})$ is given by

$$\hat{V}(\bar{y}_{sy}) = \frac{s^2}{n}\left(\frac{N-n}{N}\right)$$

DEFINITION 7.4   A population is *periodic* if the elements of the population have cyclical variation.

Suppose we are interested in determining the average daily sales volume for a chain of grocery stores. The population of daily sales is clearly periodic, with peak sales occurring toward the end of each week. The effectiveness of a 1-in-k sample depends on the value we choose for k. If we sample daily sales every Wednesday, we would probably underestimate the true average daily sales volume. Similarly, if we sample sales every Friday, we would probably overestimate the true average sales. We might sample every ninth workday to avoid consistently sampling either the low or high sales days.

Elements of a systematic sample drawn from a periodic population can be homogeneous (that is, $\rho > 0$). For example, the elements within a systematic sample of daily sales taken every Wednesday would be fairly homogeneous. It can be shown, using Equations (7.4) and (7.5), that when N is large and $\rho > 0$,

$$V(\bar{y}_{sy}) > V(\bar{y})$$

Thus in this case systematic sampling provides less information per unit cost than does simple random sampling. As in the preceding situations, $V(\bar{y}_{sy})$ cannot be estimated directly by using a single systematic sample. We can approximate its value by using $\hat{V}(\bar{y})$, as for simple random sampling. In general, this estimator should underestimate the true variance of $\bar{y}_{sy}$.

To avoid this problem that occurs with systematic sampling from a periodic population, the investigator could change the random starting point several times. This procedure would reduce the possibility of choosing observations from the same relative position in a period population. For example, when a 1-in-10 systematic sample is being drawn from a long list of file cards, a card is randomly selected from the first 10 cards (for example, no. 2) and every tenth card thereafter. This procedure can be altered by randomly selecting a card from the first 10 (for example, no. 2) and every tenth card thereafter for perhaps 15 selections to obtain the numbers

$$2, 12, 22, \ldots, 152$$

At this point another random starting point can be selected from the next 10 numbers:

$$153, 154, 155, \ldots, 162$$

If 156 is selected, we then proceed to select every tenth number thereafter for the next 15 selections. This entire process is repeated until the desired sample size is obtained.

The process of selecting a random starting point several times throughout the systematic sample has the effect of shuffling the elements of the population and then drawing a systematic sample. Hence we can assume that the sample obtained is equivalent to a systematic sample drawn from a random population. The variance of $\bar{y}_{sy}$ can then be approximated by using

$$\hat{V}(\bar{y}_{sy}) = \frac{s^2}{n}\left(\frac{N-n}{N}\right)$$

EXAMPLE 7.1

An investigaror wishes to determine the quality of maple syrup contained in the sap of trees on a Vermont farm. The total number of trees N is unknown; hence to conduct a simple random sample of trees is impossible. As an alternative procedure, the investigator decides to use a 1-in-7 systematic sample. The data from this survey are listed in the accompanying table. Entries are the percentage of sugar content (in the sap) for the trees sampled. Use these data to estimate μ, the average sugar content of maple trees on the farm. Place a bound on the error of estimation.

Tree Sampled	Sugar Content of the Sap, y	y^2
1	82	6724
2	76	5776
3	83	6889
⋮	⋮	⋮
210	84	7056
211	80	6400
212	79	6241
	$\sum\limits_{i=1}^{212} y_i = 17{,}066$	$\sum\limits_{i=1}^{212} y_i^2 = 1{,}486{,}800$

SOLUTION

An estimate of μ is given by

$$\bar{y}_{sy} = \frac{\sum\limits_{i=1}^{n} y_i}{n} = \frac{17,066}{212} = 80.5$$

To find a bound on the error of estimation, we must first compute s^2. Using the computational formula, we obtain

$$s^2 = \frac{\sum y_i^2 - (\sum y_i)^2/n}{n-1} = \frac{1,486,800 - (17,066)^2/212}{211} = 535.48$$

Intuitively, we can assume that the population of trees on the farm is random. Under this assumption the estimated variance of $\bar{y}_{sy}$ is given by Equation (7.2). Having conducted the 1-in-7 sample, we know N. Assuming $N = 1484$ gives

$$\hat{V}(\bar{y}_{sy}) = \frac{s^2}{n}\left(\frac{N-n}{N}\right) = \frac{535.483}{212}\left(\frac{1484-212}{1484}\right) = 2.16$$

An approximate bound on the error of estimation is given by

$$2\sqrt{\hat{V}(\bar{y}_{sy})} \approx 2\sqrt{2.165} = 2.9$$

To summarize, we estimate the average sugar content of the sap to be 80.5%. We are quite confident that the bound on the error of estimation is less than 2.9%.

You will recall that estimation of a population total requires knowledge of the total number of elements N in the population when we are using the procedures of Chapters 4 and 5. For example, we use

$$\hat{\tau} = N\bar{y}$$

as an estimator of τ from simple random sampling. Also, we use

$$\hat{\tau}_{st} = \sum_{i=1}^{L} N_i\bar{y}_i$$

where

$$\sum_{i=1}^{L} N_i = N$$

as an estimator of τ from stratified random sampling from L strata (Section 5.3). Similarly, we need to know N to estimate τ when we are using systematic sampling.

The population size is unknown in many practical situations, which suggests using systematic sampling; however, when N is known, we can estimate τ by using Equations (7.6), (7.7), and (7.8).

Estimator of the population total τ:

$$\hat{\tau} = N\bar{y}_{sy} \tag{7.6}$$

Estimated variance of $\hat{\tau}$:

$$\hat{V}(N\bar{y}_{sy}) = N^2\hat{V}(\bar{y}_{sy}) = N^2\left(\frac{s^2}{n}\right)\left(\frac{N-n}{N}\right) \tag{7.7}$$

Bound on the error of estimation:

$$2\sqrt{\hat{V}(N\bar{y}_{sy})} = 2\sqrt{N^2\left(\frac{s^2}{n}\right)\left(\frac{N-n}{N}\right)} \tag{7.8}$$

Note that the results presented in Equations (7.6), (7.7), and (7.8) are identical to those presented for estimating a population total under simple random sampling. This result does not imply that the variance of $N\bar{y}_{sy}$ is the same as the variance of $N\bar{y}$. Again, we cannot obtain an unbiased estimator of $V(N\bar{y}_{sy})$ from the data in a single systematic sample. However, in certain circumstances, as noted earlier, systematic sampling is equivalent to simple random sampling, and we can use the results presented in Section 4.3.

EXAMPLE 7.2

A Virginia horticulturist has an experimental orchard of $N = 1300$ apple trees of a new variety under study. The investigator wishes to estimate the total yield (in bushels) from the orchard, from a 1-in-10 systematic sample of trees. The sample mean and variance for the sampled trees are found to be $\bar{y}_{sy} = 3.52$ bushels and $s^2 = .48$ bushel. Use these data to estimate τ, and place a bound on the error of estimation.

SOLUTION

A reasonable assumption is that the population is random; hence systematic and simple random sampling are equivalent. If the population were periodic, the experimenter could choose several random starting points in selecting the trees to be included in the sample.

An estimate of τ is given by

$$N\bar{y}_{sy} = 1300(3.52) = 4576$$

A bound on the error of estimation can be found by using Equation (7.8) with $n = 130$:

$$2\sqrt{\hat{V}(N\bar{y}_{sy})} = 2\sqrt{N^2\left(\frac{s^2}{n}\right)\left(\frac{N-n}{N}\right)}$$

$$= 2\sqrt{1300^2\left(\frac{.48}{130}\right)\left(\frac{1300-130}{1300}\right)} = 150$$

Thus we estimate that the total yield from the apple orchard is 4576 bushels, with a bound on the error of estimation of 150 bushels.

If stratifying the populations is advantageous, systematic sampling can be used within each stratum in place of simple random sampling. Using the estimator of Equation (7.1) with its estimated variance (7.2) within each stratum, the resulting estimator of the population mean will look similar to Equation (5.1), with an estimated variance given by Equation (5.2). Such a situation might arise if we were to stratify an industry by plants and then take a systematic sample of the records within each plant to estimate average accounts receivable, time lost to accidents, and so on.

7.4
ESTIMATION OF A POPULATION PROPORTION

An investigator frequently wishes to use data from a systematic sample to estimate a population proportion. For example, to determine the proportion of registered voters in favor of an upcoming bond issue, the investigator might use a 1-in-k systematic sample from the voter registration list.

The estimator of the population proportion p obtained from systematic sampling is denoted by $\hat{p}_{sy}$. As in the simple random sampling (Section 4.5), the properties of $\hat{p}_{sy}$ parallel those of the sample mean $\bar{y}_{sy}$ if the response measurements are defined as follows: Let $y_i = 0$ if the ith element sampled does not possess the specified characteristic and $y_i = 1$ if it does. The estimator $\hat{p}_{sy}$ is then the average of the 0 and 1 values from the sample.

Estimator of the population proportion p:

$$\hat{p}_{sy} = \bar{y}_{sy} = \frac{\sum\limits_{i=1}^{n} y_i}{n} \tag{7.9}$$

Estimated variance of $\hat{p}_{sy}$:

$$\hat{V}(\hat{p}_{sy}) = \frac{\hat{p}_{sy}\hat{q}_{sy}}{n-1}\left(\frac{N-n}{N}\right) \tag{7.10}$$

where

$$\hat{q}_{sy} = 1 - \hat{p}_{sy}$$

Bound on the error of estimation:

$$2\sqrt{\hat{V}(\hat{p}_{sy})} = 2\sqrt{\frac{\hat{p}_{sy}\hat{q}_{sy}}{n-1}\left(\frac{N-n}{N}\right)} \tag{7.11}$$

We can ignore the fpc, $(N - n)/N$, in Equations (7.10) and (7.11) if the population size N is unknown but can be assumed large relative to n.

We again note that the estimated variance of $\hat{p}_{sy}$ (or $\bar{y}_{sy}$) is identical to the estimated variance of $\hat{p}$ (or $\bar{y}$), using simple random sampling (Section 4.5). This result does not imply that the corresponding population variances are equal; however, if N is large, and if the observations within a systematic sample are unrelated (that is, $\rho = 0$), the two population variances will be equal.

EXAMPLE 7.3

A 1-in-6 systematic sample is obtained from a voter registration list to estimate the proportion of voters in favor of the proposed bond issue. Several different random starting points are used to ensure that the results of the sample are not affected by periodic variation in the population. The coded results of this preelection survey are as shown in the accompanying table. Estimate p, the proportion of the 5775 registered voters in favor of the proposed bond issue ($N = 5775$). Place a bound on the error of estimation.

Voter	Response
4	1
10	0
16	1
⋮	⋮
5760	0
5766	0
5772	1

$$\sum_{i=1}^{962} y_i = 652$$

SOLUTION

The sample proportion is given by

$$\hat{p}_{sy} = \frac{\sum\limits_{i=1}^{962} y_i}{962} = \frac{652}{962} = .678$$

Since N is large and several random starting points were chosen in drawing the systematic sample, we can assume that

$$\hat{V}(\hat{p}_{sy}) = \frac{\hat{p}_{sy}\hat{q}_{sy}}{n - 1}\left(\frac{N - n}{N}\right)$$

provides a good estimate of $V(\hat{p}_{sy})$.

The bound on the error of estimation is

$$2\sqrt{\hat{V}(\hat{p}_{sy})} = 2\sqrt{\frac{\hat{p}_{sy}\hat{q}_{sy}}{n-1}\left(\frac{N-n}{N}\right)}$$

$$= 2\sqrt{\frac{(.678)(.322)}{961}\left(\frac{5775-962}{5775}\right)} \approx .028$$

Thus we estimate .678 (67.8%) of the registered voters favor the proposed bond issue. We are relatively confident that the error of estimation is less than .028 (2.8%).

7.5
SELECTING THE SAMPLE SIZE

Now let us determine the number of observations necessary to estimate μ to within B units. The required sample size is found by solving the following equation for n:

$$2\sqrt{V(\bar{y}_{sy})} = B \tag{7.12}$$

The solution to Equation (7.12) involves both σ^2 and ρ, which must be known (at least approximately) in order to solve for n. Although these parameters sometimes can be estimated if data from a prior survey are available, we do not discuss this method in this text. Instead, we use the formula for n for simple random sampling. This formula could give an extra large sample for ordered populations and too small a sample for periodic populations. As noted earlier, the variances of $\bar{y}_{sy}$ and $\bar{y}$ are equivalent if the population is random.

Sample size required to estimate μ with a bound B on the error of estimation:

$$n = \frac{N\sigma^2}{(N-1)D + \sigma^2} \tag{7.13}$$

where

$$D = \frac{B^2}{4}$$

EXAMPLE 7.4

The management of a large utility company is interested in the average amount of time delinquent bills are overdue. A systematic sample will be drawn from an alphabetical list of $N = 2500$ overdue customer accounts. In a similar survey conducted the previous year, the sample variance was found to be $s^2 = 100$ days. Determine the sample size required to estimate μ, the average amount of time utility bills are overdue, with a bound on the error of estimation of $B = 2$ days.

SOLUTION

A reasonable assumption is that the population is random; hence $\rho \approx 0$. Then we can use Equation (7.3) to find the approximate sample size. Replacing σ^2 by s^2 and setting

$$D = \frac{B^2}{4} = \frac{4}{4} = 1$$

we have

$$n = \frac{N\sigma^2}{(N-1)D + \sigma^2} = \frac{2500(100)}{2499(1) + 100} = 96.19$$

Thus management must sample approximately 97 accounts to estimate the average amount of time delinquent bills are overdue, to within 2 days.

To determine the sample size required to estimate τ with a bound on the error of estimation of magnitude B, we use the corresponding method presented in Section 4.4.

The sample size required to estimate p to within B units is found by using the sample size formula for estimating p under simple random sampling.

Sample size required to estimate p with a bound B on the error of estimation:

$$n = \frac{Npq}{(N-1)D + pq} \tag{7.14}$$

where $\qquad q = 1 - p \qquad$ and $\qquad D = \frac{B^2}{4}$

In a practical situation we do not know p. We can find an approximate sample size by replacing p with an estimated value. If no prior information is available to estimate p, we can obtain a conservative sample size by setting $p = .5$.

EXAMPLE 7.5

An advertising firm is starting a promotion campaign for a new product. The firm wants to sample potential customers in a small community to determine customer acceptance.

To eliminate some of the costs associated with personal interviews, the investigators decide to run a systematic sample from $N = 5000$ names listed in a community registry and collect the data via telephone interviews. Determine the sample size required to estimate p, the proportion of people who consider the product "acceptable," with a bound on the error of estimation of magnitude $B = .03$ (that is, 3%).

SOLUTION

The required sample size can be found by using Equation (7.14). Although no previous data are available on this new product, we can still find an approximate sample size. Set $p = .5$ in Equation (7.14) and

$$D = \frac{B^2}{4} = \frac{(.03)^2}{4} = .000225$$

Then the required sample size is

$$n = \frac{Npq}{(N-1)D + pq} = \frac{5000(.5)(.5)}{4999(.000225) + (.5)(.5)} = 909.240$$

Hence the firm must interview 910 people to determine consumer acceptance to within 3%.

7.6
REPEATED SYSTEMATIC SAMPLING

We stated in Section 7.3 that we cannot estimate the variance of $\bar{y}_{sy}$ from information contained in a single systematic sample unless the systematic sampling generates, for all practical purposes, a random sample. When this result occurs, we can use the random sampling estimation procedures outlined in Section 4.3. However, in most cases systematic random sampling is not equivalent to simple random sampling. An alternate method must be used to estimate $V(\bar{y}_{sy})$. Repeated systematic sampling is one such method.

As the name implies, repeated systematic sampling requires the selection of more than one systematic sample. For example, ten 1–in–50 systematic samples, each containing six measurements, could be acquired in approximately the same time as a 1–in–5 systematic sample containing 60 measurements. Both procedures yield 60 measurements for estimating the population mean μ, but the repeated sampling procedure allows us to estimate $V(\bar{y}_{sy})$ by using the square of the deviations of the $n_s = 10$ individual sample means about their mean. The average $\hat{\mu}$ of the 10 sample means will estimate the population mean μ.

To select n_s repeated systematic samples, we must space the elements of each sample further apart. Thus ten 1–in–50 samples ($n_s = 10$, $k' = 50$) of six measurements each contain the same number of measurements as does a single 1-in-5 sample ($k = 5$) containing $n = 60$ measurements. The starting point for each of the n_s systematic samples is randomly selected from the first k' elements. The remaining elements in each sample are acquired by adding k', $2k'$, and so forth, to the starting point until the total number per sample, n/n_s, is obtained.

A population consists of $N = 960$ elements, which we can number consecutively. To select a systematic sample of size $n = 60$, we choose $k = N/n = 16$ and a random number between 1 and 16 as a starting point. What procedure do we follow to select 10 repeated systematic samples in place of the one systematic sample? First, we choose $k' = 10k = 10(16) = 160$. Next, we select 10 random numbers between 1 and 160. Finally, the constant 160 is added to each of these random starting points to obtain 10 numbers between 161 and 320; the process of adding the constant is continued until 10 samples of size 6 are obtained.

A random selection of ten integers between 1 and 160 gives the following:

$$73, 42, 81, 145, 6, 21, 86, 17, 112, 102$$

These numbers form the random starting points for 10 systematic samples, as shown in Table 7.1. The second element in each sample is found by adding 160 to the first, the third by adding 160 to the second, and so forth.

TABLE 7.1 Selection of repeated systematic samples

Random Starting Point	Second Element in Sample	Third Element in Sample		Sixth Element in Sample
6	166	326	...	806
17	177	337	...	817
21	181	341	...	821
42	202	362	...	842
73	233	393	...	873
81	241	401	...	881
86	246	406	...	886
102	262	422	...	902
112	272	432	...	912
145	305	465	...	945

We frequently select $n_s = 10$ to allow us to obtain enough sample means to acquire a satisfactory estimate of $V(\hat{\mu})$. We choose k' to give the same number of measurements as would be obtained in a single 1-in-k systematic sample; thus

$$k' = kn_s$$

The formulas for estimating μ from n_s systematic samples are shown in Equations (7.15), (7.16), and (7.17).

Estimator of the population mean μ using n_s 1-in-k' systematic samples:

$$\hat{\mu} = \sum_{i=1}^{n_s} \frac{\bar{y}_i}{n_s} \tag{7.15}$$

where $\bar{y}_i$ represents the average of the ith systematic sample.

Estimated variance of $\hat{\mu}$:

$$\hat{V}(\hat{\mu}) = \left(\frac{N-n}{N}\right)\frac{\sum\limits_{i=1}^{n_s}(\bar{y}_i - \hat{\mu})^2}{n_s(n_s-1)} \tag{7.16}$$

Bound on the error of estimation:

$$2\sqrt{\hat{V}(\hat{\mu})} = 2\sqrt{\left(\frac{N-n}{N}\right)\frac{\sum\limits_{i=1}^{n_s}(\bar{y}_i - \hat{\mu})^2}{n_s(n_s-1)}} \tag{7.17}$$

We can also use repeated systematic sampling to estimate a population total τ, if N is known. The necessary formulas are given in Equations (7.18), (7.19), and (7.20).

Estimator of the population total τ using n_s 1-in-k' systematic samples:

$$\hat{\tau} = N\hat{\mu} = N\sum_{i=1}^{n_s} \frac{\bar{y}_i}{n_s} \tag{7.18}$$

Estimated variance of $\hat{\tau}$:

$$\hat{V}(\hat{\tau}) = N^2\hat{V}(\hat{\mu}) = N^2\left(\frac{N-n}{N}\right)\frac{\sum\limits_{i=1}^{n_s}(\bar{y}_i - \hat{\mu})^2}{n_s(n_s-1)} \tag{7.19}$$

Bound on the error of estimation:

$$2\sqrt{\hat{V}(\hat{\tau})} = 2\sqrt{N^2\left(\frac{N-n}{N}\right)\frac{\sum\limits_{i=1}^{n_s}(\bar{y}_i - \hat{\mu})^2}{n_s(n_s-1)}} \tag{7.20}$$

EXAMPLE 7.6

A state park charges admission by carload rather than by person, and a park official wants to estimate the average number of persons per car for a particular summer holiday. She knows from past experience that there should be about

400 cars entering the park, and she wants to sample 80 cars. To obtain an estimate of the variance, she uses repeated systematic sampling with 10 samples of 8 cars each. Using the data given in Table 7.2. estimate the average number of persons per car, and place a bound on the error of estimation.

TABLE 7.2 Data on number of persons per car (the responses y_i are in parentheses)

Random Starting Point	Second Element	Third Element	Fourth Element	Fifth Element	Sixth Element	Seventh Element	Eighth Element	$\bar{y}_i$
2 (3)	52 (4)	102 (5)	152 (3)	202 (6)	252 (1)	302 (4)	352 (4)	3.75
5 (5)	55 (3)	105 (4)	155 (2)	205 (4)	255 (2)	305 (3)	355 (4)	3.38
7 (2)	57 (4)	107 (6)	157 (2)	207 (3)	257 (2)	307 (1)	357 (3)	2.88
13 (6)	63 (4)	113 (6)	163 (7)	213 (2)	263 (3)	313 (2)	363 (7)	4.62
26 (4)	76 (5)	126 (7)	176 (4)	226 (2)	276 (6)	326 (2)	376 (6)	4.50
31 (7)	81 (6)	131 (4)	181 (4)	231 (3)	281 (6)	331 (7)	381 (5)	5.25
35 (3)	85 (3)	135 (2)	185 (3)	235 (6)	285 (5)	335 (6)	385 (8)	4.50
40 (2)	90 (6)	140 (2)	190 (5)	240 (5)	290 (4)	340 (4)	390 (5)	4.12
45 (2)	95 (6)	145 (3)	195 (6)	245 (4)	295 (4)	345 (5)	395 (4)	4.25
46 (6)	96 (5)	146 (4)	196 (6)	246 (3)	296 (3)	346 (5)	396 (3)	4.38

SOLUTION

For one systematic sample,

$$k = \frac{N}{n} = \frac{400}{80} = 5$$

Hence for $n_s = 10$ samples,

$$k' = 10k = 10(5) = 50$$

The following ten random numbers between 1 and 50 are drawn:

$$13, 35, 2, 40, 26, 7, 31, 45, 5, 46$$

Cars with these numbers form the random starting points for the systematic samples.

For Table 7.2 the quantity $\bar{y}_1$ is the average for the first row, $\bar{y}_2$ is the average for the second row, and so forth. The estimate of μ is

$$\hat{\mu} = \frac{1}{n_s} \sum_{i=1}^{n_s} \bar{y}_i = \frac{1}{10}(3.75 + 3.38 + \cdots + 4.38) = 4.16$$

The following identity can be established:

$$\sum_{i=1}^{n_s} (\bar{y}_i - \hat{\mu})^2 = \sum_{i=1}^{n_s} \bar{y}_i^2 - \frac{1}{n_s}\left(\sum_{i=1}^{n_s} \bar{y}_i\right)^2$$

Substituting, we obtain

$$\sum_{i=1}^{10} (\bar{y}_i - \hat{\mu})^2 = 177.410 - \frac{1}{10}(1733.06) = 4.104$$

Thus the estimated variance of $\hat{\mu}$ is

$$\hat{V}(\hat{\mu}) = \left(\frac{N-n}{N}\right)\frac{\sum_{i=1}^{n_s}(\bar{y}_i - \hat{\mu})^2}{n_s(n_s-1)} = \left(\frac{400-80}{400}\right)\left[\frac{4.104}{10(9)}\right] = .0365$$

The estimate of μ with a bound on the error of estimation is

$$\hat{\mu} \pm 2\sqrt{\hat{V}(\hat{\mu})}, \qquad \text{or} \qquad 4.163 \pm 2\sqrt{.0365}, \qquad \text{or} \qquad 4.16 \pm .38$$

Therefore our best estimate of the average of persons per car is 4.16. The error of estimation should be less than .38 with probability approximately .95.

7.7
SUMMARY

Systematic sampling is presented as an alternative to simple random sampling. Systematic sampling is easier to perform and, therefore, is less subject to interviewer errors than simple random sampling. In addition, systematic sampling often provides more information per unit cost than does simple random sampling.

We considered estimation of a population mean, total, and proportion using the estimators $\bar{y}_{sy}$, $N\bar{y}_{sy}$, and $\hat{p}_{sy}$, respectively. The corresponding bounds on the errors of estimation were given for these estimators.

We must first consider the type of population under investigation in order to choose between systematic and simple random sampling. For example, when N is large and $\rho < 0$, the variance of $\bar{y}_{sy}$ is smaller than the corresponding variance of $\bar{y}$ based on simple random sampling. A systematic sample is preferable when the population is ordered and N is large. When the population is random, the two sampling procedures are equivalent and either design can be used. Care must be used in applying systematic sampling to periodic populations.

Sample size requirements for estimating μ, τ, and p are determined by using formulas presented for simple random sampling.

Repeated systematic sampling was discussed in Section 7.6; it allows the experimenter to estimate the population mean or total and the variance of the estimator without making any assumptions about the nature of the population.

CASE STUDY REVISITED

THE ASSESSMENT OF PRODUCT QUALITY

THE quality control problem involving percentage of copper in bronze castings, given at the beginning of this chapter, showed a systematic sample of 16 measurements with $\bar{y} = 87$ and $s^2 = 18$. Even though the sample was selected systematically, we can estimate the population mean by

$$\bar{y} \pm 2\sqrt{\frac{s^2}{n}}$$

assuming N is large compared with n. Thus we have

$$87 \pm 2\sqrt{\frac{18}{16}}$$

$$87 \pm 2$$

or 85 to 89 as the best estimate of the true mean of the production process. Since the standard is 90, apparently the process is not performing up to the advertised standard on this day. The foreman in charge will want to look into possible causes for this failure.

In this case systematic sampling is reasonable since it forces the sample to cover the entire day of production. If quality tends to decrease (or increase) during the day, this sampling plan may detect it. A simple random sample could concentrate all sampled items in the morning (or afternoon) hours.

EXERCISES

7.1   Suppose that a home mortgage company has N mortgages numbered serially in the order that they were granted over a period of 20 years. There is a generally increasing trend in the unpaid balances because of the rising cost of housing over the years. The company wishes to estimate the total amount of unpaid balances. Would you employ a systematic or a simple random sample? Why?

7.2   A corporation lists employees by income brackets (alphabetically within brackets) from highest to lowest. If the objective is to estimate average income per employee, should systematic, stratified, or simple random sampling be used? Assume that costs are equivalent for the three methods and that you can stratify on income brackets. Discuss the advantages and disadvantages of the three methods.

7.3   A retail store with four departments has charge accounts arranged by department, with past-due accounts at the front of each departmental list. Suppose the departments average around 10 accounts each, with approximately 40% past due. On a given day the accounts might appear as shown in the accompanying table (with account numbers 1 through 40). The store wishes to estimate the proportion of past-due accounts by systematic sampling.

	Department			
Account numbers	1–11	12–20	21–28	29–40
Delinquent accounts	1, 2, 3, 4	12, 13,14	21,22, 23, 24, 25	29, 30, 31, 32

(a) List all possible 1-in-10 systematic samples, and compute the exact variance of the sample proportion. (Note that there are 10 possible values, not all distinct, for the sample proportion, each with probability $\frac{1}{10}$ of occurring.)

(b) List all possible 1-in-5 systematic samples, and compute the exact variance of the sample proportion.

(c) Compare the result in part (a) with an approximate variance that would have been obtained in a simple random sample of size $n = 4$ from this population. Similarly, compare the result in part (b) with what would have been obtained from a simple random sample with $n = 8$. What general conclusions can you make?

7.4   The management of a particular company is interested in estimating the proportion of employees favoring a new investment policy. A 1-in-10 systematic sample is obtained from employees leaving the building at the end of a particular workday. Use the data in the accompanying table to estimate p, the proportion in favor of the new policy, and place a bound on the error of estimation. Assume $N = 2000$.

Employee Sampled	Response
3	1
13	0
23	1
$\vdots$	$\vdots$
1993	1
	$\sum\limits_{i=1}^{200} y_i = 132$

7.5   For the situation outlined in Exercise 7.4, determine the sample size required to estimate p to within .01 unit. What type of systematic sample should be run?

7.6   The quality control section of an industrial firm uses systematic sampling to estimate the average amount of fill in 12-ounce cans coming off an assembly line. The data in the accompanying table represent a 1-in-50 systematic sample of the production in one day. Estimate μ, and place a bound on the error of estimation. Assume $N = 1800$.

Amount of Fill (in ounces)					
12.00	11.97	12.01	12.03	12.01	11.80
11.91	11.98	12.03	11.98	12.00	11.83
11.87	12.01	11.98	11.87	11.90	11.88
12.05	11.87	11.91	11.93	11.94	11.89
11.72	11.93	11.95	11.97	11.93	12.05
11.85	11.98	11.87	12.05	12.02	12.04

7.7   Use the data of Exercise 7.6 to determine the sample size required to estimate μ to within .03 unit.

7.8   Soil experts want to determine the amount of exchangeable calcium (in parts per million) in a plot of ground. So that the sampling scheme is simplified, a rectangular grid is superimposed on the field. Soil samples are taken at each point of intersection on the grid (see the diagram). Use the following data to determine the average amount of exchangeable calcium on the plot of ground. Place a bound on the error of estimation.

$$n = 45$$

$$\sum y_i = 90,320 \qquad \text{exchangeable calcium}$$

$$\sum y_i^2 = 184,030,000$$

7.9   The highway patrol of a particular state is concerned about the proportion of motorists who carry their licenses. A checkpoint is set up on a major highway, and the driver of every seventh car is questioned. Use the data in the accompanying table to estimate the proportion of drivers carrying their licenses. Place a bound on the error of estimation. Assume that $N = 2800$ cars pass the checkpoint during the sampling period.

Car	Response, y_i
1	1
2	1
3	0
⋮	⋮
400	1
$\sum y_i = 324$	

7.10   The highway patrol expects at least $N = 3000$ cars to pass the checkpoint. Determine the sample size required to estimate p to within $B = .015$ unit.

7.11   A college is concerned about improving its relations with a neighboring community. A 1-in-150 systematic sample of the $N = 4500$ students listed in the directory is taken to estimate the total amount of money spent on clothing during one quarter of the school year. The results of the sample are listed in the accompanying table. Use these data to estimate τ, and place a bound on the error of estimation.

Student	Amount Spent (in dollars)	Student	Amount Spent (in dollars)
1	30	16	32
2	22	17	14
3	10	18	29
4	62	19	48
5	28	20	50
6	31	21	9
7	40	22	15
8	29	23	6

9	17	24	93
10	51	25	21
11	29	26	20
12	21	27	13
13	13	28	12
14	15	29	29
15	23	30	38

7.12   What sample size is needed to estimate τ in Exercise 7.11 with a bound on the error of estimation approximately equal to \$10,000? What systematic sampling scheme would you recommend?

7.13   A census is conducted in a community. In addition to obtaining the usual population information, the surveyors question the occupants of every twentieth household to determine how long they have occupied their present home. These results are summarized next.

$$n = 115 \qquad \sum y_i^2 = 2011.15$$

$$\sum y_i = 407.1 \text{ (years)} \qquad N = 2300$$

Use these data to estimate the average amount of time people have lived in their present home. Place a bound on the error of estimation.

7.14   A group of guidance counselors is concerned about the average yearly tuition for out-of-state students in 371 junior colleges. From an alphabetical list of these colleges, a 1-in-7 systematic sample is drawn. Data concerning out-of-state tuition expenses for an academic year (September to June) are obtained for each college sampled. Let y_i be the amount of tuition required for the ith college sampled. Use the following data to estimate μ, and place a bound on the error of estimation.

$$\sum_{i=1}^{53} y_i = \$11,950 \qquad \sum_{i=1}^{53} y_i^2 = \$2,731,037$$

7.15   Museum officials are interested in the total number of persons who visit their museum during a 180-day period when an expensive antique collection is on display. Since monitoring the museum traffic each day is too costly, officials decide to obtain these data every tenth day. The information from this 1-in-10 systematic sample is summarized in the accompanying table. Use these data to estimate τ, the total number of persons visiting the museum during the specified period. Place a bound on the error of estimation.

Day	Number of People Visiting the Museum
3	160
13	350
23	225
⋮	⋮
173	290

$$\sum_{i=1}^{18} y_i = 4868$$

$$\sum_{i=1}^{18} y_i^2 = 1,321,450$$

7.16   Foresters are interested in determining the mean timber volume per acre for 520 one-acre plots ($N = 520$). A 1-in-25 systematic sample is conducted. Using the data presented in the accompanying table, estimate μ, the average timber volume per plot, and place a bound on the error of estimation.

Plot Sampled	Volume (in board feet)	Plot Sampled	Volume (in board feet)
4	7030	279	7540
29	6720	304	6720
54	6850	329	6900
79	7210	354	7200
104	7150	379	7100
129	7370	404	6860
154	7000	429	6800
179	6930	454	7050
204	6570	479	7420
229	6910	504	7090
254	7380		

7.17   The officers of a certain professional society wish to determine the proportion of the membership that favors several proposed revisions in refereeing practices. They conduct a 1-in-10 systematic sample from an alphabetical list of the $N = 650$ registered members. Let $y_i = 1$ if the ith person sampled favors the proposed changes and $y_i = 0$ if he opposes the changes. Use the following sample data to estimate p, the proportion of members in favor of the proposed changes. Place a bound on the error of estimation.

$$\sum_{i=1}^{65} y_i = 48$$

7.18   In a sociological survey a 1-in-50 systematic sample is drawn from city tax records to determine the total number of families in the city who rent their homes. Let $y_i = 1$ if the family in the ith household sampled rents and let $y_i = 0$ if the family does not. There are $N = 15{,}200$ households in the community. Use the following to estimate τ, the total number of families who rent. Place a bound on the error of estimation.

$$\sum_{i=1}^{304} y_i = 88$$

[Hint: If $\hat{p} =$ estimated fraction who rent, then $N\hat{p}$ is an estimate of the total number who rent; $\hat{V}(N\hat{p}) = N^2 \hat{V}(\hat{p})$.]

7.19   A farmer wishes to estimate the total weight of fruit to be produced in a field of zucchini (squash) by sampling just prior to harvest. The plot consists of 20 rows with 400 plants per row. The manufacturer of the seeds says that each plant can yield up to 8 pounds of fruit. Outline an appropriate systematic sampling plan for this problem so as to estimate the total weight of fruit to within 2000 pounds.

7.20   The accompanying table shows the number of live births and the birth rate per 1000 population for the United States during six systematically selected years.

(a) Estimate the average number of male births per year for the 1955–1980 period, and place a bound on the error of estimation.

(b) Estimate the average birth rate per year for the 1955–1980 period, and place a bound on the error of estimation.

(c) Do you think systematic sampling is better than simple random sampling for the problems in parts (a) and (b)? Why?

Year	Male Births	Female Births	Total Births	Birth Rate
1955	2,073,719	1,973,576	4,047,295	26.0
1960	2,179,708	2,078,142	4,257,850	23.7
1965	1,927,054	1,833,304	3,760,358	19.4
1970	1,915,378	1,816,008	3,731,386	18.4
1975	1,613,135	1,531,063	3,144,198	14.6
1980	1,852,616	1,759,642	3,612,258	15.9

Source: *The World Almanac & Book of Facts*, 1984 edition, copyright © Newspaper Enterprise Association, Inc., 1983, New York, NY 10166.

7.21 Data on divorce rates (per 1000 population) in the United States for a systematic sample of years from 1900 are given in the accompanying table. Estimate the average annual divorce rate for that time period, and place a bound on the error of estimation. Is systematic sampling better or worse than simple random sampling in this case? Why?

Year	Rate	Year	Rate
1900	0.7	1945	3.5
1905	0.8	1950	2.6
1910	0.9	1955	2.3
1915	1.0	1960	2.2
1920	1.6	1965	2.5
1925	1.5	1970	3.5
1930	1.6	1975	4.8
1935	1.7	1980	5.2
1940	2.0		

Source: *The World Almanac & Book of Facts*, 1984 edition, copyright © Newspaper Enterprise Association, Inc., 1983, New York, NY 10166.

7.22 A quality control inspector must sample silicon wafers, from which computer chips will be made, after they are baked in an oven. Slotted trays containing many wafers are put through the oven, one after another, all day long. Position on the tray and time of day may have important bearings on the quality of the wafer. Suggest a sampling plan, with the goal being to estimate the proportion of defective wafers.

7.23 A warehouse contains stacks of automobile batteries that must be sampled for quality inspection. Each stack has a different production date code, and the stacks are arranged chronologically. The stacks are of approximately equal size. Suggest a sampling plan for estimating the proportion of defective batteries.

7.24   An auditor is confronted with a long list of accounts receivable for a firm. She must verify the amounts on 10% of these accounts and estimate the average difference between the audited and book values.

(a)   Suppose the accounts are arranged chronologically, with the older accounts tending to have smaller values. Would you choose a systematic or a simple random sampling design to select the sample?

(b)   Suppose the accounts are arranged randomly. Would you choose a systematic or a simple random sampling design to select the sample?

(c)   Suppose the accounts are grouped by department and then listed chronologically within departments. The older accounts again tend to have smaller values. Would you choose a systematic or a simple random sampling design to select the sample?

7.25   The market share for a certain food product is to be estimated by recording store purchases of the product for certain weeks selected throughout the year. Discuss the advantages and disadvantages of a systematic selection of the weeks for this study.

7.26   Crop yield for a large field of wheat is to be estimated by sampling small plots within the field while the grain is ripening. The field is on sloping land, with higher fertility toward the lower side.

(a)   Suggest a systematic sampling design for the small plots.

(b)   Could other sampling designs be used effectively in this case?

EXPERIENCES WITH REAL DATA

7.1   Locate the stock price summaries for the week in your local weekend newspaper. These summaries usually list the high and low prices of each stock for the week, along with the difference between the closing price for the current week and that for the previous week.

(a)   Select a systematic sample of stocks, and estimate the proportion that have a lower closing price this week than they had the previous week. Place a bound on the error of estimation.

(b)   How do you think systematic sampling would compare with simple or stratified random sampling in this case?

7.2   Table 4.7 lists data for National Basketball Association teams for 1982–1983. Suppose a sportswriter wishes to estimate the total points scored in the league by selecting a systematic sample of teams from the list.

(a)   Should a 1–in–6 systematic sample be used? Why?

(b)   Would a 1–in–4 systematic sample be better than a 1–in–6? Why?

7.3   The data in Table 5.4 shows heights of tall buildings for selected United States cities. Discuss the strengths and weaknesses of using a systematic sample of buildings from this list to estimate average building height. Could systematic sampling be better than simple random sampling for this problem?

7.4   From a list of names, like those in a student directory, select a systematic sample, and interview the selected persons to find out whether they favor a certain issue of current importance (such as a proposed government action or a pending campus decision). Estimate the population proportion favoring the issue, and place a bound on the error of estimation.

Repeat the procedure just outlined three more times so that four independent systematic samples are available. Compare the results from the individual samples with the combined result of the four samples analyzed according to the methods of Section 7.6. If you prefer to work with something other than lists of people, use other listed records in a similar way. For example, you could systematically sample names of employees from a file and estimate average age, income, and so on.

Case Study

What Are the Characteristics of the People Living in Your Neighborhood?

SUPPOSE a firm wants to locate a business in your neighborhood. How can it find information on the characteristics of the people living there without conducting its own survey? One way is to consult the block statistics data from the United State Census Bureau. *Block statistics* give demographic information—such as total number of residents, number in certain minority groups, number over the age of 65, and number of owners and renters—on very small regions that often conform to city blocks. These data are used by market researchers, housing and transportation planners, and community associations, among others.

The business that is considering a location in your neighborhood caters to those aged 65 and over. Thus it wants to estimate the proportion of residents in this age category who live in a 40-block area. The firm decides to sample 5 of the 40 blocks and obtain the data from block statistics. The sampled blocks form clusters of people, and hence the techniques of cluster sampling must be used. (This problem is a scaled-down version of a real problem. Usually, the number of blocks and the sample size would be much larger.)

8
CLUSTER
SAMPLING

8.1
INTRODUCTION

You will recall that the objective of sample survey design is to obtain a specified amount of information about a population parameter at minimum cost. Stratified random sampling is often better suited for this than is simple random sampling for the three reasons indicated in Section 5.1. Systematic sampling often gives results at least as accurate as those from simple random sampling, and it is easier to perform, as discussed in Section 7.1. This chapter introduces a fourth design, cluster sampling, which sometimes gives more information per unit cost than do any of the other three designs discussed previously.

DEFINITION 8.1   A *cluster sample* is a simple random sample in which each sampling unit is a collection, or cluster, of elements.

Cluster sampling is less costly than simple or stratified random sampling if the cost of obtaining a frame that lists all population elements is very high or if the cost of obtaining observations increases as the distance separating the elements increases.

To illustrate, suppose we wish to estimate the average income per household in a large city. How should we choose the sample? If we use simple random sampling, we will need a frame listing all households (elements) in the city, and this frame may be very costly or impossible to obtain. We cannot avoid this problem by using stratified random sampling because a frame is still required for each stratum in the population. Rather than draw a simple random sample of *elements*, we could divide the city into regions such as blocks (or clusters of elements) and select a simple random sample of blocks from the population. This task is easily accomplished by using a frame that lists all city blocks. Then the income of every household within each sampled block could be measured.

To illustrate the second reason for using cluster sampling, suppose that a list of households in the city is available. We could select a simple random sample of households, which probably would be scattered throughout the city. The cost of conducting interviews in the scattered households would be large owing to the interviewer travel time and other related expenses. Stratified random sampling could lower these expenses, but using cluster sampling is a more effective method of reducing travel costs. Elements within a cluster should be close to each other geographically, and hence travel expenses should be reduced. Obviously, travel within a city block would be minimal when compared with the travel associated with simple random sampling of households within the city.

To summarize, cluster sampling is an effective design for obtaining a specified amount of information at minimum cost under the following conditions:

1. A good frame listing population elements either is not available or is very costly to obtain, while a frame listing clusters is easily obtained.

2. The cost of obtaining observations increases as the distance separating the elements increases.

City blocks are frequently used as clusters of households or people because the United States Census Bureau reports very detailed block statistics. In census data a block may be a standard city block or an irregularly shaped area with identifiable political or geographic boundaries. Block statistics are reported for all urban areas and for all other places with concentrations of 10,000 or more people. In total, block statistics cover 77% of the nation's population. Data reported for each block includes total population, racial mix, and number of housing units, and they may include the dollar value of the property, whether the unit is owned or rented by the inhabitants, and whether the unit has complete plumbing facilities.

Block statistics from the Census Bureau are widely used in cluster sampling by market research firms, which may want to estimate the potential market for a product, the potential sales if a new store were to open in the area, or the potential number of clients for a new service, such as an emergency medical facility.

State and local governments sample blocks (clusters of housing units or people) to plan new transportation methods and facilities and to plan housing developments. Similarly, community organizations, such as churches, use block statistics to determine optimal sites for expansion.

There are many other common examples of the uses of cluster sampling. Housing units themselves are clusters of people and may form convenient sampling units when sampling college students, for example. Hospitals form convenient clusters of patients with certain illnesses for studies of the average length of time hospitalized or average number of recurrences of these illnesses.

Elements other than people are often sampled in clusters. An automobile forms a nice cluster of four tires for studies of tire wear and safety. A circuit board manufactured for a computer forms a cluster of semiconductors for testing. An orange tree forms a cluster of oranges for investigating an insect infestation. A plot in a forest contains a cluster of trees for estimating timber volume or proportions of diseased trees. As you can see, the list of possible clusters that are convenient units for sampling is endless.

We will now discuss the details of selecting a cluster sample.

8.2
How to Draw a Cluster Sample

The first task in cluster sampling is to specify appropriate clusters. Elements within a cluster are often physically close together and hence tend to have similar characteristics. Stated another way, the measurement on one element in a cluster may be highly correlated with the measurement on another. Thus the amount of information about a population parameter may not be increased substantially as new measurements are taken within a cluster. Since measurements cost money, an experimenter would waste money by choosing too large a cluster size. However, situations may arise in which elements within a cluster are very different from one another. In such cases a sample containing a few large clusters could produce a very good estimate of a population parameter, such as the mean.

For example, suppose clusters are formed by boxes of components coming off production lines, one cluster of components per line. If all lines have approximately the same rate of defects, then the components in each cluster (box) are about as variable with respect to quality as the population as a whole. In this situation a good estimate of the proportion of defectives produced could be obtained from one or two clusters.

In contrast, suppose that school districts are specified as clusters of households for estimating the proportion of households that favor a rezoning plan. Since the clusters contain many households, resources allow only a small number of clusters, say two or three, to be sampled. In this case most of the households in one district may be happy with their schools and not favor rezoning, while most households in another district may be unhappy with their schools and strongly favor rezoning. A small sample of school districts may miss one or the other of these groups entirely, thereby yielding a very poor estimate. More information may be obtained by sampling a larger number of clusters of smaller size.

The problem of choosing an appropriate cluster size can be even more difficult when an infinite number of possible cluster sizes are available, as in the selection of forest plots to estimate the proportion of diseased trees. If there is variability in the density of diseased trees across the forest, then many small plots (clusters), randomly or systematically located, would be desirable. However, to randomly locate a plot in a forest is quite time-consuming, and once it is located, sampling many trees in that one plot is economically desirable. Thus many small plots are advantageous for controlling variability, but a few large plots are advantageous economically. A balance between size and number of plots must be achieved. There are no good rules that always hold for making this decision. Each problem must be studied on its own, but pilot surveys with various plot sizes might help point the experimenter in the correct direction.

Notice the main difference between the optimal construction of strata (Chapter 5) and the construction of clusters. Strata are to be as homogeneous (alike) as possible within, but one stratum should differ as much as possible from another with respect to the characteristic being measured. Clusters, on the other hand, should be as heterogeneous (different) as possible within, and one cluster should look very much like another in order for the economic advantages of cluster sampling to pay off.

Once appropriate clusters have been specified, a frame that lists all clusters in the population must be composed. A simple random sample of clusters is then selected from this frame by using the methods of Section 4.2. We illustrate with the following example.

EXAMPLE 8.1

A sociologist wants to estimate the average per capita income in a certain small city. No list of resident adults is available. How should he design the sample survey?

SOLUTION

Cluster sampling seems to be the logical choice for the survey design because no lsits of elements are available. The city is marked off into rectangular blocks, except for two industrial areas and three parks that contain only a few houses. The sociologist decides that each of the city blocks will be considered one cluster, the two industrial areas will be considered one cluster, and, finally, the three parks will be considered one cluster. The clusters are numbered on a city map, with the numbers from 1 to 415. The experimenter has enough time and money to sample $n = 25$ clusters and to interview every household within each cluster. Hence 25 random numbers between 1 and 415 are selected from Table 2 of the Appendix, and the clusters having these numbers are marked on the map. Interviewers are then assigned to each of the sampled clusters.

8.3
ESTIMATION OF A POPULATION MEAN AND TOTAL

Cluster sampling is simple random sampling with each sampling unit containing a number of elements. Hence the estimators of the population mean μ and total τ are similar to those for simple random sampling. In particular, the sample mean $\bar{y}$ is a good estimator of the population mean μ. An estimator of μ and two estimators of τ are discussed in this section.

The following notation is used in this chapter:

N = the number of *clusters* in the population

n = the number of clusters selected in a simple random sample

m_i = the number of elements in cluster i, $i = 1, \ldots, N$

$\bar{m} = \dfrac{1}{n} \sum\limits_{i=1}^{n} m_i$ = the average cluster size for the sample

$M = \sum\limits_{i=1}^{N} m_i$ = the number of elements in the population

$\bar{M} = \dfrac{M}{N}$ = the average cluster size for the population

y_i = the total of all observations in the ith cluster

The estimator of the population mean μ is the sample mean $\bar{y}$, which is given by

$$\bar{y} = \frac{\sum\limits_{i=1}^{n} y_i}{\sum\limits_{i=1}^{n} m_i}$$

Thus $\bar{y}$ takes the form of a ratio estimator, as developed in Chapter 6, with m_i taking the place of x_i. Then the estimated variance of $\bar{y}$ has the form of the variance of a ratio estimator, given by Equation (6.2).

Estimator of the population mean μ:

$$\bar{y} = \frac{\sum\limits_{i=1}^{n} y_i}{\sum\limits_{i=1}^{n} m_i} \tag{8.1}$$

Estimated variance of $\bar{y}$:

$$\hat{V}(\bar{y}) = \left(\frac{N-n}{Nn\bar{M}^2}\right) \frac{\sum\limits_{i=1}^{n} (y_i - \bar{y}m_i)^2}{n-1} \tag{8.2}$$

Bound on the error of estimation:

$$2\sqrt{\hat{V}(\bar{y})} = 2\sqrt{\left(\frac{N-n}{Nn\bar{M}^2}\right)\frac{\sum_{i=1}^{n}(y_i - \bar{y}m_i)^2}{n-1}} \tag{8.3}$$

Here $\bar{M}$ can be estimated by $\bar{m}$ if M is unknown.

The estimated variance in Equation (8.2) is biased and a good estimator of $V(\bar{y})$ only if n is large, say $n \geq 20$. The bias disappears if the cluster sizes $m_1, m_2, \ldots, m_N$ are equal.

Let us illustrate the use of the formulas with an example.

EXAMPLE 8.2

Interviews are conducted in each of the 25 blocks sampled in Example 8.1. The data on incomes are presented in Table 8.1. Use the data to estimate the average per capita income in the city, and place a bound on the error of estimation.

TABLE 8.1 Per capita income

Cluster i	Number of Residents, m_i	Total Income Per Cluster, y_i	Cluster i	Number of Residents, m_i	Total Income per Cluster, y_i
1	8	$ 96,000	14	10	$49,000
2	12	121,000	15	9	53,000
3	4	42,000	16	3	50,000
4	5	65,000	17	6	32,000
5	6	52,000	18	5	22,000
6	6	40,000	19	5	45,000
7	7	75,000	20	4	37,000
8	5	65,000	21	6	51,000
9	8	45,000	22	8	30,000
10	3	50,000	23	7	39,000
11	2	85,000	24	3	47,000
12	6	43,000	25	8	41,000
13	5	54,000			
				$\sum_{i=1}^{25} m_i =$	$\sum_{i=1}^{25} y_i =$
				151	$1,329,000

SOLUTION

The best estimate of the population mean μ is given by Equation (8.1) and calculated as follows:

$$\bar{y} = \frac{\sum\limits_{i=1}^{n} y_i}{\sum\limits_{i=1}^{n} m_i} = \frac{\$1,329,000}{151} = \$8801$$

In order to calculate $\hat{V}(\bar{y})$, we need the following quantities:

$$\sum_{i=1}^{25} y_i^2 = y_1^2 + y_2^2 + \cdots + y_{25}^2$$

$$= (96,000)^2 + (121,000)^2 + \cdots + (41,000)^2$$

$$= 82,039,000,000$$

$$\sum_{i=1}^{25} m_i^2 = m_1^2 + m_2^2 + \cdots + m_{25}^2$$

$$= (8)^2 + (12)^2 + \cdots + (8)^2 = 1,047$$

$$\sum_{i=1}^{25} y_i m_i = y_1 m_1 + y_2 m_2 + \cdots + y_{25} m_{25}$$

$$= (96,000)(8) + (121,000)(12) + \cdots + (41,000)(8)$$

$$= 8,403,000$$

The following equality is easily established:

$$\sum_{i=1}^{n} (y_i - \bar{y} m_i)^2 = \sum_{i=1}^{n} y_i^2 - 2\bar{y} \sum_{i=1}^{n} y_i m_i + \bar{y}^2 \sum_{i=1}^{n} m_i^2$$

Substituting into this equation from Table 8.1 yields

$$\sum_{i=1}^{25} (y_i - \bar{y} m_i)^2 = 82,039,000,000 - 2(8801)(8,403,000)$$

$$+ (8801)^2 (1047)$$

$$= 15,227,502,247$$

Since M is not known, the $\bar{M}$ appearing in Equation (8.2) must be estimated by $\bar{m}$, where

$$\bar{m} = \frac{\sum\limits_{i=1}^{n} m_i}{n} = \frac{151}{25} = 6.04$$

Example 8.1 gives $N = 415$. Then from Equation (8.2)

$$\hat{V}(\bar{y}) = \left(\frac{N - n}{N n \bar{M}^2}\right) \frac{\sum\limits_{i=1}^{n} (y_i - \bar{y} m_i)^2}{n - 1}$$

$$= \left[\frac{415 - 25}{(415)(25)(6.04)^2}\right] \left(\frac{15,227,502,247}{24}\right) = 653,785$$

Thus the estimate of μ with a bound on the error of estimation is given by

$$\bar{y} \pm 2\sqrt{\hat{V}(\bar{y})}, \qquad \text{or} \qquad 8801 \pm 2\sqrt{653{,}785}, \qquad \text{or} \qquad 8801 \pm 1617$$

The best estimate of the average per capita income is \$8801, and the error of estimation should be less than \$1617 with probability close to .95. This bound on the error of estimation is rather large; it could be reduced by sampling more clusters and, consequently, increasing the sample size.

The population total τ is now $M\mu$ because M denotes the total number of elements in the population. Consequently, as in simple random sampling, $M\bar{y}$ provides an estimator of τ.

Estimator of the population total τ:

$$M\bar{y} = M \frac{\displaystyle\sum_{i=1}^{n} y_i}{\displaystyle\sum_{i=1}^{n} m_i} \tag{8.4}$$

Estimated variance of $M\bar{y}$:

$$\hat{V}(M\bar{y}) = M^2 \hat{V}(\bar{y}) = N^2 \left(\frac{N-n}{Nn}\right) \frac{\displaystyle\sum_{i=1}^{n}(y_i - \bar{y}m_i)^2}{n-1} \tag{8.5}$$

Bound on the error of estimation:

$$2\sqrt{\hat{V}(M\bar{y})} = 2\sqrt{N^2\left(\frac{N-n}{Nn}\right) \frac{\displaystyle\sum_{i=1}^{n}(y_i - \bar{y}m_i)^2}{n-1}} \tag{8.6}$$

Note that the estimator $M\bar{y}$ is useful only if the number of elements in the population, M, is known.

EXAMPLE 8.3

Use the data in Table 8.1 to estimate the total income of all residents of the city, and place a bound on the error of estimation. There are 2500 residents of the city.

SOLUTION

The sample mean $\bar{y}$ is calculated to be \$8801 in Example 8.2. Thus the estimate of τ is

$$M\bar{y} = 2500(8801) = \$22{,}002{,}500$$

The quantity $\hat{V}(\bar{y})$ is calculated by the method used in Example 8.2, except that M can now be used in place of $\bar{m}$. The estimate of τ with a bound on the error of estimation is

$$M\bar{y} \pm 2\sqrt{\hat{V}(M\bar{y})} = M\bar{y} \pm 2\sqrt{M^2 \hat{V}(\bar{y})}$$

$$22{,}002{,}500 \pm 2\sqrt{(2500)^2(653{,}785)}$$

$$22{,}002{,}500 \pm 4{,}042{,}848$$

Again, this bound on the error of estimation is large, and it could be reduced by increasing the sample size.

Often the number of elements in the population is not known in problems for which cluster sampling is appropriate. Thus we cannot use the estimator $M\bar{y}$, but we can form another estimator of the population total that does not depend on M. The quantity $\bar{y}_t$, given by

$$\bar{y}_t = \frac{1}{n} \sum_{i=1}^{n} y_i \tag{8.7}$$

is the average of the cluster totals for the n sampled clusters. Hence $\bar{y}_t$ is an unbiased estimator of the average of the N cluster totals in the population. By the same reasoning as employed in Chapter 4, $N\bar{y}_t$ is an unbiased estimator of the sum of the cluster totals or, equivalently, of the population total τ.

For example, it is highly unlikely that the number of adult males in a city would be known, and hence the estimator $N\bar{y}_t$, rather than $M\bar{y}$, would have to be used to estimate τ.

Estimator of the population total τ, which does not depend on M:

$$N\bar{y}_t = \frac{N}{n} \sum_{i=1}^{n} y_i \tag{8.8}$$

Estimated variance of $N\bar{y}$:

$$\hat{V}(N\bar{y}_t) = N^2 \hat{V}(\bar{y}_t) = N^2 \left(\frac{N-n}{Nn}\right) \frac{\sum_{i=1}^{n} (y_i - \bar{y}_t)^2}{n-1} \tag{8.9}$$

Bound on the error of estimation:

$$2\sqrt{\hat{V}(N\bar{y}_t)} = 2\sqrt{N^2 \left(\frac{N-n}{Nn}\right) \frac{\sum_{i=1}^{n} (y_i - \bar{y}_t)^2}{n-1}} \tag{8.10}$$

If there is a large amount of variation among the cluster sizes and if cluster sizes are highly correlated with cluster totals, the variance of $N\bar{y}_t$

[Equation (8.9)] is generally larger than the variance of $M\bar{y}$ [Equation (8.5)]. The estimator $N\bar{y}_t$ does not use the information provided by the cluster sizes $m_1, m_2, \ldots, m_n$ and hence may be less precise.

EXAMPLE 8.4

Use the data of Table 8.1 to estimate the total income of all residents of the city if M is not known. Place a bound on the error of estimation.

SOLUTION

Example 8.1 gives $N = 415$. From Equation (8.8) and Table 8.1, the estimate of the total income τ is

$$N\bar{y}_t = \frac{N}{n} \sum_{i=1}^{n} y_i = \frac{415}{25}(1,329,000) = \$22,061,400$$

This figure is fairly close to the estimate given in Example 8.3.
    To place a bound on the error of estimation, we first calculate

$$\sum_{i=1}^{n} (y_i - \bar{y}_t)^2 = \sum_{i=1}^{n} y_i^2 - \frac{1}{n}\left(\sum_{i=1}^{n} y_i\right)^2$$

$$= 82,039,000,000 - \tfrac{1}{25}(1,329,000)^2$$

$$= 11,389,360,000$$

Then the estimate of the total income of all residents of the city, with a bound on the error of estimation, is

$$N\bar{y}_t \pm 2\sqrt{\hat{V}(N\bar{y}_t)}$$

Substituting into Equation (8.10), we calculate

$$N\bar{y}_t \pm 2 \sqrt{N^2\left(\frac{N-n}{Nn}\right)\frac{\sum_{i=1}^{n}(y_i - \bar{y}_t)^2}{n-1}}$$

$$22,061,400 \pm 2\sqrt{(415)^2\left[\frac{415-25}{(415)(25)}\right]\left(\frac{11,389,360,000}{24}\right)}$$

$$22,061,400 \pm 3,505,920$$

The bound on the error of estimation is slightly smaller than the bound for the estimator $M\bar{y}$ (Example 8.3), partly because the cluster sizes are not highly correlated with the cluster total in this example. In other words, the cluster sizes are providing little information on cluster totals; hence the unbiased estimator $N\bar{y}_t$ appears to be better than the estimator $M\bar{y}$.

The estimators of μ and τ possess special properties when all cluster sizes are equal (that is, $m_1 = m_2 = \cdots = m_N$). First, the estimator $\bar{y}$, given by Equation (8.1), is an unbiased estimator of the population mean μ. Second, $\hat{V}(\bar{y})$, given by Equation (8.2), is an unbiased estimator of the variance of $\bar{y}$. Finally, the two estimators, $M\bar{y}$ and $N\bar{y}_t$, of the population total τ are equivalent.

EXAMPLE 8.5

The circulation manager of a newspaper wishes to estimate the average number of newspapers purchased per household in a given community. Travel costs from household to household are substantial. Therefore the 4000 households in the community are listed in 400 geographical clusters of 10 households each, and a simple random sample of 4 clusters is selected. Interviews are conducted, with the results as shown in the accompanying table. Estimate the average number of newspapers per household for the community, and place a bound on the error of estimation.

Cluster	Number of Newspapers										Total
1	1	2	1	3	3	2	1	4	1	1	19
2	1	3	2	2	3	1	4	1	1	2	20
3	2	1	1	1	1	3	2	1	3	1	16
4	1	1	3	2	1	5	1	2	3	1	20

SOLUTION

From Equation (8.1)

$$\bar{y} = \frac{\sum\limits_{i=1}^{n} y_i}{\sum\limits_{i=1}^{n} m_i}$$

When $m_1 = m_2 = \cdots = m_n = m$, the equation becomes

$$\bar{y} = \frac{\sum\limits_{i=1}^{n} y_i}{nm} = \frac{19 + 20 + 16 + 20}{4(10)} = 1.875$$

Also, it can be shown that

$$\sum_{i=1}^{n} (y_i - \bar{y}m_i)^2 = \sum_{i=1}^{n} y_i^2 - 2\bar{y} \sum_{i=1}^{n} y_i m_i + \bar{y}^2 \sum_{i=1}^{n} m_i^2$$

$$= \sum_{i=1}^{n} y_i^2 - nm^2 \bar{y}^2$$

Substituting, we obtain

$$\sum_{i=1}^{n} (y_i - \bar{y}m_i)^2 = (19)^2 + (20)^2 + (16)^2 + (20)^2 - 4(10)^2(1.875)^2$$

$$= 10.75$$

Thus from Equation (8.2),

$$\hat{V}(\bar{y}) = \left(\frac{N-n}{Nn\bar{M}^2}\right) \frac{\sum\limits_{i=1}^{m} (y_i - \bar{y}m_i)^2}{n-1} = \frac{(400-4)(10.75)}{400(4)(10)^2(3)} = .0089$$

Therefore the best estimate of the average number of newspapers per household, with a bound on the error of estimation, is

$$\bar{y} \pm 2\sqrt{\hat{V}(\bar{y})}, \qquad \text{or} \qquad 1.88 \pm 2\sqrt{.0089}, \qquad \text{or} \qquad 1.88 \pm .19$$

Thus the estimate of the average number of newspapers per household is 1.88, with a high probability that the error of estimation is less than .19.

8.4
SELECTING THE SAMPLE SIZE FOR ESTIMATING POPULATION MEANS AND TOTALS

The quantity of information in a cluster sample is affected by two factors, the number of clusters and the relative cluster size. We have not encountered the latter factor in any of the sampling procedures discussed previously. In the problem of estimating the number of homes with inadequate fire insurance in a state, the clusters could be counties, voting districts, school districts, communities, or any other convenient grouping of homes. As we have already seen, the size of the bound on the error of estimation depends crucially upon the variation among the cluster *totals*. Thus in attempting to achieve small bounds on the error of estimation, one must select clusters with as little variation as possible among these totals. We will now assume that the cluster size (sampling unit) has been chosen and will consider only the problem of choosing the number of clusters, n.

From Equation (8.2) the estimated variance of $\bar{y}$ is

$$\hat{V}(\bar{y}) = \frac{N-n}{Nn\bar{M}^2}(s_c^2)$$

where $$s_c^2 = \frac{\sum\limits_{i=1}^{n}(y_i - \bar{y}m_i)^2}{n-1}$$ \hfill (8.11)

The actual variance of $\bar{y}$ is approximately

$$V(\bar{y}) = \frac{N-n}{Nn\bar{M}^2}(\sigma_c^2)$$ \hfill (8.12)

where σ_c^2 is the population quantity estimated by s_c^2.

Because we do not know σ_c^2 or the average cluster size $\bar{M}$, choice of the sample size, that is, the number of clusters necessary to purchase a specified quantity of information concerning a population parameter, is difficult. We overcome this difficulty by using the same method we used for ratio estimation. That is, we use an esitmate of σ_c^2 and $\bar{M}$ available from a prior survey, or we select a preliminary sample containing n' elements. Estimates of σ_c^2 and $\bar{M}$ can be computed from the preliminary sample and used to acquire an approximate total sample size n. Thus, as in all problems of selecting a sample size, we equate two standard deviations of our estimator to a bound on the error of estimation, B. This bound is chosen by the experimenter and represents the maximum error that he or she is willing to tolerate. That is,

$$2\sqrt{V(\bar{y})} = B$$

Using Equation (8.12), we can solve for n.

We obtain similar results when using $M\bar{y}$ to estimate the population total τ, because $V(M\bar{y}) = M^2 V(\bar{y})$.

Approximate sample size required to estimate μ with a bound B on the error of estimation:

$$n = \frac{N\sigma_c^2}{ND + \sigma_c^2} \tag{8.13}$$

where σ_c^2 is estimated by s_c^2 and

$$D = \frac{B^2 \bar{M}^2}{4}$$

EXAMPLE 8.6

Suppose the data in Table 8.1 represent a preliminary sample of incomes in the city. How large a sample should be taken in a future survey in order to estimate the average per capita income μ with a bound of $500 on the error of estimation?

SOLUTION

To use Equation (8.13), we must estimate σ_c^2; the best estimate available is s_c^2, which can be calculated by using the data in Table 8.1. Using the calculations in Example 8.2, we have

$$s_c^2 = \frac{\sum\limits_{i=1}^{n} (y_i - \bar{y}m_i)^2}{n - 1} = \frac{15,227,502,247}{24} = 634,479,260$$

Quantity $\bar{M}$ can be estimated by $\bar{m} = 6.04$ calculated from Table 8.1. Then

D is approximately

$$\frac{B^2\bar{m}^2}{4} = \frac{(500)^2(6.04)^2}{4} = (62,500)(6.04)^2$$

Using Equation (8.13) yields

$$n = \frac{N\sigma_c^2}{ND + \sigma_c^2} = \frac{415(634,479,260)}{415(6.04)^2(62,500) + 634,479,260} = 166.58$$

Thus 167 clusters should be sampled.

Approximate size required to estimate τ, using $M\bar{y}$, with a bound B on the error of estimation:

$$n = \frac{N\sigma_c^2}{ND + \sigma_c^2} \tag{8.14}$$

where σ_c^2 is estimated by s_c^2 and

$$D = \frac{B^2}{4N^2}$$

EXAMPLE 8.7

Again using the data in Table 8.1 as a preliminary sample of incomes in the city, how large a sample is necessary to estimate the total income of all residents, τ, with a bound of $1,000,000 on the error of estimation? There are 2500 residents of the city ($M = 2500$).

SOLUTION

We use Equation (8.14) and estimate σ_c^2 by

$$s_c^2 = 634,479,260$$

as in Example 8.6. When estimating τ, we use

$$D = \frac{B^2}{4N^2} = \frac{(1,000,000)^2}{4(415)^2}$$

$$ND = \frac{(1,000,000)^2}{4(415)} = 602,409,000$$

Then using Equation (8.14) gives

$$n = \frac{N\sigma_c^2}{ND + \sigma_c^2} = \frac{415(634,479,260)}{602,409,000 + 634,479,260} = 212.88$$

Thus 213 clusters should be sampled to estimate the total income with a bound of $1,000,000 on the error of estimation.

The estimator $N\bar{y}_t$, shown in Equation (8.8), is used to estimate τ when M is unknown. The estimated variance of $N\bar{y}_t$, shown in Equation (8.9), is

$$\hat{V}(N\bar{y}_t) = N^2 \left(\frac{N - n}{Nn} \right) s_t^2$$

where

$$s_t^2 = \frac{\sum\limits_{i=1}^{n} (y_i - \bar{y}_t)^2}{n - 1} \qquad (8.15)$$

Thus the population variance of $N\bar{y}_t$ is

$$V(N\bar{y}_t) = N^2 V(\bar{y}_t) = N^2 \left(\frac{N - n}{Nn} \right) \sigma_t^2 \qquad (8.16)$$

where σ_t^2 is the population quantity estimated by s_t^2.

Estimation of τ with a bound of B units on the error of estimation leads to the following equation:

$$2\sqrt{V(N\bar{y}_t)} = B$$

Using Equation (8.16), we can solve for n.

Approximate sample size required to estimate τ, using $N\bar{y}_t$ with a bound B on the error of estimation:

$$n = \frac{N\sigma_t^2}{ND + \sigma_t^2} \qquad (8.17)$$

where σ_t^2 is estimated by s_t^2, and

$$D = \frac{B^2}{4N^2}$$

EXAMPLE 8.8

Assume the data of Table 8.1 are from a preliminary study of incomes in the city and M is not known. How large a sample must be taken to estimate the total income of all residents, τ, with a bound of $1,000,000 on the error of estimation?

SOLUTION

The quantity σ_t^2 must be estimated by s_t^2, which is calculated from the data of Table 8.1. Using the calculations of Example 8.4 gives

$$s_t^2 = \frac{\sum\limits_{i=1}^{n} (y_i - \bar{y}_t)^2}{n - 1} = \frac{11,389,360,000}{24} = 474,556,667$$

The bound on the error of estimation is $B = \$1{,}000{,}000$. Hence

$$D = \frac{B^2}{4N^2} = \frac{(1{,}000{,}000)^2}{4(415)^2}$$

From Equation (8.17)

$$n = \frac{N\sigma_t^2}{ND + \sigma_t^2} = \frac{415(474{,}556{,}667)}{415(1{,}000{,}000)^2/4(415)^2 + 474{,}556{,}667} = 182.88$$

Thus a sample of 183 clusters must be taken to have a bound of $1,000,000 on the error of estimation.

8.5
ESTIMATION OF A POPULATION PROPORTION

Suppose an experimenter wishes to estimate a population proportion, or fraction, such as the proportion of houses in a state with inadequate plumbing or the proportion of corporation presidents who are college graduates. The best estimator of the population proportion p is the sample proportion $\hat{p}$. Let a_i denote the total number of elements in cluster i that possess the characteristic of interest. Then the proportion of elements in the sample of n clusters possessing the characteristic is given by

$$\hat{p} = \frac{\sum\limits_{i=1}^{n} a_i}{\sum\limits_{i=1}^{n} m_i}$$

where m_i is the number of elements in the ith cluster, $i = 1, 2, \ldots, n$. Note that $\hat{p}$ has the same form as $\bar{y}$ [see Equation (8.1)], except that y_i is replaced by a_i. The estimated variance of $\hat{p}$ is similar to that of $\bar{y}$.

Estimator of the population proportion p:

$$\hat{p} = \frac{\sum\limits_{i=1}^{n} a_i}{\sum\limits_{i=1}^{n} m_i} \tag{8.18}$$

Estimated variance of $\hat{p}$:

$$\hat{V}(\hat{p}) = \left(\frac{N-n}{Nn\bar{M}^2}\right) \frac{\sum\limits_{i=1}^{n} (a_i - \hat{p}m_i)^2}{n-1} \tag{8.19}$$

Bound on the error of estimation:

$$2\sqrt{\hat{V}(\hat{p})} = 2\sqrt{\left(\frac{N-n}{Nn\bar{M}^2}\right)\frac{\sum\limits_{i=1}^{n}(a_i - \hat{p}m_i)^2}{n-1}} \qquad (8.20)$$

The variance formula, (8.19), is a good estimator only when the sample size n is large, say $n \geq 20$. If $m_1 = m_2 = \cdots = m_N$, then $\hat{p}$ is an unbiased estimator of p, and $\hat{V}(\hat{p})$, shown in Equation (8.19), is an unbiased estimator of the actual variance of $\hat{p}$ for any sample size.

EXAMPLE 8.9

In addition to being asked about their income, the residents of the sample survey of Example 8.2 are asked whether they rent or own their homes. The results are given in Table 8.2. Use the data in Table 8.2 to estimate the proportion of residents who live in rented housing. Place a bound on the error of estimation.

TABLE 8.2 Number of renters

Cluster	Number of Residents, m_i	Number of Renters, a_i	Cluster	Number of Residents, m_i	Number of Renters, a_i
1	8	4	14	10	5
2	12	7	15	9	4
3	4	1	16	3	1
4	5	3	17	6	4
5	6	3	18	5	2
6	6	4	19	5	3
7	7	4	20	4	1
8	5	2	21	6	3
9	8	3	22	8	3
10	3	2	23	7	4
11	2	1	24	3	0
12	6	3	25	8	3
13	5	2			

$$\sum_{i=1}^{25} m_i = 151 \qquad \sum_{i=1}^{25} a_i = 72$$

$$\sum_{i=1}^{25} a_i^2 = 262 \qquad \sum_{i=1}^{25} m_i^2 = 1047 \qquad \sum_{i=1}^{25} a_i m_i = 511$$

SOLUTION

The best estimate of the population proportion of renters is $\hat{p}$, shown in Equation (8.18), where

$$\hat{p} = \frac{\sum\limits_{i=1}^{m} a_i}{\sum\limits_{i=1}^{n} m_i} = \frac{72}{151} = .48$$

To estimate the variance of $\hat{p}$, we must calculate

$$\sum_{i=1}^{n} (a_i - \hat{p}m_i)^2 = \sum_{i=1}^{n} a_i^2 - 2\hat{p} \sum_{i=1}^{n} a_i m_i + \hat{p}^2 \sum_{i=1}^{n} m_i^2$$

and from Table 8.2

$$\sum_{i=1}^{n} (a_i - \hat{p}m_i)^2 = 262 - 2(.477)(511) + (.477)^2(1047) = 12.729$$

Quantity $\bar{M}$ is estimated by $\bar{m}$, where

$$\bar{m} = \frac{\sum\limits_{i=1}^{n} m_i}{n} = \frac{151}{25} = 6.04$$

Then from Equation (8.19),

$$\hat{V}(\hat{p}) = \left(\frac{N-n}{Nn\bar{M}^2}\right) \frac{\sum\limits_{i=1}^{n} (a_i - \hat{p}m_i)^2}{n-1}$$

$$= \frac{(415 - 25)(12.729)}{415(25)(6.04)^2(24)} = .00055$$

The estimate of p with a bound on the error is

$$\hat{p} \pm 2\sqrt{\hat{V}(\hat{p})}, \quad \text{or} \quad .48 \pm 2\sqrt{.00055}, \quad \text{or} \quad .48 \pm .05$$

Thus the best estimate of the proportion of people who rent homes is .48. The error of estimation should be less than .05 with probability of approximately .95.

8.6
SELECTING THE SAMPLE SIZE FOR ESTIMATING PROPORTIONS

Estimation of the population proportion p with a bound of B units on the error of estimation implies that the experimenter wants

$$2\sqrt{V(\hat{p})} = B$$

This equation can be solved for n, and the solution is similar to Equation (8.13). That is,

$$n = \frac{N\sigma_c^2}{ND + \sigma_c^2}$$

where $D = B^2 \bar{M}^2/4$, and σ_c^2 is estimated by

$$s_c^2 = \frac{\sum_{i=1}^{n} (a_i - \hat{p}m_i)^2}{n - 1} \qquad (8.21)$$

Equation (8.21) is Equation (8.11) with y_i replaced by a_i and $\bar{y}$ by $\hat{p}$.

EXAMPLE 8.10

The data in Table 8.2 are out of date. A new study will be conducted in the same city for the purpose of estimating the proportion p of residents who rent their homes. How large a sample should be taken to estimate p with a bound of .04 on the error of estimation?

SOLUTION

The best estimate of σ_c^2 is s_c^2, which is calculated by using data from Table 8.2:

$$s_c^2 = \frac{\sum_{i=1}^{n} (a_i - \hat{p}m_i)^2}{n - 1} = \frac{12.729}{24} = .530$$

Quantity $\bar{M}$ is estimated by $\bar{m} = 6.04$. Also, D is approximated by

$$\frac{B^2 \bar{m}^2}{4} = \frac{(.04)^2 (6.04)^2}{4} = .0146$$

Then $\qquad n = \dfrac{N\sigma_c^2}{ND + \sigma_c^2} = \dfrac{(415)(.530)}{(415)(.0146) + .530} = 33.40$

Thus 34 clusters should be sampled to estimate p with a bound of .04 on the error of estimation.

8.7
CLUSTER SAMPLING COMBINED WITH STRATIFICATION

As is the case with all other sampling methods, cluster sampling can be combined wtih stratified sampling, in the sense that the population may be divided into L strata and a cluster sample can then be selected from each stratum.

Recall that Equation (8.1) has the form of a ratio estimator and can be thought of as the ratio of an estimator of the average cluster total to an estimator of the average cluster size. Thinking in terms of ratio estimators, then, we have two ways to form the estimator of a population mean across strata, the separate estimator and the combined estimator. A little investigation will show that if the separate estimator is employed, the total number of elements in each stratum must be known in order to assign proper stratum weights. Since these quantities are usually unknown, we will investigate only the combined form of the ratio estimator in the context of cluster sampling.

Instead of presenting formidable-looking general formulas, we will illustrate the technique with a numerical example.

EXAMPLE 8.11

Let the data of Table 8.1 form the sample of stratum 1, with, as in Example 8.2, $N_1 = 415$ and $n_1 = 25$. A smaller neighboring city is taken to be stratum 2. For stratum 2, $n_2 = 10$ blocks are to be sampled from $N_2 = 168$. Estimate the average per capita income in the two cities combined, and place a bound on the error of estimation, given the additional data shown in the accompanying table.

Cluster i	Number of Residents, m_i	Total Income per Cluster, y_i
1	2	$ 18,000
2	5	52,000
3	7	68,000
4	4	36,000
5	3	45,000
6	8	96,000
7	6	64,000
8	10	115,000
9	3	41,000
10	1	12,000

SOLUTION

The average cluster totals in the respective samples are $\bar{y}_{t1} = 53,160$ and $\bar{y}_{t2} = 54,700$. The average cluster sizes in the respective samples are $\bar{m}_1 = 6.04$ and $\bar{m}_2 = 4.90$. The estimate of the population average cluster total is then

$$\frac{1}{N}(N_1 \bar{y}_{t1} + N_2 \bar{y}_{t2})$$

while the estimate of the average cluster size is

$$\frac{1}{N}(N_1 \bar{m}_1 + N_2 \bar{m}_2)$$

An estimate of the population mean per element is then

$$\bar{y}^* = \frac{N_1 \bar{y}_{t1} + N_2 \bar{y}_{t2}}{N_1 \bar{m}_1 + N_2 \bar{m}_2}$$

and this equation does have the form of a combined ratio estimate. Analogous to the variance used in Section 6.6, the variance of $\bar{y}^*$ can be estimated by

$$\hat{V}(\bar{y}^*) = \frac{1}{M^2} \left\{ \frac{N_1(N_1 - n_1)}{n_1(n_1 - 1)} \sum_{i=1}^{n_1} [(y_i - \bar{y}_{t1}) - \bar{y}^*(m_i - \bar{m}_1)]^2 \right.$$

$$\left. + \frac{N_2(N_2 - n_2)}{n_2(n_2 - 1)} \sum_{i=1}^{n_2} [(y_i - \bar{y}_{t2}) - \bar{y}^*(m_i - \bar{m}_2)]^2 \right\}$$

where M is the total number of elements in the population and can be estimated by $N_1 \bar{m}_1 + N_2 \bar{m}_2$ if it is not known. The first sum in the variance expression is over all the sample observations from stratum 1, and the second sum is over all the observations from stratum 2.

For the data given in the table,

$$\bar{y}^* = \frac{415(53,160) + 168(54,700)}{415(6.04) + 168(4.90)} = 9385$$

For stratum 1

$$\left(\frac{1}{n_1 - 1}\right) \sum_{i=1}^{n_1} [(y_i - \bar{y}_{t1}) - \bar{y}^*(m_i - \bar{m}_1)]^2 = 675,930,246$$

and for stratum 2

$$\left(\frac{1}{n_2 - 1}\right) \sum_{i=1}^{n_2} [(y_i - \bar{y}_{t2}) - \bar{y}^*(m_i - \bar{m}_2)]^2 = 74,934,600$$

Since $N_1 \bar{m}_1 + N_2 \bar{m}_2 = 3329.8$

it follows that $\hat{V}(\bar{y}^*) = 412,563.8$

and $2\sqrt{\hat{V}(\bar{y}^*)} = 1285$

Thus the average per capita income for the two cities combined is

$$\$9385 \pm \$1285$$

We see that the bound on the error of estimation is slightly smaller than the bound for stratum 1 alone, as found in Example 8.2.

8.8
CLUSTER SAMPLING WITH PROBABILITIES PROPORTIONAL TO SIZE

We saw in Section 4.6 that we can sometimes reduce the variance of an esitmator by sampling units with probabilities proportional to a measure of the size of the unit. Cluster sampling often provides an ideal situation in which to use pps sampling since the number of elements in a cluster, m_i, forms a natural measure of the size of the cluster. Sampling with probabilities proportional to m_i pays big dividends in terms of reducing the bound on the error of estimation when the cluster total y_i is highly correlated with the number of elements in the cluster, which is often the case.

In the notation of Section 4.6, let π_i, the probability that the ith sampling unit appears in the sample, be given by

$$\pi_i = \frac{m_i}{M} \tag{8.22}$$

Thus the estimator of a population total $\hat{\tau}_{pps}$ becomes [see Equation (4.20)]

$$\hat{\tau}_{pps} = \frac{1}{n} \sum_{i=1}^{n} \frac{y_i}{\pi_i} = \frac{1}{n} \sum_{i=1}^{n} \frac{y_i}{(m_i/M)}$$

$$= \frac{M}{n} \sum_{i=1}^{n} \frac{y_i}{m_i} = \frac{M}{n} \sum_{i=1}^{n} \bar{y}_i$$

where $\bar{y}_i$ is the average of the observations in the ith cluster. The estimated variance of $\hat{\tau}_{pps}$ has a particularly simple form, as given later.

Since there are now M elements in the population, the estimator of the population mean, $\hat{\mu}_{pps}$, is simply

$$\hat{\mu}_{pps} = \frac{1}{M} \hat{\tau}_{pps} = \frac{1}{n} \sum_{i=1}^{n} \bar{y}_i$$

The estimated variance of $\hat{\mu}_{pps}$ is also easy to calculate.

Estimator of the population mean μ:

$$\hat{\mu}_{pps} = \frac{1}{n} \sum_{i=1}^{n} \bar{y}_i \tag{8.23}$$

where $\bar{y}_i$ is the mean for the ith cluster.

Estimated variance of $\hat{\mu}_{pps}$:

$$\hat{V}(\hat{\mu}_{pps}) = \frac{1}{n(n-1)} \sum_{i=1}^{n} (\bar{y}_i - \hat{\mu}_{pps})^2 \tag{8.24}$$

Bound on the error of estimation:

$$2\sqrt{\hat{V}(\hat{\mu}_{pps})} = 2\sqrt{\frac{1}{n(n-1)} \sum_{i=1}^{n} (\bar{y}_i - \hat{\mu}_{pps})^2} \tag{8.25}$$

Estimator of the population total τ:

$$\hat{\tau}_{pps} = \frac{M}{n} \sum_{i=1}^{n} \bar{y}_i \tag{8.26}$$

Estimated variance of $\hat{\tau}_{pps}$:

$$\hat{V}(\hat{\tau}_{pps}) = \frac{M^2}{n(n-1)} \sum_{i=1}^{n} (\bar{y}_i - \hat{\mu}_{pps})^2 \tag{8.27}$$

Bound on the error of estimation:

$$2\sqrt{\hat{V}(\hat{\tau}_{pps})} = 2\sqrt{\frac{M^2}{n(n-1)} \sum_{i=1}^{n} (\bar{y}_i - \hat{\mu}_{pps})^2} \tag{8.28}$$

We illustrate the technique of sampling with probabilities proportional to cluster sizes and the use of the formulas just given in the next two examples.

EXAMPLE 8.12

An auditor wishes to sample sick-leave records of a large firm in order to estimate the average number of days of sick leave per employee over the past quarter. The firm has eight divisions, with varying numbers of employees per division. Since number of days of sick leave used within each division should be highly correlated with the number of employees, the auditor decides to sample $n = 3$ divisions with probabilities proportional to number of employees. Show how to select the sample if the respective numbers of employees are 1200, 450, 2100, 860, 2840, 1910, 290, 3200.

SOLUTION

We first list the number of employees and the cumulative range for each division, as follows:

Division	Number of Employees	Cumulative Range
1	1,200	1–1200
2	450	1201–1650
3	2,100	1651–3750
4	860	3751–4610
5	2,840	4611–7450
6	1,910	7451–9360
7	390	9361–9750
8	3,200	9751–12,950
	12,950	

Since $n = 3$ divisions are to be sampled, we must select three random numbers between 00001 and 12,500. We can make this selection by starting anywhere on a random number table and selecting five-digit numbers, but we chose to start on line 1, column 4 of Table 2 in the Appendix. The first three numbers between 00001 and 12,950, as we proceed down the column, are 02011, 07972, and 10281. The first appears in the cumulative range of division 3, the second appears in the range of division 6, and the third apppears in the range of division 8. Thus divisions 3, 6, and 8 constitute the sample. (Note that one division can be selected more than once. In that event we treat the resulting data as two separate but equal sample values.)

EXAMPLE 8.13

Suppose the total number of sick days used by the three sampled divisions during the past quarter are, respectively,

$$y_1 = 4320, \qquad y_2 = 4160, \qquad y_3 = 5790$$

Estimate the average number of sick days used per person for the entire firm, and place a bound on the error of estimation.

SOLUTION

We must first compute the cluster means for the sampled clusters, which are

$$\bar{y}_1 = \frac{4320}{2100} = 2.06, \qquad \bar{y}_2 = \frac{4160}{1910} = 2.18, \qquad \bar{y}_3 = \frac{5790}{3200} = 1.81$$

(Note that the numbers of employees per sampled firm come from the data in Example 8.12.)

Now by Equation (8.23)

$$\hat{\mu}_{pps} = \frac{1}{n} \sum_{i=1}^{n} \bar{y}_i = \frac{1}{3}(2.06 + 2.18 + 1.81) = 2.02$$

Also, by Equation (8.24)

$$\hat{V}(\hat{\mu}_{pps}) = \frac{1}{n(n-1)} \sum_{i=1}^{n} (\bar{y}_i - \hat{\mu}_{pps})^2$$

$$= \frac{1}{3(2)} [(2.06 - 2.02)^2 + (2.18 - 2.02)^2 + (1.81 - 2.02)^2]$$

$$= .0119$$

Thus the bound on the error of estimation is

$$2\sqrt{.0119} = .22$$

Our estimate of the average number of sick days used by employees of the firm is

$$2.02 \pm .22$$

We now have three estimators of the population total in cluster sampling, the ratio estimator (8.4), the unbiased estimator (8.8), and the pps estimator (8.26). How do we know which is best? Here are some guidelines about how to answer this question: If y_i is uncorrelated with m_i, then the unbiased estimator is better than either of the other two. If y_i is correlated with m_i, then the ratio and pps estimators are more precise than the unbiased estimator. The pps estimator is better than the ratio estimator if the within–cluster variation does not change with changing m_i. The ratio estimator is better than the pps estimator if the within–cluster variation increases with increasing m_i.

In Examples 8.12 and 8.13, the number of sick-leave days used should increase as the number of employees increases. Thus the unbiased estimator is a poor choice here. But the variation of sick-leave days within divisions may remain relatively constant across divisions. In that case the pps estimator is the best choice.

8.9
SUMMARY

This chapter introduces a third sample survey design, cluster sampling. In this design each sampling unit is a group, or cluster, of elements. Cluster sampling may provide maximum information at minimum cost when a frame listing population elements is not available or when the cost of obtaining observations increases with increasing distance between elements.

The estimator of the population mean μ is the sample mean $\bar{y}$, given by Equation (8.1). The estimated variance of $\bar{y}$ is given by Equation (8.2). Two estimators of the population total τ were given with their estimated variances. The estimator $M\bar{y}$ is presented in Equation (8.4); it is used when the number of elements M in the population is known. Estimator $N\bar{y}_t$ [see Equation (8.8)] is used when M is unknown.

In Section 8.4 we discussed an appropriate sample size for estimating μ or τ with a specified bound on the error of estimation.

In cluster sampling the estimator of a population proportion p is the sample proportion $\hat{p}$, given by Equation (8.18). The estimated variance of $\hat{p}$ is given by Equation (8.19). The problem of selecting a sample size for estimating a proportion is similar to the problem for estimating a mean.

Cluster sampling can also be used within strata in a stratified population, and an example was given in Section 8.7.

CASE STUDY REVISITED

THE NEIGHBORHOOD CHARACTERISTICS PROBLEM

AT the beginning of this chapter we suggested using U.S. Census data on block statistics to estimate the proportion of residents aged 65 and over in a 40-block area. The $n = 5$ blocks were randomly sampled from the 40 and the following data obtained:

Number of Residents, m_i	Number Aged 65 and Over, a_i	$\hat{p}m_i$	$a_i - \hat{p}m_i$	$(a_i - \hat{p}m_i)^2$
90	15	21.60	−6.60	43.5600
32	8	7.68	0.32	0.1024
47	14	11.28	2.72	7.3984
25	9	6.00	3.00	9.0000
16	4	3.84	0.16	0.0256
$\overline{210}$	$\overline{50}$			60.0864

$$\hat{p} = \frac{\sum\limits_{i=1}^{n} a_i}{\sum\limits_{i=1}^{n} m_i} = \frac{50}{210} = .24$$

So the best estimate of the proportion of people aged 65 or over is .24. The bound on the error of estimation is

$$2\sqrt{\hat{V}(\hat{p})} = 2\sqrt{\left(\frac{N-n}{Nn\bar{m}^2}\right)\left(\frac{1}{n-1}\right) \sum_{i=1}^{n} (a_i - pm_i)^2}$$

$$= 2\sqrt{\left[\frac{35}{(40)(5)(42)^2}\right]\left(\frac{1}{4}\right)(60.0864)}$$

$$= .08$$

Thus the estimate of the true proportion for the 40-block area is $.24 \pm .08$, or .16 to .32. We are confident that over 16% of the residents have ages of 65 or more.

EXERCISES

8.1 An experimenter working in an urban area desires to estimate the average value of a variable highly correlated with race. She thinks she should use cluster sampling, with city blocks as clusters and adults within blocks as elements. Explain why you would, or would not, use cluster sampling in each of the following situations.

(a) Most of the adults in certain blocks are white and most in other blocks are nonwhite.

(b) The proportion of nonwhites is the same in every block and is not close to zero or one.

(c) The proportion of nonwhites differs from block to block in the manner that would be expected if the clusters were made up by randomly assigning adults in the population to clusters.

8.2 A manufacturer of band saws wants to estimate the average repair cost per month for the saws he has sold to certain industries. He cannot obtain a repair cost for each saw, but he can obtain the total amount spent for saw repairs and the number of saws owned by each industry. Thus he decides to use cluster sampling, with each industry as a cluster. The manufacturer selects a simple random sample of $n = 20$ from the $N = 96$ industries he services. The data on total cost of repairs per industry and number of saws per industry are as given in the accompanying table. Estimate the average repair cost per saw for the past month, and place a bound on the error of estimation.

Industry	Number of Saws	Total Repair Cost for Past Month (in dollars)	Industry	Number of Saws	Total Repair Cost for Past Month (in dollars)
1	3	50	11	8	140
2	7	110	12	6	130
3	11	230	13	3	70
4	9	140	14	2	50
5	2	60	15	1	10
6	12	280	16	4	60
7	14	240	17	12	280
8	3	45	18	6	150
9	5	60	19	5	110
10	9	230	20	8	120

8.3 For the data in Exercise 8.2, estimate the total amount spent by the 96 industries on band saw repairs. Place a bound on the error of estimation.

8.4 After checking his sales records, the manufacturer of Exercise 8.2 finds that he sold a total of 710 band saws to these industries. Using this additional information, estimate the total amount spent on saw repairs by these industries, and place a bound on the error of estimation.

8.5 The same manufacturer (Exercise 8.2) wants to estimate the average repair cost per saw for next month. How many clusters should he select for his sample if he wants the bound on the error of estimation to be less than $2.00?

8.6 A political scientist developed a test designed to measure the degree of awareness of current events. She wánts to estimate the average score that would be achieved on this test by all students in a certain high school. The administration at the school will not allow the experimenter to randomly select students out of classes in session, but it will allow her to interrupt a small number of classes for the purpose of giving the test to every member of the class. Thus the experimenter selects 25 classes at random from the 108 classes in session at a particular hour. The test is given to each member of the sampled classes, with results as shown in the accompanying table. Estimate the average score that would be achieved on this test by all students in the school. Place a bound on the error of estimation.

Class	Number of Students	Total Score	Class	Number of Students	Total Score
1	31	1590	14	40	1980
2	29	1510	15	38	1990
3	25	1490	16	28	1420
4	35	1610	17	17	900
5	15	800	18	22	1080
6	31	1720	19	41	2010
7	22	1310	20	32	1740
8	27	1427	21	35	1750
9	25	1290	22	19	890
10	19	860	23	29	1470
11	30	1620	24	18	910
12	18	710	25	31	1740
13	21	1140			

8.7 The political scientist of Exercise 8.6 wants to estimate the average test score for a similar high school. She wants the bound on the error of estimation to be less than 2 points. How many classes should she sample? Assume the school has 100 classes in session during each hour.

8.8 An industry is considering revision of its retirement policy and wants to estimate the proportion of employees that favor the new policy. The industry consists of 87 separate plants located throughout the United States. Since results must be obtained quickly and with little cost, the industry decides to use cluster sampling with each plant as a cluster. A simple random sample of 15 plants is selected, and the opinions of the employees in these plants are obtained by questionnaire. The results are as shown in the accompanying table. Estimate the proportion of employees in the industry who favor the new retirement policy, and place a bound on the error of estimation.

Plant	Number of Employees	Number Favoring New Policy	Plant	Number of Employees	Number Favoring New Policy
1	51	42	9	73	54
2	62	53	10	61	45
3	49	40	11	58	51
4	73	45	12	52	29
5	101	63	13	65	46
6	48	31	14	49	37
7	65	38	15	55	42
8	49	30			

8.9 The industry of Exercise 8.8 modified its retirement policy after obtaining the results of the survey. It now wants to estimate the proportion of employees in favor of the modified policy. How many plants should be sampled to have a bound of .08 on the error of estimation? Use the data from Exercise 8.8 to approximate the results of the new survey.

8.10 An economic survey is designed to estimate the average amount spent on utilities for households in a city. Since no list of households is available, cluster sampling is used, with divisions (wards) forming the clusters. A simple random sample of 20 wards is selected from the 60 wards of the city. Interviewers then obtain the cost of utilities from each household within the sampled wards; the total costs are shown in the accompanying table. Estimate the average amount a household in the city spends on utilities, and place a bound on the error of estimation.

Sampled Ward	Number of Households	Total Amount Spent on Utilities	Sampled Ward	Number of Households	Total Amount Spent on Utilities
1	55	$2210	11	73	$2930
2	60	2390	12	64	2470
3	63	2430	13	69	2830
4	58	2380	14	58	2370
5	71	2760	15	63	2390
6	78	3110	16	75	2870
7	69	2780	17	78	3210
8	58	2370	18	51	2430
9	52	1990	19	67	2730
10	71	2810	20	70	2880

8.11 In the survey of Exercise 8.10 the number of households in the city is not known. Estimate the total amount spent on utilities for all households in the city, and place a bound on the error of estimation.

8.12 The economic survey of Exercise 8.10 is to be performed in a neighboring city of similar structure. The objective is to estimate the total amount spent on utilities by households in the city, with a bound of $5000 on the error of estimation. Use the data in Exercise 8.10 to find the approximate number of clusters needed to achieve this bound.

8.13 An inspector wants to estimate the average weight of fill for cereal boxes packaged in a certain factory. The cereal is available to him in cartons containing 12 boxes each. The inspector randomly selects 5 cartons and measures the weight of fill for every box in the sampled cartons, with the results (in ounces) as shown in the accompanying table. Estimate the average weight of fill for boxes packaged by this factory, and place a bound on the error of estimation. Assume that the total number of cartons packaged by the factory is large enough for the finite population correction to be ignored.

Carton	Ounces of Fill
1	16.1 15.9 16.1 16.2 15.9 15.8 16.1 16.2 16.0 15.9 15.8 16.0
2	15.9 16.2 15.8 16.0 16.3 16.1 15.8 15.9 16.0 16.1 16.1 15.9
3	16.2 16.0 15.7 16.3 15.8 16.0 15.9 16.0 16.1 16.0 15.9 16.1
4	15.9 16.1 16.2 16.1 16.1 16.3 15.9 16.1 15.9 15.9 16.0 16.0
5	16.0 15.8 16.3 15.7 16.1 15.9 16.0 16.1 15.8 16.0 16.1 15.9

8.14 A newspaper wants to estimate the proportion of voters favoring a certain candidate, candidate A, in a statewide election. Since selecting and interviewing a simple random sample of registered voters is very expensive, cluster sampling is used, with precincts as clusters. A simple random sample of 50 precincts is selected from the 497 precincts in the state. The newspaper wants to make the estimation on election day but before final returns are tallied. Therefore reporters are sent to the polls of each sample precinct to obtain the pertinent information directly from the voters. The results are shown in the accompanying table. Estimate the proportion of voters favoring candidate A, and place a bound on the error of estimation.

Number of Voters	Number Favoring A	Number of Voters	Number Favoring A	Number of Voters	Number Favoring A
1290	680	1893	1143	843	321
1170	631	1942	1187	1066	487
840	475	971	542	1171	596
1620	935	1143	973	1213	782
1381	472	2041	1541	1741	980
1492	820	2530	1679	983	693
1785	933	1567	982	1865	1033
2010	1171	1493	863	1888	987
974	542	1271	742	1947	872
832	457	1873	1010	2021	1093
1247	983	2142	1092	2001	1461
1896	1462	2380	1242	1493	1301
1943	873	1693	973	1783	1167
798	372	1661	652	1461	932
1020	621	1555	523	1237	481
1141	642	1492	831	1843	999
1820	975	1957	932		

8.15 The newspaper of Exercise 8.14 wants to conduct a similar survey during the next election. How large a sample size will be needed to esitmate the proportion of voters favoring a similar candidate with a bound of .05 on the error of estimation? Use the data in Exercise 8.14.

8.16 A forester wishes to estimate the average height of trees on a plantation. The plantation is divided into quarter-acre plots. A simple random sample of 20 plots is selected from the 386 plots on the plantation. All trees on the sampled plots are measured, with the results as shown in the accompanying table. Estimate the average height of trees on the plantation, and place a bound on the error of estimation. (*Hint:* The total for cluster i can be found by taking m_i times the cluster average.)

Number of Trees	Average Height (in feet)	Number of Trees	Average Height (in feet)
42	6.2	60	6.3
51	5.8	52	6.7
49	6.7	61	5.9
55	4.9	49	6.1

47	5.2	57	6.0
58	6.9	63	4.9
43	4.3	45	5.3
59	5.2	46	6.7
48	5.7	62	6.1
41	6.1	58	7.0

8.17 To emphasize safety, a taxicab company wants to estimate the proportion of unsafe tires on their 175 cabs. (Ignore spare tires.) Selecting a simple random sample of tires is impractial, so cluster sampling is used, with each cab as a cluster. A random sample of 25 cabs gives the following number of unsafe tires per cab:

$$2, 4, 0, 1, 2, 0, 4, 1, 3, 1, 2, 0, 1,$$
$$1, 2, 2, 4, 1, 0, 0, 3, 1, 2, 2, 1$$

Estimate the proportion of unsafe tires being used on the company's cabs, and place a bound on the error of estimation.

8.18 Accountants frequently require their business clients to provide cost inventories. Since a complete inventory is costly, quarterly inventories can conveniently be accomplished by sampling. Suppose a plumbing supply firm desires a cost inventory for many small items in stock. To obtain a simple random sample of items is difficult. However, the items are arranged on shelves, and selecting a simple random sample of shelves is relatively easy, treating each shelf as a cluster of items. Sampling 10 of the 48 shelves gave the results shown in the accompanying table. Estimate the total dollar amount of the items on the shelves, and place a bound on the error of estimation.

Cluster	Number of Items, m_i	Total Dollar Amount, y_i
1	42	83
2	27	62
3	38	45
4	63	112
5	72	96
6	12	58
7	24	75
8	14	58
9	32	67
10	41	80

8.19 A certain firm specializing in the manufacture and sale of leisure clothing has 80 retail stores in Florida and 140 in California. With each state as a stratum, the firm wishes to estimate average sick-leave time per employee for the past year. Each outlet can be viewed as a cluster of employees, and total sick leave time for each store can be determined from records. Simple random samples of 8 stores from Florida and 10 stores from California gave the results shown in the accompanying table (m_i denotes the number of employees and y_i denotes total days sick leave for the ith store). Estimate the average amount of sick leave per employee, and calculate an estimate of the variance of your estimator.

Florida		California	
m_i	y_i	m_i	y_i
12	40	16	51
20	52	8	32
8	30	4	11
14	36	3	10
24	71	12	33
15	48	17	39
10	39	24	61
6	21	30	37
		21	40
		9	41

8.20 Block statistics report the number of housing units, the number of residents, and the total number of rooms within housing units for a random sample of eight blocks selected from a large city. (Assume the number of blocks in the city is very large.) The data are given in the accompanying table.

Block	Number of Housing Units	Number of Residents	Number of Rooms
1	12	40	58
2	14	39	72
3	3	12	26
4	20	52	98
5	12	37	74
6	8	33	57
7	10	41	76
8	6	14	48

(a) Estimate the average number of residents per housing unit, and place a bound on the error of estimation.

(b) Estimate the average number of rooms per resident, and place a bound on the error of estimation.

8.21 A certain type of circuit board manufactured for installation in computers has 12 microchips per board. During the quality control inspection of 10 of these boards, the numbers of defective microchips per board were as follows:

$$2, 0, 1, 3, 2, 0, 0, 1, 3, 4$$

Estimate the proportion of defective microchips in the population from which this sample was drawn, and place a bound on the error of estimation.

8.22 Refer to the setting of Exercise 8.21. Suppose the sample of 10 boards used there came from a shipment of 50 such boards. Estimate the total number of defective microchips in the shipment, and place a bound on the error of estimation.

8.23 A large firm has its equipment inventories listed separately by department. From the 15 departments in the firm, 5 are to be randomly sampled by an auditor, who will then check to make sure that all equipment is properly identified and located.

The proportion of equipment items not properly identified is of interest to the auditor. The data are given in the accompanying table. Estimate the proportion of equipment items in the firm not properly identified, and place a bound on the error of estimation.

Department	Number of Equipment Items	Number of Items Not Properly Identified
1	15	2
2	27	3
3	9	1
4	31	1
5	16	2

8.24 Suppose that for the firm of Exercise 8.23 the 15 departments have the number of equipment items given in the accompanying table. Select a sample of 3 departments with probabilities proportional to number of equipment items.

Department	Number of Items	Department	Number of Items
1	12	9	31
2	9	10	26
3	27	11	22
4	40	12	19
5	35	13	16
6	15	14	33
7	18	15	6
8	10		

8.25 Suppose the three departments selected in Exercise 8.24 each have two improperly identified equipment items. Estimate the total number of improperly identified items in the firm, and place a bound on the error of estimation.

8.26 A large shipment of frozen seafood is packaged in cartons, each containing twenty-four 5-pound packages. There are 100 cartons in the shipment. The total weight (in pounds) of spoiled seafood is determined by a government inspector for each of a sample of 5 cartons. These data are as follows:

$$9, 6, 3, 10, 2$$

Estimate the total weight of spoiled seafood in the shipment, and place a bound on the error of estimation.

8.27 Using the data of Exercise 8.26, estimate the average amount of spoiled seafood per 5-pound package, and place a bound on the error of estimation.

8.28 A political scientist wishes to sample resident students on a large-university campus. Individual housing units can be conveniently used as clusters of students, or collections of housing units (freshman dormitories, fraternity houses, and so on) can be used as strata. Discuss the merits of cluster versus stratified random sampling if the goal is to estimate the proportion of students favoring a certain candidate in the following types of elections:

(a) A student government election,
(b) A national presidential election.

8.29 Under what conditions does cluster sampling produce a smaller bound on the error of estimation for a mean than simple random sampling?

8.30 Disregarding costs of sampling, what criteria would you use for selecting appropriate clusters in a cluster sampling problem?

EXPERIENCES WITH REAL DATA

8.1 Per capita income in the United States (as of 1977) is shown in Table 3 in the Appendix. Population figures for 1980 are also given. Treating each state as a cluster of people, select a random sample of states, and estimate the total personal income for the United States. Place a bound on the error of estimation.

8.2 Try an economic study, perhaps by treating households in a certain fixed geographic area (perhaps a few city blocks) as clusters of people. Sample n households and, upon gaining permission for an interview, record the total weekly amount spent on food by all individuals in the household and the number of individuals. Then estimate the average amount spent on food per person among the households in this population. Even if all the money is actually spent by one person (say, the mother), that total amount is the same as would have been recorded if each individual had purchased his or her own food. Thus the cluster total is available even though the observations per element may not be.

CASE STUDY

HOW MUCH DO STUDENTS SPEND FOR ENTERTAINMENT?

THE entertainment dollars are important to the businesses in a town containing a university. How can we estimate the average monthly amount spent on entertainment per student? To locate students who may be randomly selected from a directory is difficult, but to locate randomly selected classrooms, all of which should contain students at a prime class hour, like 10:00 A.M. on Monday, is relatively easy. Since classes may be large, sufficient information can be obtained by sampling a subset of those students in each sampled class. The result is a two-stage cluster sample.

A certain mid-sized university has 12,000 students divided into 150 classes at 10:00 A.M. on Mondays. Almost all the students should be in class at this hour. For the purpose of estimating the average monthly amount spent on entertainment, 4 classes are randomly selected, and approximately 10% of the students in each class are interviewed. The methods of two-stage cluster sampling are used in the analysis.

9

TWO-STAGE CLUSTER SAMPLING

9.1
INTRODUCTION

Two-stage cluster sampling is an extension of the concept of cluster sampling. You will recall from the discussion of cluster sampling in Chapter 8 that a cluster is usually a convenient or natural collection of elements, such as blocks of households or cartons of flashbulbs. A cluster often contains too many elements to obtain a measurement on each, or it contains elements so nearly alike that measurement of only a few elements provides information on an entire cluster. When either situation occurs, the experimenter can select a simple random sample of clusters and then take a simple random sample of elements within each cluster. The result is a two-stage cluster sample.

DEFINITION 9.1   A *two-stage cluster sample* is obtained by first selecting a simple random sample of clusters and then selecting a simple random sample of elements from each sampled cluster.

For example, a national survey of university students' opinions can be conducted by selecting a simple random sample of universities from all those in the country and then selecting a simple random sample of students from each university. Thus a university corresponds to a cluster of students. Similarly, the total amount of accounts receivable for a chain store can be estimated by first taking a simple random sample of stores and then selecting a simple random sample of accounts from each. Thus each chain store provides a cluster of accounts.

Two-stage cluster sampling is commonly used in large surveys involving the sampling of housing units. We mentioned in Chapter 4 that the Gallup poll samples approximately three hundred election districts from around the United States. At the second stage this poll randomly (or systematically) selects approximately five households per district, for a total sample size of about fifteen hundred households. In other polls block statistics from the U.S. Census Bureau form clusters of households, as discussed in Chapter 8, which are then subsampled before interviews are conducted.

Sampling for quality control purposes often involves two (or more) stages of sampling. For example, when an inspector samples packaged products, such as frozen food, he or she commonly samples cartons and then samples packages from within cartons. When one is sampling products turned out at various workstations, one might sample workstations and then sample items produced at each sampled station. When sampling requires detailed investigation of components of products, such as measuring plate thicknesses in automobile batteries, a quite natural procedure is to sample some of the products (batteries) and then sample components (plates) within these products.

There is a certain similarity between cluster sampling and stratified random sampling. Think of a population being divided into nonoverlapped groups of elements. If these groups are considered to be strata, then a simple random sample is selected from *each* group. If these groups are considered to be clusters, then a simple random sample of *groups* is selected, and the sampled groups are then subsampled. Stratified random sampling provides estimators with small variance when there is little variation among elements within each group. Cluster sampling does well when the elements within each group are highly variable, and all groups are quite similar to one another.

The advantages of two-stage cluster sampling over other designs are the same as those listed in Chapter 8 for cluster sampling. First, a frame listing all elements in the population may be impossible or costly to obtain, whereas to obtain a list of all clusters may be easy. For example, to compile a list of all university students in the country would be expensive and time–consuming, but a list of universities could be readily acquired. Second, the cost of obtaining data may be inflated by travel costs if the sampled elements are spread over a large geographic area. Thus to sample clusters of elements that are physically close together is often economical.

9.2
How to Draw a Two-Stage Cluster Sample

The first problem in selecting a two-stage cluster sample is the choice of appropriate clusters. Two conditions are desirable: (1) geographic proximity of the elements within a cluster and (2) cluster sizes that are convenient to administer.

The selection of appropriate clusters also depends on whether we want to sample a few clusters and many elements from each or many clusters and a few elements from each. Ultimately, the choice is based on costs. Large clusters tend to possess heterogeneous elements, and hence a large sample is required from each in order to acquire accurate estimates of population parameters. In contrast, small clusters frequently contain relatively homogeneous elements, in which case accurate information on the characteristics of a cluster can be obtained by selecting a small sample from each cluster.

Consider the problem of sampling personal incomes in a large city. The city can be divided into large clusters, for example precincts, which contain a heterogeneous assortment of incomes. Thus a small number of precincts may yield a representative cross section of incomes within the city, but a fairly large sample of elements from each cluster will be required in order to accurately estimate its mean (because of the heterogeneity of incomes within the cluster). In contrast, the city can be divided into small, relatively homogeneous clusters, say city blocks. Then a small sample of people from each block will give adequate information on each cluster's mean, but to obtain accurate information on the mean income for the entire city will require many blocks.

As another example, consider the university student opinion poll. If students within a university hold similar opinions on the question of interest but opinions differ widely from university to university, then the sample should contain a few representatives from many different universities. If the opinions vary greatly within each university, then the survey should include many representatives from each of a few universities.

To select the sample, we first obtain a frame listing all clusters in the population. We then draw a simple random sample of clusters, using the random sampling procedures presented in Chapter 4. Third, we obtain frames that list all elements in each of the sampled clusters. Finally, we select a simple random sample of elements from each of these frames.

9.3
UNBIASED ESTIMATION OF A POPULATION MEAN AND TOTAL

As in previous chapters, we are interested in estimating a population mean μ or a population total τ and placing a bound on the error of estimation. The following notation is used:

N = the number of clusters in the population

n = the number of clusters selected in a simple random sample

M_i = the number of elements in cluster i

m_i = the number of elements selected in a simple random sample from cluster i

$M = \sum_{i=1}^{N} M_i$ = the number of elements in the population

$$\bar{M} = \frac{M}{N} = \text{the average cluster size for the population}$$

y_{ij} = the jth observation in the sample from the ith cluster

$$\bar{y}_i = \frac{1}{m_i} \sum_{j=1}^{m_i} y_{ij} = \text{the sample mean for the } i\text{th cluster}$$

In constructing an estimator of the population mean μ, we might try to parallel what was done in Chapter 8 on single-stage cluster sampling. Equation (8.8) gives

$$\frac{N}{n} \sum_{i=1}^{n} y_i$$

as an unbiased estimator of τ. Thus if we divide by M,

$$\frac{N}{Mn} \sum_{i=1}^{n} y_i$$

becomes an unbiased estimator of μ. But we cannot evaluate this estimator now since we no longer know the cluster totals, y_i. We can, however, estimate y_i by $M_i\bar{y}_i$, and, on substituting $M_i\bar{y}_i$ for y_i, we have an unbiased estimator of μ, which we can calculate from our sample data.

Unbiased estimator of the population mean μ:

$$\hat{\mu} = \left(\frac{N}{M}\right) \frac{\sum_{i=1}^{n} M_i\bar{y}_i}{n} \tag{9.1}$$

Estimated variance of $\hat{\mu}$:

$$\hat{V}(\hat{\mu}) = \left(\frac{N-n}{N}\right)\left(\frac{1}{n\bar{M}^2}\right)s_b^2 + \frac{1}{nN\bar{M}^2} \sum_{i=1}^{n} M_i^2 \left(\frac{M_i - m_i}{M_i}\right)\left(\frac{s_i^2}{m_i}\right) \tag{9.2}$$

where

$$s_b^2 = \frac{\sum_{i=1}^{n} (M_i\bar{y}_i - \bar{M}\hat{\mu})^2}{n-1} \tag{9.3}$$

and

$$s_i^2 = \frac{\sum_{j=1}^{m_i} (y_{ij} - \bar{y}_i)^2}{m_i - 1} \qquad i = 1, 2, \ldots, n \tag{9.4}$$

Bound on the error of estimation:

$$2\sqrt{\hat{V}(\hat{\mu})} \tag{9.5}$$

The estimator $\hat{\mu}$ shown in Equation (9.1) depends on M, the number of elements in the population. A method of estimating μ when M is unknown is given in the next section.

Note that s_i^2 is the sample variance for the sample selected from cluster i.

EXAMPLE 9.1

A garment manufacturer has 90 plants located throughout the United States and wants to estimate the average number of hours that the sewing machines were down for repairs in the past months. Because the plants are widely scattered, she decides to use cluster sampling, specifying each plant as a cluster of machines. Each plant contains many machines, and checking the repair record for each machine would be time-consuming. Therefore she uses two-stage sampling. Enough time and money are available to sample $n = 10$ plants and approximately 20% of the machines in each plant.

TABLE 9.1 Downtime for sewing machines

Plant	M_i	m_i	Downtime (in hours)	$\bar{y}_i$	s_i^2
1	50	10	5, 7, 9, 0, 11, 2, 8, 4, 3, 5	5.40	11.38
2	65	13	4, 3, 7, 2, 11, 0, 1, 9, 4, 3, 2, 1, 5	4.00	10.67
3	45	9	5, 6, 4, 11, 12, 0, 1, 8, 4	5.67	16.75
4	48	10	6, 4, 0, 1, 0, 9, 8, 4, 6, 10	4.80	13.29
5	52	10	11, 4, 3, 1, 0, 2, 8, 6, 5, 3	4.30	11.12
6	58	12	12, 11, 3, 4, 2, 0, 0, 1, 4, 3, 2, 4	3.83	14.88
7	42	8	3, 7, 6, 7, 8, 4, 3, 2	5.00	5.14
8	66	13	3, 6, 4, 3, 2, 2, 8, 4, 0, 4, 5, 6, 3	3.85	4.31
9	40	8	6, 4, 7, 3, 9, 1, 4, 5	4.88	6.13
10	56	11	6, 7, 5, 10, 11, 2, 1, 4, 0, 5, 4	5.00	11.80

Using the data in Table 9.1, estimate the average downtime per machine, and place a bound on the error of estimation. The manufacturer knows she has a combined total of 4500 machines in all plants.

SOLUTION

The best estimate of μ is $\hat{\mu}$, shown in Equation (9.1), which yields

$$\hat{\mu} = \frac{N}{Mn} \sum_{i=1}^{n} M_i \bar{y}_i$$

$$= \frac{90}{(4500)(10)}[(50)(5.40) + (65)(4.00) + \cdots + (56)(5.00)]$$

$$= \frac{90}{(4500)(10)}(2400.59) = 4.80$$

To estimate the variance of $\hat{\mu}$, we must calculate

$$s_b^2 = \frac{1}{n-1} \sum_{i=1}^{n} (M_i \bar{y}_i - \bar{M}\hat{\mu})^2$$

$$= \frac{1}{n-1} \left[\sum_{i=1}^{n} (M_i \bar{y}_i)^2 - 2\bar{M}\hat{\mu} \sum_{i=1}^{n} M_i \bar{y}_i + n(\bar{M}\hat{\mu})^2 \right]$$

$$= \tfrac{1}{9}[583{,}198.6721 - 2(50)(4.80)(2400.59) + 10(240)^2]$$

$$= 768.38$$

$$\sum_{i=1}^{n} M_i^2 \left(\frac{M_i - m_i}{M_i} \right) \left(\frac{s_i^2}{m_i} \right) = (50)^2 \left(\frac{50 - 10}{50} \right) \left(\frac{11.38}{10} \right)$$

$$+ \cdots + (56)^2 \left(\frac{56 - 11}{56} \right) \left(\frac{11.80}{11} \right)$$

$$= 21{,}990.96$$

Then from Equation (9.2)

$$\hat{V}(\hat{\mu}) = \left(\frac{N-n}{N} \right) \left(\frac{1}{n\bar{M}^2} \right) s_b^2 + \frac{1}{nN\bar{M}^2} \sum_{i=1}^{n} M_i^2 \left(\frac{M_i - m_i}{M_i} \right) \left(\frac{s_i^2}{m_i} \right)$$

$$= \left(\frac{90 - 10}{90} \right) \left[\frac{1}{(10)(50)^2} \right] (768.38) + \frac{1}{(10)(90)(50)^2} (21{,}990.96)$$

$$= .037094$$

The estimate of μ with a bound on the error of estimation is given by

$$\hat{\mu} \pm 2\sqrt{\hat{V}(\hat{\mu})}, \qquad \text{or} \qquad 4.80 \pm 2\sqrt{.037094}, \qquad \text{or} \qquad 4.80 \pm .38$$

Thus the average downtime is estimated to be 4.80 hours. The error of estimation should be less than .38 hour with a probability of approximately .95.

An unbiased estimator of a population total can be found by taking an unbiased estimator of the population mean and multiplying by the number of elements in the population in a manner similar to that used in simple random sampling. Thus $M\hat{\mu}$ is an unbiased estimator of τ for two-stage cluster sampling.

Estimation of the population total τ:

$$\hat{\tau} = M\hat{\mu} = N \frac{\sum_{i=1}^{n} M_i \bar{y}_i}{n} \tag{9.6}$$

Estimated variance of $\hat{\tau}$:

$$\hat{V}(\hat{\tau}) = M^2 \hat{V}(\hat{\mu})$$

$$= \left(\frac{N-n}{N}\right)\left(\frac{N^2}{n}\right)s_b^2 + \frac{N}{n}\sum_{i=1}^{n} M_i^2\left(\frac{M_i - m_i}{M_i}\right)\left(\frac{s_i^2}{m_i}\right) \tag{9.7}$$

where s_b^2 is given by Equation (9.3) and s_i^2 is given by Equation (9.4).

Bound on the error of estimation:

$$2\sqrt{\hat{V}(\hat{\tau})} = 2\sqrt{M^2\hat{V}(\hat{\mu})} \tag{9.8}$$

Note that we do not need to know M in order to calculate $\hat{\tau}$ or the estimated variance of $\hat{\tau}$, since the M's cancel in the formulas for $\hat{\tau}$ and $\hat{V}(\hat{\tau})$ [see Equations (9.6) and (9.7)].

EXAMPLE 9.2

Estimate the total amount of downtime during the past month for all machines owned by the manufacturer in Example 9.1. Place a bound on the error of estimation.

SOLUTION

The best estimate of τ is

$$\hat{\tau} = M\hat{\mu} = \frac{N}{n}\sum_{i=1}^{n} M_i\bar{y}_i = \frac{90}{10}(2400.59) = 21{,}605.31$$

The estimated variance of $\hat{\tau}$ is found by using the value of $\hat{V}(\hat{\mu})$ calculated in Example 9.1 and substituting as follows:

$$\hat{V}(\hat{\tau}) = M^2\hat{V}(\hat{\mu}) = (4500)^2(.037094)$$

The estimate of τ with a bound on the error of estimation is

$$\hat{\tau} \pm 2\sqrt{\hat{V}(\hat{\tau})}, \quad \text{or} \quad 21{,}605.31 \pm 2\sqrt{(4500)^2(.037094)},$$

or

$$21{,}605.31 \pm 1733.4$$

Thus the estimate of total downtime is 21,605.31 hours. We are fairly confident that the error of estimation is less than 1733.4 hours.

9.4
RATIO ESTIMATION OF A POPULATION MEAN

The estimator $\hat{\mu}$, given by Equation (9.1), depends on the total number of elements in the population, M. When M is unknown, as is frequently the case, it must be estimated from the sample data. We obtain an estimator of M by multiplying the average cluster size, $\sum_{i=1}^{n} M_i/n$, by the number of clusters in the population, N. If we replace M by its estimator, we obtain a ratio estimator, denoted by $\hat{\mu}_r$, because the numerator and denominator are both random variables.

Ratio estimator of the population mean μ:

$$\hat{\mu}_r = \frac{\sum\limits_{i=1}^{n} M_i \bar{y}_i}{\sum\limits_{i=1}^{n} M_i} \tag{9.9}$$

Estimated variance of $\hat{\mu}_r$:

$$\hat{V}(\hat{\mu}_r) = \left(\frac{N-n}{N}\right)\left(\frac{1}{n\bar{M}^2}\right)s_r^2 + \frac{1}{nN\bar{M}^2}\sum_{i=1}^{n} M_i^2 \left(\frac{M_i - m_i}{M_i}\right)\left(\frac{s_i^2}{m_i}\right) \tag{9.10}$$

where

$$s_r^2 = \frac{\sum\limits_{i=1}^{n} M_i^2(\bar{y}_i - \hat{\mu}_r)^2}{n-1} \tag{9.11}$$

and

$$s_i^2 = \frac{\sum\limits_{i=1}^{m_i} (y_{ij} - \bar{y}_i)^2}{m_i - 1} \qquad i = 1, 2, \ldots, n \tag{9.12}$$

Bound on the error of estimation:

$$2\sqrt{\hat{V}(\hat{\mu}_r)} \tag{9.13}$$

The estimator $\hat{\mu}_r$ is biased, but the bias is negligible when n is large.

EXAMPLE 9.3

Using the data in Table 9.1, estimate the average downtime per machine, and place a bound on the error of estimation. Assume the manufacturer does not know how many machines there are in all plants combined.

SOLUTION

Because M is unknown, we must use $\hat{\mu}_r$, given by Equation (9.9), to estimate μ. Our calculations yields

$$\hat{\mu}_r = \frac{\sum\limits_{i=1}^{n} M_i \bar{y}_i}{\sum\limits_{i=1}^{n} M_i} = \frac{(50)(5.40) + (65)(4.00) + \cdots + (56)(5.00)}{50 + 65 + \cdots + 56}$$

$$= \frac{2400.59}{522} = 4.60$$

To find the estimated variance of $\hat{\mu}_r$, we must calculate

$$s_r^2 = \frac{1}{n-1} \sum_{i=1}^{n} M_i^2(\bar{y}_i - \hat{\mu}_r)^2$$

$$= \frac{1}{n-1}\left[\sum_{i=1}^{n} (M_i\bar{y}_i)^2 - 2\hat{\mu}_r \sum_{i=1}^{n} M_i^2\bar{y}_i + (\hat{\mu}_r)^2 \sum_{i=1}^{n} M_i^2\right]$$

$$= \tfrac{1}{9}[583{,}198.6721 - 2(4.60)(126{,}530.87) + (4.6)^2(27{,}978)]$$

$$= 1236.57$$

Note that as in Example 9.1,

$$\sum_{i=1}^{n} M_i^2\left(\frac{M_i - m_i}{M_i}\right)\left(\frac{s_i^2}{m_i}\right) = 21{,}990.96$$

We can estimate $\bar{M}$ by using the average cluster size for the sample:

$$\frac{\sum_{i=1}^{n} M_i}{n} = \frac{522}{10} = 52.2$$

Substituting into Equation (9.10) yields the estimated variance of $\hat{\mu}_r$:

$$\hat{V}(\hat{\mu}_r) = \left(\frac{N-n}{N}\right)\left(\frac{1}{n\bar{M}^2}\right)s_r^2 + \frac{1}{nN\bar{M}^2} \sum_{i=1}^{n} M_i^2\left(\frac{M_i - m_i}{M_i}\right)\left(\frac{s_i^2}{m_i}\right)$$

$$= \left(\frac{90-10}{90}\right)\left[\frac{1}{(10)(52.2)^2}\right](1236.57) + \frac{1}{(10)(90)(52.2)^2}(21{,}990.96)$$

$$= .049306$$

The estimate of the average downtime with a bound on the error of estimation is

$$\hat{\mu}_r \pm 2\sqrt{\hat{V}(\hat{\mu}_r)}, \qquad \text{or} \qquad 4.60 \pm 2\sqrt{.049306}, \qquad \text{or} \qquad 4.60 \pm .44$$

Thus the estimated mean downtime per machine is 4.60 hours with a bound on the error of estimation of .44 hour.

9.5
ESTIMATION OF A POPULATION PROPORTION

Consider the problem of estimating a population proportion p such as the proportion of university students in favor of a certain law or the proportion of machines that have had no downtime for the past month. An estimate of p can be obtained by using $\hat{\mu}$, given in Equation (9.1), or $\hat{\mu}_r$, given in Equation (9.9), and letting $y_{ij} = 1$ or 0 depending on whether or not the jth element in the ith cluster falls into the category of interest.

Because M is usually unknown, we present the formulas for estimating p with a ratio estimator analogous to $\hat{\mu}_r$, given in Equation (9.9). Let $\hat{p}_i$ denote the proportion of sampled elements from cluster i that fall into the category of interest.

Estimator of a population proportion p:

$$\hat{p} = \frac{\sum\limits_{i=1}^{n} M_i \hat{p}_i}{\sum\limits_{i=1}^{n} M_i} \tag{9.14}$$

Estimated variance of $\hat{p}$:

$$\hat{V}(\hat{p}) = \left(\frac{N-n}{N}\right)\left(\frac{1}{n\bar{M}^2}\right)s_r^2 + \frac{1}{nN\bar{M}^2}\sum_{i=1}^{n} M_i^2\left(\frac{M_i - m_i}{M_i}\right)\left(\frac{\hat{p}_i\hat{q}_i}{m_i - 1}\right) \tag{9.15}$$

where

$$s_r^2 = \frac{\sum\limits_{i=1}^{n} M_i^2(\hat{p}_i - \hat{p})^2}{n-1} \tag{9.16}$$

and

$$\hat{q}_i = 1 - \hat{p}_i$$

Bound on the error of estimation:

$$2\sqrt{\hat{V}(\hat{p})} \tag{9.17}$$

EXAMPLE 9.4

The manufacturer in Example 9.1 wants to estimate the proportion of machines that have been shut down for major repairs (those requiring parts from stock outside the factory). The sample proportions of machines requiring

TABLE 9.2 Proportion of sewing machines requiring major repairs

Plant	M_i	m_i	Proportion of Machines Requiring Major Repairs, $\hat{p}_i$
1	50	10	.40
2	65	13	.38
3	45	9	.22
4	48	10	.30
5	52	10	.50
6	58	12	.25
7	42	8	.38
8	66	13	.31
9	40	8	.25
10	56	11	.36

major repairs are given in Table 9.2. The data are for the machines sampled in Example 9.1. Estimate p, the proportion of machines involved in major repairs for all plants combined, and place a bound on the error of estimation.

SOLUTION

The best estimate of p is given by

$$\hat{p} = \frac{\sum\limits_{i=1}^{n} M_i \hat{p}_i}{\sum\limits_{i=1}^{n} M_i} = \frac{(50)(.40) + (65)(.38) + \cdots + (56)(.36)}{50 + 65 + \cdots + 56}$$

$$= \frac{176.08}{522} = .34$$

To estimate the variance of $\hat{p}$, we calculate

$$s_r^2 = \frac{1}{n-1} \sum_{i=1}^{n} M_i^2 (\hat{p}_i - \hat{p})^2$$

$$= \frac{1}{n-1} \left[\sum_{i=1}^{n} (M_i \hat{p}_i)^2 - 2\hat{p} \sum_{i=1}^{n} M_i^2 \hat{p}_i + (\hat{p})^2 \sum_{i=1}^{n} M_i^2 \right]$$

$$= \tfrac{1}{9}[3381.4688 - 2(.34)(9484.84) + (.34)^2(27{,}978)]$$

$$= 18.4482$$

$$\sum_{i=1}^{n} M_i^2 \left(\frac{M_i - m_i}{M_i} \right) \left(\frac{\hat{p}_i \hat{q}_i}{m_i - 1} \right) = (50)^2 \left(\frac{50 - 10}{50} \right) \left[\frac{(.4)(.6)}{9} \right]$$

$$+ \cdots + (56)^2 \left(\frac{56 - 11}{56} \right) \left[\frac{(.36)(.64)}{10} \right]$$

$$= 509.4881$$

Then the estimated variance of $\hat{p}$ when $\bar{M}$ is estimated by the sample average, 52.2, is

$$\hat{V}(\hat{p}) = \left(\frac{N - n}{N} \right) \left(\frac{1}{n\bar{M}^2} \right) s_r^2 + \frac{1}{nN\bar{M}^2} \sum_{i=1}^{n} M_i^2 \left(\frac{M_i - m_i}{M_i} \right) \left(\frac{\hat{p}_i \hat{q}_i}{m_i - 1} \right)$$

$$= \left(\frac{90 - 10}{90} \right) \left[\frac{1}{(10)(52.2)^2} \right](18.4482) + \frac{1}{(10)(90)(52.2)^2}(509.4881)$$

$$= .00081$$

The best estimate of the proportion of machines that have undergone major repairs is

$$\hat{p} \pm 2\sqrt{\hat{V}(\hat{p})}, \quad \text{or} \quad .34 \pm 2\sqrt{.00081}, \quad \text{or} \quad .34 \pm .056$$

We estimate the proportion of machines involved in major repairs to be .34, with a bound of .056 on the error of estimation.

9.6
SELECTING SAMPLE SIZES

The problem of choosing the sample sizes in two-stage sampling is much more difficult than in previous chapters, which only involved one stage of sampling. We have to select values for n and all the m_i's; in addition, the best choices for these values depend on two sources of variation, variation between the clusters and variation among elements within the clusters. The general principle is to allocate the sample resources to the component of variation that is larger. That is, if measurements are homogeneous within clusters, but cluster means vary greatly from cluster to cluster, then we sample many clusters with few measurements from each. However, if measurements within clusters vary greatly, but cluster means are homogeneous, then we sample few clusters and many measurements from each. We will make this statement more precise for a simplified sampling situation.

Suppose that all clusters contain $\bar{M}$ elements, and m elements are to be subsampled from each of the n sampled clusters. That is,

$$M_1 = M_2 = \cdots = M_N = \bar{M} \quad \text{and} \quad m_1 = m_2 = \cdots = m_n = m$$

Under these conditions Equation (9.1) gives

$$\hat{\mu} = \frac{1}{n} \sum_{i=1}^{n} \bar{y}_i \tag{9.18}$$

which is equivalent to the overall average of all sample measurements. Also, under these conditions and the assumption that all fpc's can be ignored, the theoretical variance of $\hat{\mu}$ is of the form

$$V(\hat{\mu}) = \frac{\sigma_b^2}{n} + \frac{\sigma_w^2}{nm} \tag{9.19}$$

where σ_b^2 = variance among (between) the true cluster means

and σ_w^2 = variance among the elements within the clusters

As in the case of stratified random sampling, we now want to find sample sizes, m and n, that either minimize $V(\hat{\mu})$ for a fixed cost or minimize total cost of sampling for a fixed $V(\hat{\mu})$. To accomplish this minimization, we must introduce a cost function. Suppose the cost associated with sampling each cluster is c_1 and the cost associated with sampling each element within a cluster is c_2. Then the total cost C is

$$C = nc_1 + nmc_2 \tag{9.20}$$

The value of m that minimizes $V(\hat{\mu})$ for fixed C, or minimizes C for fixed $V(\hat{\mu})$, is given by

$$m = \sqrt{\frac{\sigma_w^2 c_1}{\sigma_b^2 c_2}} \tag{9.21}$$

After m is determined, n is found through (9.19) if $V(\hat{\mu})$ is fixed or (9.20) if C is fixed.

Note that m increases as σ_w^2 increases, and m decreases as σ_b^2 increases. Thus more and more elements are sampled from within clusters as σ_w^2 gets large compared with σ_b^2.

One problem still remains. How can σ_w^2 and σ_b^2 be estimated from sample data? Equation (9.4) gives an expression for an estimate of the within-cluster variance for a single cluster. When we consider all n sampled clusters,

$$s_w^2 = \frac{1}{n} \sum_{i=1}^{n} s_i^2 \tag{9.22}$$

becomes an unbiased estimator of the within-cluster variance σ_w^2.

Since σ_b^2 is the variance of the cluster means, it may seem natural to attempt to estimate this variance by

$$s_1^2 = \left(\frac{1}{n-1}\right) \sum_{i=1}^{n} (\bar{y}_i - \hat{\mu})^2 \tag{9.23}$$

the sample variance calculated from the observed estimates of cluster means, $\bar{y}_i$. Since each $\bar{y}_i$ is only an estimate of the true cluster mean, Equation (9.23) measures a combination of cluster-to-cluster variation and element-to-element variation. Actually, s_1^2 is an unbiased estimator of

$$\sigma_b^2 + \frac{\sigma_w^2}{m}$$

Since s_w^2 estimates σ_w^2, an estimator of σ_b^2 is given by

$$\hat{\sigma}_b^2 = s_1^2 - \frac{s_w^2}{m} \tag{9.24}$$

Thus if we have values for s_1^2 and s_w^2, perhaps from a pilot study, then σ_w^2 and σ_b^2 can each be estimated. These estimates can then be used in (9.21) to find the optimal value of m, and then in (9.19) to find the optimal value of n for fixed $V(\hat{\mu})$.

EXAMPLE 9.5

A quality assurance plan in an automobile battery factor calls for sampling n batteries and then sampling m positive plates from each. The key measurement is the thickness of the positive plates, in thousandths of an inch. The investigator wishes to choose n and m so that the variance of the estimate of mean plate thickness is .5. The cost of selecting a battery and cutting it open is six times the cost of measuring a plate.

Preliminary studies of similar batteries manufactured in this factory yield, for $n = 40$ and $m = 5$, $s_w^2 = 3.0$ and $s_1^2 = 3.4$. Use these data to determine an m and an n that will meet the variance condition given.

SOLUTION

Before we can use Equation (9.21), we must estimate σ_w^2 and σ_b^2. Now σ_w^2 is estimated by

$$\hat{\sigma}_w^2 = s_w^2 = 3.0$$

and σ_b^2 is estimated by [see Equation (9.24)]

$$\hat{\sigma}_b^2 = s_1^2 - \frac{s_w^2}{m} = 3.4 - \frac{3.0}{5} = 2.8$$

Since c_1 is six times c_2, then $c_1/c_2 = 6$. (Note that only the ratio of costs is needed in choosing m.)

From Equation (9.21) we have

$$m = \sqrt{\frac{\hat{\sigma}_w^2}{\hat{\sigma}_b^2}\left(\frac{c_1}{c_2}\right)} = \sqrt{\left(\frac{3.0}{2.8}\right)(6)} = 2.53 \text{ or } 3$$

Hence 3 positive plates should be sampled from each selected battery.

To find n, we substitute the estimates $\hat{\sigma}_w^2$ and $\hat{\sigma}_b^2$, and m, into the variance function, Equation (9.19), which is to equal .5. This substitution yields

$$.5 = \frac{2.8}{n} + \frac{3.0}{3n}$$

or

$$n = \frac{1}{.5}(2.8 + 1.0) = 7.6 \text{ or } 8$$

Thus the quality assurance plan should call for sampling $n = 8$ batteries and $m = 3$ positive plates from each. The estimate $\hat{\mu}$ should then have a variance of approximately .5.

9.7
TWO-STAGE CLUSTER SAMPLING WITH PROBABILITIES PROPORTIONAL TO SIZE

Since the number of elements in a cluster may vary greatly from cluster to cluster, a technique often advantageous is to sample clusters with probabilities proportional to their sizes, as was discussed in Section 8.8. Generally, pps sampling is only used at the first stage of a two-stage sampling procedure because the elements within clusters tend to be somewhat similar in size. Hence we will present estimators of μ and τ for two-stage cluster sampling in which the first-stage sampling is carried out with probabilities proportional to size.

Equation (8.23) provides an estimator of μ, in the case of single-stage cluster sampling, of the form

$$\frac{1}{n}\sum_{i=1}^{n}\bar{y}_i \tag{9.25}$$

In Chapter 8 $\bar{y}_i$ was calculated from all the elements in cluster i and was *exactly* the cluster mean. In this chapter $\bar{y}_i$ is calculated from a sample of elements from cluster i and is only an *estimate* of the cluster mean. Nevertheless, Equation (9.25) still forms an unbiased estimate of μ, with an estimated variance as given in Equation (8.24).

To form an unbiased estimate of τ, one merely multiplies Equation (9.25) by M, the number of elements in the population.

Estimator of the population mean μ:

$$\hat{\mu}_{pps} = \frac{1}{n} \sum_{i=1}^{n} \bar{y}_i \qquad (9.26)$$

Estimated variance of $\hat{\mu}_{pps}$:

$$\hat{V}(\hat{\mu}_{pps}) = \frac{1}{n(n-1)} \sum_{i=1}^{n} (\bar{y}_i - \hat{\mu}_{pps})^2 \qquad (9.27)$$

Bound on the error estimation:

$$2\sqrt{\hat{V}(\hat{\mu}_{pps})} = 2\sqrt{\frac{1}{n(n-1)} \sum_{i=1}^{n} (\bar{y}_i - \hat{\mu}_{pps})^2} \qquad (9.28)$$

Estimator of the population total τ:

$$\hat{\tau}_{pps} = \frac{M}{n} \sum_{i=1}^{n} \bar{y}_i \qquad (9.29)$$

Estimated variance of $\hat{\tau}_{pps}$:

$$\hat{V}(\hat{\tau}_{pps}) = \frac{M^2}{n(n-1)} \sum_{i=1}^{n} (\bar{y}_i - \hat{\mu}_{pps})^2 \qquad (9.30)$$

Bound on the error of estimation:

$$2\sqrt{\hat{V}(\hat{\tau}_{pps})} = 2\sqrt{\frac{M^2}{n(n-1)} \sum_{i=1}^{n} (\bar{y}_i - \hat{\mu}_{pps})^2} \qquad (9.31)$$

We illustrate this pps procedure with the following examples.

EXAMPLE 9.6

From the six hospitals in a city a researcher desires to sample three hospitals for the purpose of estimating the proportion of current patients who have been (or will be) in the hospital for more than two consecutive days. Since the hospitals vary in size, they will be sampled with probabilities proportional to their numbers of patients. For the three sampled hospitals 10% of the records of current patients will be examined to determine how many patients will stay in the hospital for more than two days. Given the information on hospital sizes in the accompanying table, select a sample of three hospitals with probabilities proportional to size.

Hospital	Number of Patients	Cumulative Range
1	328	1–328
2	109	329–437
3	432	438–869
4	220	870–1089
5	280	1090–1369
6	190	1370–1559

SOLUTION

Since three hospitals are to be selected, three random numbers between 0001 and 1559 must be chosen from the random number table. Our numbers turned out to be 1505, 1256, and 0827. Locating these numbers in the cumulative range column leads to the selection of hospitals 3, 5, and 6.

EXAMPLE 9.7

Suppose the sampled hospitals in Example 9.6 yielded the following data on number of patients staying more than two days:

Hospital	Number of Patients Sampled	Number Staying More than Two Days
3	43	25
5	28	15
6	19	8

Estimate the proportion of patients staying more than two days, for all six hospitals, and place a bound on the error of estimation.

SOLUTION

The proportion of interest for each hospital is simply the sample mean and, by Equation (9.26), the best estimate of the population proportion is the average of the three sample means. Thus

$$\hat{\mu}_{pps} = \tfrac{1}{3}(\tfrac{25}{43} + \tfrac{15}{28} + \tfrac{8}{19}) = \tfrac{1}{3}(.58 + .54 + .42) = .51$$

By Equation (9.27),

$$\hat{V}(\hat{\mu}_{pps}) = \frac{1}{3(2)}[(.58 - .51)^2 + (.54 - .51)^2 + (.42 - .51)^2]$$

$$= .0025$$

Thus the bound on the error of estimation is

$$2\sqrt{\hat{V}(\hat{\mu}_{pps})} = 2\sqrt{.0025} = .10$$

and our estimate of the true population proportion is

$$.51 \pm .10$$

One further comment about when one uses pps sampling is in order. If the variation as measured by s_b^2 is small in comparison with that measured by s_i^2 [that is, if the second term dominates the variance expression (9.2)], then we would want to select few clusters and many elements from within each sampled cluster. In that case any sampling plan for clusters will work well.

If, however, the s_i^2 terms are small compared with s_b^2 [Equation (9.2) is dominated by the first term], then great care should be taken in planning the selection of clusters. In this case the comments made at the end of Section 8.8 still hold, and the pps method works well if the cluster sizes vary appreciably.

9.8
SUMMARY

The concept of cluster sampling can be extended to two-stage sampling by taking a simple random sample of elements from each sampled cluster. Two-stage cluster sampling is advantageous when one wishes to have sample elements in geographic proximity because of travel costs.

Two-stage cluster sampling eliminates the need to sample all elements in each sampled cluster. Thus the cost of sampling can often be reduced with little loss of information.

An unbiased estimator of μ was presented for the case when M, the total number of elements in the population, is known. When M is unknown, a ratio estimator is employed. Estimators were also given for a population total τ and for a population proportion p.

CASE STUDY REVISITED

THE ESTIMATION OF ENTERTAINMENT EXPENSES

THE sampling plan for estimating average monthly entertainment expenses among students, outlined at the beginning of this chapter, involved the selection of $n = 4$

classrooms from $N = 150$ and subsampling students from each. The data are as follows (averages in dollars):

No. of Students per class, M_i	No. of Students Sampled, m_i	$\bar{y}_i$	s_i^2	$M_i\bar{y}_i$	$[M_i(\bar{y}_i - \hat{\mu}_r)]^2$
80	8	16	9	1280	518,400
47	5	30	15	1410	55,225
62	6	21	22	1302	61,504
39	4	45	18	1755	608,400
228				5747	1,243,529

$$\hat{\mu}_r = \frac{\sum\limits_{i=1}^{n} M_i\bar{y}_i}{\sum\limits_{i=1}^{n} M_i} = \frac{5747}{228} = 25$$

$$\bar{M} = 57$$

$$s_r^2 = \left(\frac{1}{n-1}\right) \sum\limits_{i=1}^{n} [M_i(\bar{y}_i - \hat{\mu}_r)]^2 = 414,510$$

$$s_w^2 = \sum\limits_{i=1}^{n} M_i(M_i - m_i)\left(\frac{s_i^2}{m_i}\right) = 31,275$$

The best estimate of the average monthly amount spent on entertainment among all students at the university is $\hat{\mu}_r = \$25$. The bound on the error of estimation is

$$2\sqrt{\hat{V}(\hat{\mu}_r)} = \left(\frac{N-n}{N}\right)\left(\frac{1}{n\bar{M}^2}\right)(s_r^2) + \left(\frac{1}{nN\bar{M}^2}\right)(s_w^2)$$

$$= \left(\frac{146}{150}\right)\left[\frac{1}{4(57)^2}\right](414,510) + \left[\frac{1}{4(150)(57)^2}\right](31,275) = 11$$

Hence we are confident that the true average monthly entertainment amount is between $25 - 11 = \$14$ and $25 + 11 = \$36$. (This interval could be reduced in size by sampling more classrooms or more students per classroom. Which procedure would you recommend?)

EXERCISES

9.1 Suppose a large retail store has its accounts receivable listed by department. The firm wishes to estimate the total accounts receivable on a given day by sampling. Discuss the relative merits of stratified random sampling, single-stage cluster sampling, systematic sampling, and two-stage cluster sampling. What extra information would you like to have on these accounts before selecting the sampling design?

9.2 A nurseryman wants to estimate the average height of seedlings in a large field that is divided into 50 plots that vary slightly in size. He believes the heights are fairly constant throughout each plot but may vary considerably from plot to plot. Therefore he decides to sample 10% of the trees within each of 10 plots using a two-stage cluster sample. The data are as given in the accompanying table. Estimate the average height of seedlings in the field, and place a bound on the error of estimation.

Plot	Number of Seedlings	Number of Seedlings Sampled	Heights of Seedlings (in inches)
1	52	5	12, 11, 12, 10, 13
2	56	6	10, 9, 7, 9, 8, 10
3	60	6	6, 5, 7, 5, 6, 4
4	46	5	7, 8, 7, 7, 6
5	49	5	10, 11, 13, 12, 12
6	51	5	14, 15, 13, 12, 13
7	50	5	6, 7, 6, 8, 7
8	61	6	9, 10, 8, 9, 9, 10
9	60	6	7, 10, 8, 9, 9, 10
10	45	6	12, 11, 12, 13, 12, 12

9.3 In Exercise 9.2, assume that the nurseryman knows there are approximately 2600 seedlings in the field. Use this additional information to estimate the average height, and place a bound on the error of estimation.

9.4 A supermarket chain has stores in 32 cities. A company official wants to estimate the proportion of stores in the chain that do not meet a specified cleanliness criterion. Stores within each city appear to possess similar characteristics; therefore she decides to select a two-stage cluster sample containing one-half of the stores within each of 4 cities. Cluster sampling is desirable in this situation because of travel costs. The data collected are given in the accompanying table. Estimate the proportion of stores not meeting the cleanliness criterion, and place a bound on the error of estimation.

City	Number of Stores in City	Number of Stores Sampled	Number of Stores Not Meeting Criterion
1	25	13	3
2	10	5	1
3	18	9	4
4	16	8	2

9.5 Repeat Exercise 9.4 given that the chain contains 450 stores.

9.6 To improve telephone service, an executive of a certain company wants to estimate the total number of phone calls placed by secretaries in the company during one day. The company contains 12 departments, each making approximately the same number of calls per day. Each department employs approximately 20 secretaries, and the number of calls made varies considerably from secretary to secretary. The executive decides to employ two-stage cluster sampling, using a small number of departments (clusters) and selecting a fairly large number of secretaries (elements) from each. Ten secretaries are sampled from each of 4 departments. The data are summarized in the accompanying table. Estimate the total number of calls placed by the secreataries in this company, and place a bound on the error of estimation.

Department	Number of Secretaries	Number of Secretaries Sampled	Mean, $\bar{y}_i$	Variance, s_i^2
1	21	10	15.5	2.8
2	23	10	15.8	3.1
3	20	10	17.0	3.5
4	20	10	14.9	3.4

9.7 A city–zoning commission wants to estimate the proportion of property owners in a certain section of a city who favor a proposed zoning change. The section is divided into seven distinct residential areas, each containing similar residents. Because the results must be obtained in a short period of time, two-stage cluster sampling is used. Three of the seven areas are selected at random, and 20% of the property owners in each area selected are sampled. The figure of 20% seems reasonable because the people living within each area seem to be in the same socioeconomic class and hence they tend to hold similar opinions on the zoning question. The results are given in the accompanying table. Estimate the proportion of property owners who favor the proposed zoning change, and place a bound on the error of estimation.

Area	Number of Property Owners	Number of Property Owners Sampled	Number in Favor of Zoning Change
1	46	9	1
2	67	13	2
3	93	20	2

9.8 A forester wants to estimate the total number of trees in a certain county that are infected with a particular disease. There are ten well-defined forest areas in the county; these areas can be subdivided into plots of approximately the same size. Four crews are available to conduct the survey, which must be completed in one day. Hence two-stage cluster sampling is used. Four areas (clusters) are chosen with six plots (elements) randomly selected from each. (Each crew can survey six plots

Area	Number of Plots	Number of Plots Sampled	Number of Infected Trees per Plot
1	12	6	15, 14, 21, 13, 9, 10
2	15	6	4, 6, 10, 9, 8, 5
3	14	6	10, 11, 14, 10, 9, 15
4	21	6	8, 3, 4, 1, 2, 5

in one day.) The data are given in the accompanying table. Estimate the total number of infected trees in the county, and place a bound on the error of estimation.

9.9 A new bottling machine is being tested by a company. During a test run the machine fills 24 cases, each containing 12 bottles. The company wishes to estimate the average number of ounces of fill per bottle. A two-stage cluster sample is employed using 6 cases (clusters), with 4 bottles (elements) randomly selected from each. The results are given in the accompanying table. Estimate the average number of ounces per bottle, and place a bound on the error of estimation.

Case	Average Ounces of Fill for Sample, $\bar{y}_i$	Sample Variance, s_i^2
1	7.9	.15
2	8.0	.12
3	7.8	.09
4	7.9	.11
5	8.1	.10
6	7.9	.12

9.10 A certain manufacturing plant contains 40 machines, all producing the same product (say boxes of cereal). An estimate of the proportion of defective products (say boxes underfilled) for a given day is desired. Discuss the relative merits of two-stage cluster sampling (machines as clusters of boxes) and stratified random sampling (machines as strata) as possible designs for this study.

9.11 A market research firm constructed a sampling plan to estimate the weekly sales of brand A cereal in a certain geographic area. The firm decided to sample cities within the area and then to sample supermarkets within cities. The number of boxes of brand A cereal sold in a specified week is the measurement of interest. Five cities are sampled from the 20 in the area. Using the data given in the accompanying table, estimate the average sales for the week for all supermarkets in the area. Place a bound on the error of the estimation. Is the estimator you used unbiased?

City	Number of Supermarkets	Number of Supermarkets Sampled	$\bar{y}_i$	s_i^2
1	45	9	102	20
2	36	7	90	16
3	20	4	76	22
4	18	4	94	26
5	28	6	120	12

9.12 In Exercise 9.11, do you have enough information to estimate the total number of boxes of cereal sold by all supermarkets in the area during the week? If so, explain how you would estimate this total, and place a bound on the error of estimation.

9.13 If a study like the one outlined in Exercise 9.11 is to be done again, would you recommend that cities be sampled with probabilities proportional to their numbers of supermarkets? Why?

9.14 Suppose a sociologist wants to estimate the total number of retired people residing in a certain city. She decides to sample blocks and then sample households within blocks. (Block statistics from the Census Bureau aid in determining the number of households in each block.) Four blocks are randomly selected from the 300 of the city. From the data in the accompanying table, estimate the total number of retired residents in the city, and place a bound on the error of estimation.

Block	Number of Households	Number of Households Sampled	Number of Retired Residents per Household
1	18	3	1, 0, 2
2	14	3	0, 3, 0
3	9	3	1, 1, 2
4	12	3	0, 1, 1

9.15 Using the data in Exercise 9.14, estimate the average number of retired residents per household, and place a bound on the error of estimation.

9.16 From the data in Exercise 9.14, can you estimate the average number of retired residents per block? How can you construct this estimate and place a bound on the error of estimation?

9.17 In the estimation of the amount of impurities in a bulk product like sugar, the sampling procedure may select bags of sugar from a warehouse and then select small test samples from each bag. The test samples are analyzed for amount of impurities. Discuss how one might choose the number of bags to sample and the number and size of the test samples taken from each bag.

9.18 A quality assurance program requires sampling of manufactured products as they come off assembly lines in a production facility. One could treat the assembly lines as clusters or as strata. Discuss the relative merits of these two options, and design a sampling plan for each case.

EXPERIENCES WITH REAL DATA

9.1 Refer to the United States population data in the Appendix. Construct a two-stage cluster sampling estimate of the total rural population in the United States for 1980 by first sampling divisions and then sampling states within divisions. Sample four of the nine divisions and at least two states within each sampled division. Compute an estimate of the variance attached to your estimator. Would you recommend this procedure over startified random sampling? Why?

9.2 When one is sampling people, the naturally occurring frames typically involve people grouped in clusters. Hence two-stage cluster sampling is commonly employed as a matter of economic convenience. For example, try estimating the total number of library books currently checked out by students on your campus. (Any other numerical variable of interest to you can be substituted for the number of library books.) Some naturally occurring clusters of students are those in residence halls, classrooms, fraternities and sororities, and pages of a student directly. (Can you think of others?) Estimate the total of interest, and place a bound on the error of estimation, by using the following procedures:

(a) Sampling residences and students within residences.
(b) Sampling classrooms in use and students within classrooms.
(c) Sampling pages of the student directory and students' names within pages.

Whichever method you choose, think carefully about the relative sample sizes for the first and second stages. If the experiment with students is not applicable to your situation, a simpler problem to carry out is to estimate the number of words in this (or any other) book by randomly sampling pages and then sampling lines within a page. Should the two-stage sampling scheme for a statistics book with formulas and tables differ from the scheme for a novel?

Case Study _____

How Many People Attended the Concert?

A NEWSPAPER reporter wants to estimate the number of people in attendance at a free rock concert, and she goes to a statistician for advice. The statistician observes that the newspaper is handing out free orange hats to some attendees, and he asks how many are to be given out. Armed with this knowledge, he suggests the following estimation scheme:

Suppose t hats are being worn by a subgroup of the N people in attendance. After the concert begins, n people are randomly selected, and the number s wearing the orange hats is observed. Now the sample proportion of those wearing the hats, $\hat{p} = s/n$, is an estimate of the population proportion t/N. In other words,

$$\hat{p} = \frac{s}{n} \approx \frac{t}{N}$$

or

$$N \approx \frac{nt}{s} = \hat{N}$$

The details of this estimator are developed in Chapter 10.

10

ESTIMATING THE POPULATION SIZE

10.1
INTRODUCTION

In the preceding chapters we estimated means, totals, and proportions assuming that the population size was either known or so large that it could be ignored if not expressly needed to calculate an estimate. Frequently, however, the population size is not known and is important to the goals of the study. In fact, in some studies estimation of the population size is the main goal. The study of the growth, evolution, and maintenance of wildlife populations depends crucially on accurate estimates of population sizes, and estimating the size of such populations will motivate much of our discussion in this chapter. The techniques can also be used for estimating the number of people at a concert or a sporting event, the number of defects in a bolt of material, and many similar quantities. We will present and discuss four methods for estimating population sizes.

The first method is *direct sampling*. This procedure entails drawing a random sample from a wildlife population of interest, tagging each animal sampled, and returning the tagged animals to the population. At a later date another random sample (of a fixed size) is drawn from the same population, and the number of tagged animals is observed. If N represents the total population size, t represents the number of animals tagged in the initial sample, and p represents the proportion of tagged animals in the population, then

$$\frac{t}{N} = p$$

Consequently, $N = t/p$. We can obtain an estimate of N because t is known and p can be estimated by $\hat{p}$, the proportion of tagged animals in the second sample. Thus

$$\hat{N} = \frac{\text{the number of animals tagged}}{\text{the proportion of tagged animals in the second sample}}$$

or, equivalently,

$$\hat{N} = \frac{t}{\hat{p}}$$

The second technique is *inverse sampling*. It is similar to direct sampling, but the second sample size is not fixed. That is, we sample until a fixed number of tagged animals is observed. Using this procedure, we can also obtain an estimate of N, the total population size, using

$$\hat{N} = \frac{t}{\hat{p}}$$

The third technique depends upon first estimating the density of elements in the population and then multiplying by an appropriate measure of area. If we estimate that there are $\hat{\lambda}$ animals per acre, and the area of interest contains A acres, then $A\hat{\lambda}$ provides an estimate of the population size.

The fourth method is similar to the third but depends only upon being able to identify the presence or absence of animals on the sampled plots. Then under certain conditions the density and the total number of animals can still be estimated.

10.2
ESTIMATION OF A POPULATION SIZE USING DIRECT SAMPLING

Direct sampling can be used to estimate the size of a mobile population. First, a random sample of size t is drawn from the population. At a later date a second sample of size n is drawn. For example, suppose a conservationist is concerned about the apparent decline in the number of seals in Alaska. Estimates of the population size are available from previous years. For determination of whether or not there has been a decline, a random sample of $t = 200$ seals is caught, tagged, and then released. A month later a second sample of size $n = 100$ is obtained. Using these data (often called recapture data), we can estimate N, the population size.

Let s be the number of tagged individuals observed in the second sample. The proportion of tagged individuals in the sample is

$$\hat{p} = \frac{s}{n}$$

An estimate of n is given by

$$\hat{N} = \frac{t}{\hat{p}} = \frac{nt}{s}$$

Estimator of N:

$$\hat{N} = \frac{nt}{s} \qquad (10.1)$$

Estimated variance of $\hat{N}$:

$$\hat{V}(\hat{N}) = \frac{t^2 n(n-s)}{s^3} \qquad (10.2)$$

Bound on the error of estimation:

$$2\sqrt{\hat{V}(\hat{N})} = 2\sqrt{\frac{t^2 n(n-s)}{s^3}} \qquad (10.3)$$

Note that s must be greater than 0 for Equations (10.1), (10.2), and (10.3) to hold. We will assume that n is large enough so that s is greater than 0 with high probability.

You should also realize that $\hat{N}$, which is presented in Equation (10.1), is not an unbiased estimator of N. For $s > 0$

$$E(\hat{N}) \approx N + \frac{N(N-t)}{nt}$$

Hence for fairly large sample sizes, that is, t and n large, the term

$$\frac{N(N-t)}{nt}$$

is small and the bias of the estimator $\hat{N}$ approaches 0. Estimator $\hat{N}$ tends to overestimate the true value of N. Chapman (1952) gives another estimator of N, along with its approximate variance, which is nearly unbiased for most direct sampling situations.

EXAMPLE 10.1

Before posting a schedule for the upcoming hunting season, the game commission for a particular county wishes to estimate the size of the deer population. A random sample of 300 deer is captured ($t = 300$). The deer are tagged and released. A second sample of 200 is taken two weeks later ($n = 200$). If 62 tagged deer are recaptured in the second sample ($s = 62$), estimate N, and place a bound on the error of estimation.

SOLUTION

Using Equation (10.1), we have

$$\hat{N} = \frac{nt}{s} = \frac{200(300)}{62} = 967.74$$

or $\hat{N} = 968$.

A bound on the error of estimation is given by

$$2\sqrt{\hat{V}(\hat{N})} = 2\sqrt{\frac{t^2 n(n-s)}{s^3}} = 2\sqrt{\frac{(300)^2(200)(138)}{(62)^3}} = 204.18$$

Thus the game commission estimates the total number of deer is 968, with a bound on the error of estimation of approximately 205 deer.

You may be concerned about the magnitude of the bound on the error of estimation in this example. As might be expected, we can obtain a more accurate estimator of N by increasing the two sample sizes (n and t). Further information on the choice of t and n is given in Section 10.4.

10.3
ESTIMATION OF A POPULATION SIZE USING INVERSE SAMPLING

Inverse sampling is the second method for estimating N, the total size of a population. We again assume that an initial sample of t individuals is drawn, tagged, and released. Later, random sampling is conducted until exactly s tagged animals are recaptured. If the sample contains n individuals, the proportion of tagged individuals in the sample is given by $\hat{p} = s/n$. We use this sample proportion to estimate the proportion of tagged individuals in the population.

Again, the estimator of N is given by

$$\hat{N} = \frac{t}{\hat{p}} = \frac{nt}{s}$$

but note that s is fixed and n is random.

Estimator of N:

$$\hat{N} = \frac{nt}{s} \tag{10.4}$$

Estimated variance of $\hat{N}$:

$$\hat{V}(\hat{N}) = \frac{t^2 n(n-s)}{s^2(s+1)} \tag{10.5}$$

Bound on the error of estimation:

$$2\sqrt{\hat{V}(\hat{N})} = 2\sqrt{\frac{t^2 n(n-s)}{s^2(s+1)}} \tag{10.6}$$

Note that Equations (10.4), (10.5), and (10.6) hold only for $s > 0$. This restriction offers no difficulty; we simply specify that s must be greater than 0, and we sample until s tagged individuals are recaptured. The estimator $\hat{N} = nt/s$, obtained by using inverse sampling, provides an unbiased estimator of N, and the variance given by Equation (10.5) is an unbiased estimator of the true variance of $\hat{N}$.

Variance (10.5) for the inverse case looks very much like variance (10.2) for the direct case, and the estimators $\hat{N}$ appear to be identical. However, the inverse method offers the advantages that s can be fixed in advance, $\hat{N}$ is unbiased, and an unbiased estimator of the true variance of $\hat{N}$ is available.

EXAMPLE 10.2

Authorities of a large wildlife preserve are interested in the total number of birds of a particular species that inhabit the preserve. A random sample of $t = 150$ birds is trapped, tagged, and then released. In the same month a second sample is drawn until 35 tagged birds are recaptured ($s = 35$). In total, 100 birds are recaptured in order to find 35 tagged ones ($n = 100$). Estimate N, and place a bound on the error of estimation.

SOLUTION

Using Equation (10.4), we estimate N by

$$\hat{N} = \frac{nt}{s} = \frac{100(150)}{35} = 428.57$$

A bound on the error of estimation is found by using Equation (10.6) as follows:

$$2\sqrt{\hat{V}(\hat{N})} = 2\sqrt{\frac{t^2 n(n-s)}{s^2(s+1)}} = 2\sqrt{\frac{(150)^2(100)(65)}{(35)^2(36)}} = 115.173$$

Hence we estimate that 429 birds of the particular species inhabit the preserve. We are quite confident that our estimate is within approximately 116 birds of the true population size.

10.4
CHOOSING SAMPLE SIZES FOR DIRECT AND INVERSE SAMPLING

We have been discussing direct sampling and inverse sampling techniques. You probably wonder which is the better one to use. Either method can be used. Inverse sampling yields more precise information than does direct sampling, provided the second sample size n required to recapture s tagged individuals is small relative to the population size N. However, if nothing is known about the size of N, a poor choice of t could make n quite large when inverse sampling is used. For example, if $N = 10,000$ and a first sample of $t = 50$ individuals is drawn, a large second sample would be needed to obtain exactly $s = 10$ tagged animals.

Table 10.1 is useful in determining the sample sizes (t and n) required to estimate $\hat{N}$ with a fixed bound on the error of estimation. However, to use these data requires some prior knowledge concerning the magnitude of N. Entries in Table 10.1 are $V(\hat{N})/N$ for direct sampling. If you know the approximate size of N, you can determine the variance of the estimator for fixed values of the sample sizes t and n. In Table 10.1 these sample sizes are expressed as fractions of N. These fractions, given by

$$p_1 = \frac{t}{N} \qquad \text{and} \qquad p_2 = \frac{n}{N}$$

are called sampling fractions.

TABLE 10.1 Values of $V(\hat{N})/N$ for direct sampling

	$p_1 = t/N$					
$p_2 = n/N$	.001	.01	.1	.25	.50	1.0
.001	999,000	99,000	9000	3000	1000	0
.01	99,000	9,900	900	300	100	0
.1	9,990	990	90	30	10	0
.25	3,996	396	36	12	4	0
.50	1,998	198	18	6	2	0
1.0	999	99	9	3	1	0

Having a graph of the entries in this table would be convenient. However, the numbers are so large that we can only display a portion of Table 10.1. In Figure 10.1 we display the values of $V(\hat{N})/N$ for various values of the sampling fractions $p_1 = t/N$ and $p_2 = n/N$. Note that as either p_1 or p_2 increases, the variance of $\hat{N}$ divided by N decreases; consequently, $V(\hat{N})$ decreases for a fixed value of N. Intuitively, this result makes sense, since we should obtain a more accurate estimate of N by taking large sample sizes.

EXAMPLE 10.3

The game commission in Example 10.1 believes that the size of the deer population this year is approximately the same as in the preceding year, when there were between 800 and 1000 deer. Determine the bound on the error of estimation associated with the sampling fractions of $p_1 = .25$ and $p_2 = .25$.

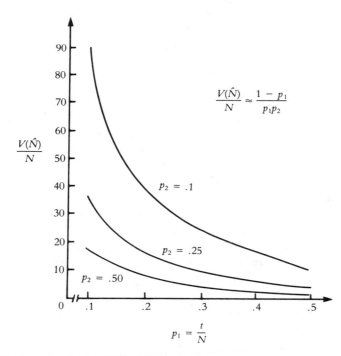

FIGURE 10.1 Graph of values of Table 10.1

SOLUTION

We take the larger of the two figures (N approximately 1000) to obtain a conservative estimate of $V(\hat{N})$ (one that is larger than would be expected). We see from Figure 10.1 (or Table 10.1) that the sampling fractions of $p_1 = t/N = .25$ and $p_2 = s/N = .25$ yield

$$\frac{V(\hat{N})}{N} = 12$$

Taking $N = 1000$, we have

$$V(\hat{N}) = 1000(12) = 12,000$$

$$\sqrt{V(\hat{N})} = \sqrt{12,000} = 109.541$$

The corresponding bound on the error of estimation is

$$2\sqrt{V(\hat{N})} = 2(109.541) = 219.082$$

An investigator could use this information to plan his survey. If this bound on the error of estimation is acceptable, he could run a survey using $p_1 = .25$ and $p_2 = .25$; that is, he could draw an initial sample of

$$t = p_1 N = (.25)(1000) = 250$$

and a second sample of

$$n = p_2 N = (.25)(1000) = 250$$

He could then estimate N by using the data from the survey. The bound on the error of estimation should be approximately equal to 220, provided the original range for N is accurate.

If the bound on the error for $\hat{N}$ is not acceptable for the sampling fractions of $p_1 = p_2 = .25$, the investigator can work with Table 10.1 (or Figure 10.1) to determine the sampling fractions required to achieve an acceptable bound on the error of estimation.

We can examine $V(\hat{N})$ for inverse sampling in the same manner as for direct sampling. Entries in Table 10.2 are the values of $V(\hat{N})/N$ for various sampling fractions $p_1 = t/N$ and $p_2 = s/N$ when inverse sampling is used. Recall that in inverse sampling we fix s rather than n; hence the second sampling fraction is in terms of s. A graphical representation of these data would be helpful, but again the numbers are too large to plot conveniently. A portion of Table 10.2 is presented in Figure 10.2.

TABLE 10.2 Values of $V(\hat{N})/N$ for inverse sampling

	$p_1 = t/N$					
$p_2 = s/N$	.001	.01	.1	.25	.5	1.0
.001	999	990	900	750	500	0
.01		99	90	75	50	0
.1			9	7.5	5	0
.25				3	2	0
.5					1	0
1.0						0

Note that $V(\hat{N})/N$ [or, equivalently, $V(\hat{N})$ for a given value of N] decreases as p_1 and p_2 increase. If the experimenter has an approximate range for N, he or she use either Figure 10.2 or Table 10.2 to determine the sampling fractions ($p_1 = t/N$, $p_2 = s/N$) necessary to achieve a reasonable bound. Then the experimenter can conduct a survey with an initial sample of

$$t = p_1 N$$

The experimenter would begin a second sample at a future time and continue until

$$s = p_2 N$$

tagged animals are recaptured. The corresponding bound on the error of estimation for N will be acceptable provided the original estimate for N was reasonable.

The preceding tag-recapture techniques can be extended to more than two stages. At the second stage the $(n - s)$ untagged animals can be tagged

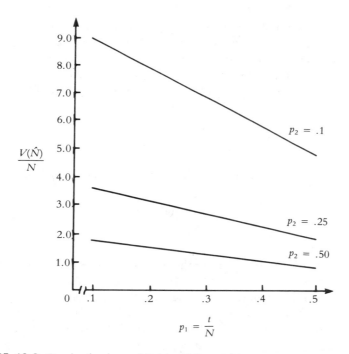

FIGURE 10.2 Graph of values of Table 10.2

and all n returned to the population. At a later time a third sample can be taken, and the counting and tagging operation repeated. This multiple-stage approach will result in an improved estimator of N and is especially useful in ongoing studies where samples might be taken every week or so.

10.5
ESTIMATING POPULATION DENSITY AND SIZE FROM QUADRAT SAMPLES

Estimation of the number of elements in a defined area can be accomplished by first estimating the number of elements per unit area (that is, the *density* of the elements) and then multiplying the estimated density by the size of the area under study. For example, if a loom produces 2 defects per square yard of material, on the average, then a bolt containing 40 square yards should contain approximately 80 defects. We will discuss estimates of both the density and the total number of elements. Our discussion, however, need not be confined to areas, because the same methods work for estimating the total number of bacteria in a fixed volume of liquid or the total number of telephone calls coming into a switchboard over a fixed interval of time. To talk in terms of areas is convenient for illustrative purposes.

Suppose a region of total area A is to be sampled by randomly selecting n plots, each of area a. For convenience, we assume $A = Na$. Each plot will be called a *quadrat* (even though they may not be square). In the terminology of earlier chapters, a quadrat can be thought of as a cluster of elements. We let m_i denote the number of elements in quadrat i, and we let M, given by

$$M = \sum_{i=1}^{N} m_i$$

denote the total number of elements in the population (having area A). Also, we let

$$\lambda = \frac{M}{A}$$

denote the density of elements, or the number of elements per unit area. Our goal is to estimate λ and then $M = \lambda A$. In this discussion, note that the m_i's are random variables, since they are the numbers of elements that happen to be located in a randomly located quadrat of fixed area.

The estimate of λ introduced here assumes that the elements themselves are randomly dispersed throughout the population. If we are speaking of defects in materials, then we are assuming that the defects are not clustered together but, rather, are spread throughout the material in no apparent order. (The reader who has studied probability theory may recognize this assumption as equivalent to the assumption that m_i has a Poisson distribution.)

The n randomly sampled quadrats are examined closely, and an accurate count of elements m_i is determined for each. Letting

$$\bar{m} = \frac{1}{n} \sum_{i=1}^{n} m_i$$

we have the following estimators of λ and M.

Estimator of the density λ:

$$\hat{\lambda} = \frac{\bar{m}}{a} \tag{10.7}$$

Estimated variance of $\hat{\lambda}$:

$$\hat{V}(\hat{\lambda}) = \frac{\hat{\lambda}}{an} \tag{10.8}$$

Bound on the error of estimation:

$$2\sqrt{\hat{V}(\hat{\lambda})} = 2\sqrt{\frac{\hat{\lambda}}{an}} \tag{10.9}$$

Estimator of the total M:

$$\hat{M} = \hat{\lambda} A \tag{10.10}$$

Estimated variance of $\hat{M}$:

$$\hat{V}(\hat{M}) = A^2 \hat{V}(\hat{\lambda}) = A^2\left(\frac{\hat{\lambda}}{an}\right) \tag{10.11}$$

Bound on the error of estimation:

$$2\sqrt{\hat{V}(\hat{M})} = 2A\sqrt{\frac{\hat{\lambda}}{an}} \tag{10.12}$$

We illustrate the use of these estimators in the following examples.

EXAMPLE 10.4

The density of trees having fusiform rust on a southern-pine plantation of 200 acres is to be estimated from a sample of $n = 10$ quadrats of .5 acre each. The 10 sampled plots had an average $\bar{m}$ of 2.8 infected trees per quadrat. Estimate the density of infected trees, and place a bound on the error of estimation.

SOLUTION

Using Equation (10.7) with $a = .5$, we determine the estimated density as

$$\hat{\lambda} = \frac{\bar{m}}{a} = \frac{2.8}{.5} = 5.6 \text{ trees per acre}$$

The bound on the error, from Equation (10.9), is

$$2\sqrt{\frac{\hat{\lambda}}{an}} = 2\sqrt{\frac{5.6}{(.5)(10)}} = 2.1$$

Thus we estimate the density as 5.6 ± 2.1, or from 3.5 to 7.7 infected trees per acre. This interval is a fairly large interval since the sample size is relatively small.

EXAMPLE 10.5

For the situation and data in Example 10.4, estimate the total number of infected trees in the 200-acre plantation. Place a bound on the error of estimation.

SOLUTION

Using Equation (10.10), we see that the estimated total is

$$\hat{M} = \hat{\lambda} A = (5.6)(200) = 1120 \text{ trees}$$

The bound on the error, from Equation (10.12), is

$$2\sqrt{\hat{V}(\hat{M})} = 2A\sqrt{\frac{\hat{\lambda}}{an}} = 2(200)\sqrt{\frac{5.6}{(.5)(10)}} = 420$$

Thus we estimate the total number of infected trees as 1120 ± 420, or 700 to 1540.

Notice that the bound on the error of estimation, for both λ and M, contains both a and n in the denominator. Hence this bound will decrease as a is increased or as n is increased. A useful sample size can be determined by fixing a at some desirable level for convenient fieldwork and then choosing n to produce the desired bound, assuming some preliminary knowledge of λ. A rough rule for optimally determining a will be given in the next section for a slightly different estimator of λ.

10.6
ESTIMATING POPULATION DENSITY AND SIZE FROM STOCKED QUADRATS

In quadrat sampling of plants or animals, counting the exact number of the species under investigation is often difficult. In contrast detecting the presence or absence of the species of interest is often easy. We will now show that just knowing whether or not a species is present in a sample quadrat can lead to an estimate of density and of population size.

Foresters refer to a quadrat that contains the species of interest as being *stocked*. We will adopt that terminology. For a sample of n quadrats, each of area a, from a population of area A, let y denote the number of sampled quadrats that are *not* stocked. Under the assumption of randomness of elements, introduced in Section 10.5, the proportion of unstocked quadrats in the population is approximately $e^{-\lambda a}$. We know from our discussions of estimating proportions in Chapter 4 that the sample proportion of unstocked quadrats is a good estimator of the population proportion. Thus (y/n) is an estimator of $e^{-\lambda a}$. This result leads to the following estimators of λ and M.

Estimator of the density λ:

$$\hat{\lambda} = -\left(\frac{1}{a}\right) \ln\left(\frac{y}{n}\right) \tag{10.13}$$

(ln denotes natural logarithm.)

Estimated variance of $\hat{\lambda}$:

$$\hat{V}(\hat{\lambda}) = \frac{1}{na^2}(e^{\hat{\lambda}a} - 1) \qquad (10.14)$$

Bound on the error of estimation:

$$2\sqrt{\hat{V}(\hat{\lambda})} = 2\sqrt{\frac{1}{na^2}(e^{\hat{\lambda}a} - 1)} \qquad (10.15)$$

Estimator of the total M:

$$\hat{M} = \hat{\lambda}A \qquad (10.16)$$

Estimated variance of $\hat{M}$:

$$\hat{V}(\hat{M}) = A^2\hat{V}(\hat{\lambda}) = \frac{A^2}{na^2}(e^{\hat{\lambda}a} - 1) \qquad (10.17)$$

Bound on the error of estimation:

$$2\sqrt{\hat{V}(\hat{M})} = 2A\sqrt{\frac{1}{na^2}(e^{\hat{\lambda}a} - 1)} \qquad (10.18)$$

The following example illustrates the use of these estimators.

EXAMPLE 10.6

Again, refer to the 200-acre tree plantation of Example 10.4. Now for estimation of the density of trees infected by fusiform rust, $n = 20$ quadrats of .5 acre each will be sampled, but only the presence or absence of infected trees will be noted for each sampled quadrat. (Since this task is easier than counting trees, the sample size can be increased.) Suppose $y = 4$ of the 20 quadrats show no signs of fusiform rust. Estimate the density and number of infected trees, placing bounds on the error of estimation in both cases.

SOLUTION

From Equation (10.13) we see that the density is estimated by

$$\hat{\lambda} = -\left(\frac{1}{a}\right)\ln\left(\frac{y}{n}\right)$$

$$= -\frac{1}{(.5)}\ln\left(\frac{4}{20}\right) = 3.2 \text{ trees per acre}$$

The bound on the error is, according to Equation (10.15),

$$2\sqrt{\frac{1}{na^2}(e^{\hat{\lambda}a} - 1)} = 2\sqrt{\frac{1}{(10)(.5)^2}(e^{(3.2)(.5)} - 1)} = 1.8$$

We then estimate the density as 3.2 ± 1.8, or 1.4 to 5.0 infected trees per acre. From Equation (10.16) we have

$$\hat{M} = \hat{\lambda}A = (3.2)(200) = 640$$

and the bound on the error, from Equation (10.18), is

$$2\sqrt{\hat{V}(\hat{M})} = 2(200)\sqrt{\frac{1}{(10)(.5)^2}(e^{(3.2)(.5)} - 1)} = 360$$

Our estimate of the total number of infected trees is 640 ± 360, or 280 to 1000.

Generally, the estimator based on stocked quadrats alone is less precise than the one based on actual count data. However, since making the measurements is easier when one is only looking for stocked quadrats, the sample size can usually be quite large. The stocked–quadrat estimator does not work if $y = 0$ or $y = n$. Thus choice of the quadrat size a is very important. Swindel (1983) gives a rough rule for choosing a as

$$a = \frac{1.6}{\lambda}$$

when some preliminary knowledge of λ is available. If, for example, we expect to see approximately 4 infected trees per acre, then each sampled quadrat should be $1.6/4 = .4$ acre.

The stocked–quadrat technique can also be used with volume or time samples. Cochran (1950) discusses the use of this technique, and modifications of it, for estimating bacterial densities in liquids.

Many other techniques for estimating population sizes are aviable. An excellent reference is the manual of wildlife investigational techniques (Mosby, 1969) listed in the references in the Appendix.

10.7
SUMMARY

Estimation of the size of a population is often very important, especially when one is studying plant and animal populations. This chapter presents four procedures for estimating the total population size N.

The first technique is direct sampling. A random sample of t individuals is drawn from a population and tagged. At a later date a fixed random sample of size n is drawn, and the number of tagged individuals is observed. Using these data, we can estimate N and place a bound on the error of estimation.

The second technique, inverse sampling, is similar to direct sampling, with the exception that we continue sampling until a fixed number s of tagged individuals is recaptured in the second sample. The sample data are then used to estimate N and to place a bound on the error of estimation.

When a choice is available between inverse and direct sampling procedures, the inverse procedure appears to provide more accurate results. However, in some instances, particularly when little or nothing is known concerning the relative size of N, the direct sampling procedure is the better choice.

The third and fourth methods both involve sampling quadrats, volumes, or intervals of time and then counting elements of interest within these relatively small units. This procedure leads to estimates of both the density of elements and the total number of elements in the population.

CASE STUDY REVISITED

THE ESTIMATION OF CONCERT ATTENDANCE

IN the example used to introduce this chapter, the newspaper reporter tells the statistician that $t = 500$ orange hats are given out. From their seat high in a balcony the statistician and the reporter locate $n = 200$ seats in a random fashion. In those 200 seats they observe $s = 40$ of the orange hats. Now N, the size of the crowd, is estimated to be

$$\hat{N} = \frac{nt}{s} = \frac{200(500)}{40} = 2500$$

The bound on the error of estimation is given by

$$2\sqrt{\hat{V}(\hat{N})} = 2\sqrt{\frac{t^2 n(n-s)}{s^3}}$$

$$= 2\sqrt{\frac{(500)^2(200)(160)}{(40)^3}} = 707$$

We confidently estimate the size of the crowd to be between $2500 - 707 = 1793$ and $2500 + 707 = 3207$. This interval is large and could be reduced by increasing n. (Note that this estimate assumes that the people wearing the orange hats are somewhat randomly dispersed through the crowd.)

EXERCISES

10.1 Discuss the differences between direct and inverse sampling.

10.2 Name the restriction implicit in the use of (a) direct sampling or (b) inverse sampling. How can this restriction be satisfied in each case?

10.3 Assuming the cost of sampling is not significant, how can you improve the bound on the error of estimation, using either direct or inverse sampling?

10.4 A particular sportsmen's club is concerned about the number of brook trout in a certain stream. During a period of several days, $t = 100$ trout are caught, tagged, and then returned to the stream. Note that the sample represents 100 different fish; hence any fish caught during these dates that had already been tagged was immediately released. Several weeks later a second sample of $n = 120$ trout is caught and observed. Suppose 27 in the second sample were tagged ($s = 27$). Estimate N, the total size of the population, and place a bound on the error of estimation.

10.5 Wildlife biologists wish to estimate the total size of the bobwhite quail population in a section of southern Florida. A series of 50 traps is used. In the first sample $t = 320$ quail are caught. After being captured, each bird is removed from the trap and tagged with a metal band on its left leg. All birds are then released. Several months later a second sample of $n = 515$ quail is obtained. Suppose $s = 91$ of these birds have tags. Estimate N, and place a bound on the error of estimation.

10.6 A game commission is interested in estimating the number of large-mouth bass in a reservoir. A random sample of $t = 2876$ bass is caught. Each bass is marked and released. One month later a second sample of $n = 2562$ is caught. Suppose $s = 678$ have tags in the second sample. Estimate the total population size. Place a bound on the error of estimation.

10.7 A team of conservationists is interested in estimating the size of the pheasant population in a particular area prior to the hunting season. The team believes that the true population size is between 2000 and 3000. Assuming $N \approx 3000$, the sampling fractions of p_1 and p_2 equal to .25 should give a bound on the error of estimation approximately equal to $2(189.74) = 379.48$ (Figure 10.1). The conservationists feel that this bound on the error is reasonable and so decide to choose $t = 750$ and $n = 750$. By using traps, they catch 750 pheasants for the first sample. Each of these pheasants is tagged and released. Several weeks later the second sample of $n = 750$ is obtained. Suppose 168 of these pheasants have tags ($s = 168$). Estimate the population size, and place a bound on the error of estimation.

10.8 City officials are concerned about the nuisance caused by pigeons around city hall. To emphasize the problem, they hire a team of investigators to estimate the number of pigeons occupying the building. With several different traps a sample of $t = 60$ pigeons is captured, tagged, and released. One month later the process is repeated, using $n = 60$. Suppose $s = 18$ tagged pigeons are observed in the second sample. Estimate N, and place a bound on the error of estimation.

10.9 Animal resource experts on a particular game preserve are concerned about an apparent decline in the rabbit population. In a study conducted two years ago, the population size was estimated to be $N = 2500$. Assume the population size is still of this magnitude, and use Figure 10.1 to determine the approximate sample sizes (t and n) required to estimate N with a bound equal to 356.

10.10 A zoologist wishes to estimate the size of the turtle population in a given geographical area. She believes that the turtle population size is between 500 and 1000; hence an initial sample of 100 (10%) appears to be sufficient. The $t = 100$ turtles are caught, tagged, and released. A second sampling is begun one month later, and she decides to continue sampling until $s = 15$ tagged turtles are recaptured. She catches 160 turtles before obtaining 15 tagged turtles ($n = 160$, $s = 15$). Estimate N, and place a bound on the error of estimation.

10.11 Because of a particularly harsh winter, state park officials are concerned about the number of squirrels inhabiting their parks. An initial sample of $t = 100$ squirrels is trapped, tagged, and released. As soon as the first sample is completed, the officials begin working on a second sample of $n = 75$. They trap 10 squirrels that were tagged previously. Estimate N, and place a bound on the error of estimation.

10.12 Assume the costs of taking an observation in the first sample and in the second sample are the same. Determine which is the most desirable: to have $t > n$, $t = n$, or $t < n$ for a fixed cost of conducting the two samples. (*Hint*: Consult Figures 10.1 and 10.2.)

10.13 A team of wildlife ecologists is interested in the effectiveness of an antifertility drug in controlling the growth of pigeon populations. To measure effectiveness, they will estimate the size of the population this year and compare it with the estimated size for a previous year. A large trap was constructed for the experiment. The trap was then baited with a corn feed containing a fixed amount of the drug. An initial sample of $t = 120$ pigeons is trapped and allowed to eat the medicated feed. Each bird is then tagged on its leg and released. At a later date a second sample of $n = 100$ pigeons is trapped. Suppose 48 of these birds have tags ($s = 48$). Estimate the size of the pigeon population, and place a bound on the error of estimation.

10.14 Air samples of 100 cubic centimeters each are taken periodically from an industrial section of a city. The density of a certain type of harmful particle is the parameter of interest. Suppose 15 samples gave an average particle count of 210 per sample. Estimate the particle density, per cubic centimeter, and place a bound on the error of estimation.

10.15 Suppose in the air sampling of Exercise 10.14 detecting the presence or absence of particles is easy but counting the particles is difficult. Among 500 such samples, 410 showed the particles to be present. Estimate the particle density, and place a bound on the error of estimation.

10.16 Cars passing through an intersection are counted during randomly selected 10-minute intervals throughout the working day. Twenty such samples showed an average of 40 cars per interval. Estimate, with a bound on the error, the number of cars that you expect to go through the intersection in an 8-hour period.

10.17 Would you recommend use of the stocked-quadrat method in counting cars, as in Exercise 10.16?

10.18 Discuss the problem of estimating highly mobile animal populations by using quadrat sampling.

10.19 The data in the accompanying table show the number of bacteria colonies observed in 240 microscopic fields. Estimate, with a bound on the error of estimation, the density of colonies per field. What assumptions are necessary for this procedure?

Colonies per Field	Number of Fields
0	11
1	37
2	64
3	55
4	37
5	24
6	12

Source: C. I. Bliss and R. A. Fisher, "Fitting the Negative Binomial Distribution to Biological Data," *Biometrics*, vol. 9, 1953, p. 176–200. With permission from The Biometric Society.

10.20 Outline how you can estimate the number of cars in a city during the working day. Compare four different methods for making this estimate. Which of the four do you think will work best? Why?

EXPERIENCES WITH REAL DATA

10.1 Simulate the tag-recapture technique for animal populations by conducting the following experiment. Put a known number N of beads in a jar. Mark t of them in some distinguishable manner, and thoroughly mix the beads. Then sample n beads, record the number of "tags," and estimate N by the direct method, placing a bound on the error of estimation. Does the resulting interval include your known N?

Replace the n beads and repeat the sampling, using the inverse method (sample until you have s tagged beads), and place a bound on the error. Does this interval include N?

You might try various sample sizes and various degrees of mixing. How will you choose an appropriate sample size? What do you think will happen if the marked beads are not thoroughly mixed with the others? Does this question suggest a realistic difficulty with the tag-recapture method?

10.2 The structure of the problems discussed in this chapter require that there be t marked objects (tags) randomly distributed among the N objects in a population. If t is known, a random sample of size n will supply information to estimate N, provided that some marked objects show up in the sample. The marked objects can be entered into the population without taking an initial sample.

Try the following technique for estimating the size of a crowd at a sporting event, lecture, movie, or other similar event. Obtain the names and descriptions of t people that you know will be attending the event. Ask them to distribute themselves somewhat randomly in the crowd. Then sample n people at random, perhaps as people leave the building. Count the number of original t "tagged" individuals who appear in your sample, and estimate N. (You can use the inverse sample method here also.)

10.3 Estimate the number of three-lettered words in this book by first estimating the density of three-lettered words per page. Place a bound on the error of estimation. Try two different techniques for making this estimate. Which works better in your opinion? What assumptions are necessary for either method to work?

11

SUPPLEMENTAL

TOPICS

11.1
INTRODUCTION

Four sample survey designs, simple and stratified random sampling, cluster sampling, and systematic sampling, have been discussed in preceding chapters. For each design we assumed that the data were correctly recorded and provided an accurate representation of the n elements sampled from the population. Under these assumptions we were able to estimate certain population parameters and place a bound on the error of estimation.

There are many situations in which the assumptions underlying these designs are not fulfilled. First, the recorded measurements are not always accurate representations of the desired data because of biases of the interviewers or measuring equipment. Second, the frame is not always adequate, and hence the sample might not have been selected from the complete population. Third, obtaining accurate sample data might be impossible because of the sensitive nature of the questions.

In this chapter we give some methods for analyzing data when measurement errors are present or an inadequate frame is used.

11.2
INTERPENETRATING SUBSAMPLES

An experimenter is interested in obtaining information from a simple random sample of n persons selected from a population of size N. She has k interviewers available to do the fieldwork, but the interviewers differ in their manner of interviewing, and hence they obtain slightly different responses from identical subjects. For example, suppose the interviewer is to rate the health of a respondent on a scale from 0 to 5, with 0 denoting poor health. Obtaining this type of data requires skill in interviewing and a subjective judgment by the interviewer. One interviewer might not obtain enough information and might tend to rate the health of an individual too high, while another might obtain detailed information and tend to rate the health too low.

A good estimate of the population mean can be obtained by using the following technique. Randomly divide the n sampled elements into k subsamples of m elements each, and assign one interviewer to each of the k subsamples. Note that $m = n/k$ and n can always be chosen so that m is an integer. We consider the first subsample to be a simple random sample of size m selected from the n elements in the total sample. The second subsample is then a simple random sample of size m selected from the $(n - m)$ remaining elements. This process is continued until the n elements have been randomly divided into k subsamples. The k subsamples are sometimes called *interpenetrating subsamples*.

We expect some interviewers to give measurements that are too small and some too large, but the average of all sample measurements should be close to the population mean. That is, we expect the biases of the investigators to possess an average that is very near zero. Thus the sample mean $\bar{y}$ is the best estimator of the population mean μ, even though the measurements are biased.

We use the following notation. Let y_{ij} denote the jth observation in the ith subsample, $j = 1, 2, \ldots, m$, $i = 1, 2, \ldots, k$. Then $\bar{y}_i$, given by

$$\bar{y}_i = \frac{1}{m} \sum_{j=1}^{m} y_{ij} \tag{11.1}$$

is the average of all observations in the ith subsample. The sample mean $\bar{y}$ is the average of the k subsample means.

Estimator of the population mean μ:

$$\bar{y} = \frac{1}{k} \sum_{i=1}^{k} \bar{y}_i \tag{11.2}$$

Estimated variance of $\bar{y}$:

$$\hat{V}(\bar{y}) = \left(\frac{N - n}{N}\right) \frac{\sum_{i=1}^{k} (\bar{y}_i - \bar{y})^2}{k(k - 1)} \tag{11.3}$$

Bound on the error of estimation:

$$2\sqrt{\hat{V}(\bar{y})} = 2\sqrt{\left(\frac{N - n}{n}\right) \frac{\sum_{i=1}^{k} (\bar{y}_i - \bar{y})^2}{k(k - 1)}} \tag{11.4}$$

The technique of interpenetrating subsamples gives an estimate of the variance of $\bar{y}$, given in Equation (11.3), that accounts for interviewer biases. That is, the estimated variance given in Equation (11.3) is usually larger than the standard estimate of the variance of a sample mean obtained in simple random sampling because of the biases present in the measurements.

EXAMPLE 11.1

A sociologist wants to estimate the average height of adult males in a community containing 800 men. He has 10 assistants, each with his or her own equipment, to acquire the measurements. Since the experimenter believes the assistants will produce slightly biased measurements, he decides to take a simple random sample of $n = 80$ males from the population and randomly divide the sample into 10 subsamples of 8 persons each. Each assistant is then assigned to one subsample. The measurements produce the following subsample means (measurements in feet):

$$\bar{y}_1 = 5.9 \qquad \bar{y}_6 = 5.7$$
$$\bar{y}_2 = 5.8 \qquad \bar{y}_7 = 5.8$$
$$\bar{y}_3 = 6.1 \qquad \bar{y}_8 = 5.6$$
$$\bar{y}_4 = 6.0 \qquad \bar{y}_9 = 5.9$$
$$\bar{y}_5 = 6.1 \qquad \bar{y}_{10} = 6.0$$

Estimate the mean height of adult males in the community, and place a bound on the error of estimation.

SOLUTION

The best estimate of the population mean is the sample mean $\bar{y}$. Thus from Equation (11.2),

$$\bar{y} = \frac{1}{k} \sum_{i=1}^{k} \bar{y}_i = \frac{1}{10}(5.9 + 5.8 + \cdots + 6.0) = 5.89$$

We must now estimate the variance of $\bar{y}$ by using Equation (11.3). The following identity can be established:

$$\sum_{i=1}^{k} (\bar{y}_i - \bar{y})^2 = \sum_{i=1}^{k} \bar{y}_i^2 - \frac{\left(\sum_{i=1}^{k} \bar{y}_i\right)^2}{k}$$

Substituting, we obtain

$$\sum_{i=1}^{k} (\bar{y}_i - \bar{y})^2 = 347.17 - \frac{(58.9)^2}{10} = .25$$

Then

$$\hat{V}(\bar{y}) = \left(\frac{N-n}{N}\right) \frac{\sum\limits_{i=1}^{k} (\bar{y}_i - \bar{y})^2}{k(k-1)} = \left(\frac{800-80}{800}\right) \left[\frac{.25}{10(9)}\right] = .0025$$

The estimate of the mean height of adult males, with a bound on the error of estimation, is given by

$$\bar{y} \pm 2\sqrt{\hat{V}(\bar{y})}, \quad \text{or} \quad 5.89 \pm 2\sqrt{.0025}, \quad \text{or} \quad 5.89 \pm .10$$

To summarize, the best estimate of the mean height is 5.89 feet, and we are reasonably confident that our error of estimation is less than .10 foot.

11.3
ESTIMATION OF MEANS AND TOTALS OVER SUBPOPULATIONS

Obtaining a frame that lists only those elements in the population is often impossible. For example, we may wish to sample households containing children, but the best frame available may be a list of all households in a city. We may be interested in a firm's overdue accounts, but the only frame available may list all the firm's accounts receivable. In situations of this type we wish to estimate parameters of a subpopulation of the population represented in the frame. Sampling is complicated because we do not know whether an element belongs to the subpopulation until after it has been sampled.

   The problem of estimating a subpopulation mean is solved essentially in the same manner as in Chapter 4. Let N denote the number of elements in the population and N_1 the number of elements in the subpopulation. A simple random sample of n elements is selected from the population of N elements. Let n_1 denote the number of sampled elements that are from the subpopulation. Let y_{1j} denote the jth sampled observation that falls in the subpopulation. Then the sample mean for elements from the subpopulation, denoted by $\bar{y}_1$, is given by

$$\bar{y}_1 = \frac{1}{n_1} \sum_{j=1}^{n_1} y_{1j}$$

The sample mean $\bar{y}_1$ is an unbiased estimate of the subpopulation mean μ_1.

Estimator of the subpopulation mean μ_1:

$$\bar{y}_1 = \frac{1}{n_1} \sum_{j=1}^{n_1} y_{1j} \tag{11.5}$$

Estimated variance of $\bar{y}_1$:

$$\hat{V}(\bar{y}_1) = \left(\frac{N_1 - n_1}{N_1}\right)\frac{\sum\limits_{j=1}^{n_1}(y_{1j} - \bar{y}_1)^2}{n_1(n_1 - 1)} \tag{11.6}$$

Bound on the error of estimation:

$$2\sqrt{\hat{V}(\bar{y}_1)} = 2\sqrt{\left(\frac{N_1 - n_1}{N_1}\right)\frac{\sum\limits_{j=1}^{n_1}(y_{1j} - \bar{y}_1)^2}{n_1(n_1 - 1)}} \tag{11.7}$$

The quantity $(N_1 - n_1)/N_1$ can be estimated by $(N - n)/N$ if N_1 is unknown.

EXAMPLE 11.2

An economist wants to estimate the average weekly amount spent on food by families with children in a certain county known to be a poverty area. A complete list of all the 250 families in the county is available, but identifying those families with children is impossible. The economist selects a simple random sample of $n = 50$ families and finds that $n_1 = 42$ families have at least one child. The 42 families with children are interviewed and give the following information:

$$\sum_{j=1}^{42} y_{1j} = \$1720 \qquad \sum_{j=1}^{42} y_{1j}^2 = 72,200$$

Estimate the average weekly amount spent on food by all families with children, and place a bound on the error of estimation.

SOLUTION

The estimator of the population mean is $\bar{y}_1$, given by Equation (11.5). Calculations yield

$$\bar{y}_1 = \frac{1}{n_1}\sum_{j=1}^{n_1} y_{1j} = \frac{1}{42}(1720) = 40.95$$

We have the equality

$$\sum_{j=1}^{n_1}(y_{1j} - \bar{y}_1)^2 = \sum_{j=1}^{n_1} y_{1j}^2 - \frac{1}{n_1}\left(\sum_{j=1}^{n_1} y_{1j}\right)^2$$

and substituting gives

$$\sum_{j=1}^{n_1}(y_{1j} - \bar{y}_1)^2 = 72,200 - \frac{1}{42}(1720)^2 = 1762$$

The quantity $(N_1 - n_1)/N_1$ must be estimated by $(N - n)/N$, since N_1 is unknown. The estimated variance of $\bar{y}_1$, given in Equation (11.6), then becomes

$$\hat{V}(\bar{y}_1) = \left(\frac{N-n}{N}\right) \frac{\sum\limits_{j=1}^{n_1}(y_{1j}-\bar{y}_1)^2}{n_1(n_1-1)} = \left(\frac{250-50}{250}\right)\left[\frac{1762}{42(41)}\right]$$

$$= .819$$

Thus the estimate of the population average, with a bound on the error of estimation, is given by

$$\bar{y}_1 \pm 2\sqrt{\hat{V}(\bar{y}_1)}, \qquad \text{or} \qquad 40.95 \pm 2\sqrt{.819}, \qquad \text{or} \qquad 40.95 \pm 1.81$$

Our best estimate of the average weekly amount spent on food by families with children is $40.95. The error of estimation should be less than $1.81 with probability approximately .95.

If the number of elements in the subpopulation N_1 is known, the subpopulation total τ_1 can be estimated by $N_1\bar{y}_1$.

Estimator of the subpopulation total τ_1:

$$N_1\bar{y}_1 = \frac{N_1}{n_1}\sum\limits_{j=1}^{n_1} y_{1j} \tag{11.8}$$

Estimated variance of $N_1\bar{y}_1$:

$$\hat{V}(N_1\bar{y}_1) = N_1^2\hat{V}(\bar{y}_1) = N_1^2\left(\frac{N_1-n_1}{N_1}\right)\frac{\sum\limits_{j=1}^{n_1}(y_{1j}-\bar{y}_1)^2}{n_1(n_1-1)} \tag{11.9}$$

Bound on the error of estimation:

$$2\sqrt{\hat{V}(N_1\bar{y}_1)} = 2\sqrt{N_1^2\left(\frac{N_1-n_1}{N_1}\right)\frac{\sum\limits_{j=1}^{n_1}(y_{1j}-\bar{y}_1)^2}{n_1(n_1-1)}} \tag{11.10}$$

EXAMPLE 11.3

A recent preliminary study of the county in Example 11.2 reveals $N_1 = 205$ families with children. Using this information and data given in Example 11.2, estimate the total weekly amount spent on food by families with children. (*Note:* N_1 will vary over time. We assume that the value of N_1 used in this analysis is correct.)

SOLUTION

The best estimator of the total is $N_1 \bar{y}_1$, given in Equation (11.8), which yields an estimate of

$$N_1 \bar{y}_1 = 205(40.95) = 8394.75$$

The quantity $\sum_{j=1}^{n_1} (y_{1j} - \bar{y}_1)^2$ is calculated in Example 11.2 to be 1762. The estimated variance of $N_1 \bar{y}_1$ is then [from Equation (11.9)]

$$\hat{V}(N_1 \bar{y}_1) = N_1^2 \left(\frac{N_1 - n_1}{N_1} \right) \frac{\sum\limits_{j=1}^{n_1} (y_{1j} - \bar{y}_1)^2}{n_1(n_1 - 1)}$$

$$= (205)^2 \left(\frac{205 - 42}{205} \right) \left[\frac{1762}{42(41)} \right] = 34{,}191.19$$

The estimate of the total weekly amount that families with children spend on food, given with a bound on the error of estimation, is

$$N_1 \bar{y}_1 \pm 2\sqrt{\hat{V}(N_1 \bar{y}_1)}, \quad \text{or} \quad 8394.75 \pm 2\sqrt{34{,}191.19},$$

or

$$8394.75 \pm 369.82$$

Frequently, the number of elements in the subpopulation, N_1, is unknown. For example, the exact number of households containing children in a city, may be difficult to determine, whereas the total number of households can perhaps be obtained from a city directory. An unbiased estimate of τ can still be obtained even though N_1 is unknown.

Estimator of the subpopulation total τ_1 when N_1 is unknown:

$$\hat{\tau}_1 = \frac{N}{n} \sum_{j=1}^{n_1} y_{1j} \tag{11.11}$$

Estimated variance of $\hat{\tau}_1$:

$$\hat{V}(\hat{\tau}_1) = N^2 \left(\frac{N - n}{N} \right) \frac{\sum\limits_{j=1}^{n_1} y_{1j}^2 - \left[\left(\sum\limits_{j=1}^{n_1} y_{1j} \right)^2 \Big/ n \right]}{n(n - 1)} \tag{11.12}$$

Bound on the error of estimation:

$$2\sqrt{\hat{V}(\hat{\tau}_1)} = 2\sqrt{N^2 \left(\frac{N - n}{N} \right) \frac{\sum\limits_{j=1}^{n_1} y_{1j}^2 - \left[\left(\sum\limits_{j=1}^{n_1} y_{1j} \right)^2 \Big/ n \right]}{n(n - 1)}} \tag{11.13}$$

EXAMPLE 11.4

Suppose that the experimenter in Example 11.3 doubts the accuracy of the preliminary value of N_1. Use the data of Example 11.3 to estimate the total weekly amount spent on food by families with children, without using the value given for N_1.

SOLUTION

The estimator of the total that does not depend on N_1 is $\hat{\tau}_1$, given by Equation (11.11). Thus

$$\hat{\tau}_1 = \frac{N}{n} \sum_{j=1}^{n_1} y_{1j} = \frac{250}{50}(1720) = 8600$$

Substituting into Equation (11.12) gives the estimated variance of $\hat{\tau}_1$:

$$\hat{V}(\hat{\tau}_1) = N^2 \left(\frac{N-n}{N} \right) \frac{\sum_{j=1}^{n_1} y_{1j}^2 - \frac{1}{n} \left(\sum_{j=1}^{n_1} y_{1j} \right)^2}{n(n-1)}$$

$$= (250)^2 \left(\frac{250-50}{250} \right) \frac{72{,}200 - (1/50)(1720)^2}{50(49)}$$

$$= 265{,}960$$

Thus the estimate of the total weekly amount spent on food, with a bound on the error of estimation, is

$$\hat{\tau}_1 \pm 2\sqrt{\hat{V}(\hat{\tau}_1)}, \quad \text{or} \quad 8600 \pm 2\sqrt{265{,}960},$$

or

$$8600 \pm 1031.44$$

This interval is a large bound on the error of estimation and should be reduced by increasing the sample size n.

Note that the variance of $\hat{\tau}_1$, calculated in Example 11.4, is much larger than the variance of $N_1\bar{y}_1$, calculated in Example 11.3. The variance of τ_1 is larger because the information provided by N_1 is used in $N_1\bar{y}_1$ but not in $\hat{\tau}_1$. Thus if N_1 is known, or if it can be found with little additional cost, the estimator $N_1\bar{y}_1$ should be used.

11.4
RANDOM-RESPONSE MODEL

Persons being interviewed often refuse to answer or give correct answers to sensitive questions that may embarrass them or be harmful to them in some way. For example, some persons may not respond truthfully to political questions such as, "Are you a Communist?" In this section we present a method of estimating the proportion of people who have some characteristic of interest without obtaining direct answers from the people interviewed. The method is due to S. L. Warner (1965).

Designate the people in the population who have or do not have the characteristic of interest as groups A and B, respectively. Thus each person in the population is in either group A or group B. Let p be the proportion of people in group A. The objective is to estimate p without asking each person directly whether or not he belongs to group A. We can estimate p by using a device called a *random-response model*. We start with a stack of cards that are identical except that a fraction, θ, are marked with A and the remaining fraction, $(1 - \theta)$, are marked with B. A simple random sample of n people is selected from the population. Each person in the sample is asked to randomly draw a card from the deck and to state "yes" if the letter on the card agrees with the group to which he or she belongs, or "no" if the letter on the card is different from the group to which he or she belongs. The card is replaced before the next person draws. The interviewer does not see the card and simply records whether the response is "yes" or "no". Let n_1 be the number of people in the sample who respond with "yes." An unbiased estimator $\hat{p}$ of the population proportion p is given in Equation (11.14).

Estimator of a population proportion p:

$$\hat{p} = \frac{\theta - 1}{2\theta - 1} + \frac{n_1}{(2\theta - 1)n} \qquad (11.14)$$

Estimated variance of $\hat{p}$:

$$\hat{V}(\hat{p}) = \frac{1}{n}\left[\frac{1}{16(\theta - \frac{1}{2})^2} - (\hat{p} - \frac{1}{2})^2\right] \qquad (11.15)$$

Bound on the error of estimation:

$$2\sqrt{\hat{V}(\hat{p})} = 2\sqrt{\frac{1}{n}\left[\frac{1}{16(\theta - \frac{1}{2})^2} - (\hat{p} - \frac{1}{2})^2\right]} \qquad (11.16)$$

Equations (11.14), (11.15), and (11.16) are based on the assumption that the population size is large relative to n, so the finite population correction can be ignored. The fraction θ of cards marked A may be arbitrarily chosen by the experimenter but must not equal $\frac{1}{2}$. A value $\theta = 1$ must not be used, because the respondents will then realize that they are telling whether or not they belong to group A, which is exactly what they do not wish to do. A value of θ between $\frac{1}{2}$ and 1, for example $\frac{3}{4}$, is usually adequate.

EXAMPLE 11.5

A study is designed to estimate the proportion of people in a certain district who give false information on income tax returns. Since respondents would not admit to cheating on tax returns, a random-response technique is used. The experimenter constructs a deck of cards in which $\frac{3}{4}$ of the cards are marked F, denoting a falsified return, and $\frac{1}{4}$ are marked C, denoting a correct return. A simple random sample of $n = 400$ persons is selected from the large population of taxpayers in the district. In separate interviews each sampled taxpayer is asked to draw a card from the deck and to respond "yes" if the letter agrees with the group to which he or she belongs. The experiment results in $n_1 = 120$ yes responses. Estimate p, the proportion of taxpayers in the district who have falsified returns, and place a bound on the error of estimation.

SOLUTION

From Equation (11.14)

$$\hat{p} = \frac{\theta - 1}{2\theta - 1} + \frac{n_1}{(2\theta - 1)n} = \frac{\frac{3}{4} - 1}{2(\frac{3}{4}) - 1} + \frac{120}{[2(\frac{3}{4}) - 1](400)}$$

$$= -\frac{1}{2} + \frac{3}{5} = \frac{1}{10} = .1$$

The estimated variance of $\hat{p}$ is given in Equation (11.15) as

$$\hat{V}(\hat{p}) = \frac{1}{n}\left[\frac{1}{16(\theta - \frac{1}{2})^2} - (\hat{p} - \frac{1}{2})^2\right]$$

$$= \frac{1}{400}\left[\frac{1}{16(\frac{3}{4} - \frac{1}{2})^2} - (\frac{1}{10} - \frac{1}{2})^2\right] = .0021$$

The estimate of p with a bound on the error of estimation is then

$$\hat{p} \pm 2\sqrt{\hat{V}(\hat{p})}, \quad \text{or} \quad .1 \pm 2\sqrt{.0021}, \quad \text{or} \quad .1 \pm .092$$

This method generally requires a very large sample size in order to obtain a reasonably small variance of the estimator. A large sample size is needed because each response provides little information on the population proportion p.

The randomized-response technique presented here is the very simplest of many such techniques. For further information on these techniques, see the papers by Campbell and Joiner (1973) and by Leysieffer and Warner (1976).

Randomized-response techniques can be used more widely than the simple yes or no type of response situation employed here might indicate. To see how this technique is developed, refer to the paper by Greenberg, Kuebler, Abernathy, and Horvitz (1971).

11.5
SELECTING THE NUMBER OF CALLBACKS

As discussed earlier in the book, nonresponse is an important problem to consider in any survey. If a simple random sample of size n is employed and only $n_1(n_1 < n)$ responses are obtained, then the two groups (response and nonresponse) can be thought of as constituting a stratified random sample with two strata. Note that this situation is not quite a true stratified random sampling situation since n_1 and $n_2 = n - n_1$ are random variables whose values are determined only after the initial sampling is completed. Nevertheless, thinking in terms of stratified sampling allows us to find an approximately optimal rule for allocating resources to callbacks.

Suppose that out of the n_2 nonrespondents, we decide to make intensive callbacks on r of them, where $r = n_2/k$ for some constant $k > 1$. Also, suppose that it costs c_1 dollars for a standard response and c_2 dollars $(c_2 > c_1)$ for a callback response, with c_0 denoting the initial cost of sampling each item. Then the total cost is

$$C = nc_0 + n_1 c_1 + r c_2$$

If $\bar{y}_1$ denotes the average of the initial responses and $\bar{y}_2$ the average of the r callback responses, then

$$\bar{y}^* = \frac{1}{n}(n_1\bar{y}_1 + n_2\bar{y}_2) \tag{11.17}$$

is an unbiased estimator of the population mean μ.

A theoretical variance expression for $\bar{y}^*$ can be derived, and then we can find the values of k and n that minimize the expected cost of sampling for a desired fixed value of $V(\bar{y}^*)$, say V_0. The optimal values of k and n are, for large N, approximately

$$k = \sqrt{\frac{c_2(\sigma^2 - W_2\sigma_2^2)}{\sigma_2^2(c_0 + c_1 W_1)}} \tag{11.18}$$

$$n = \frac{N[\sigma^2 + (k - 1)W_2\sigma_2^2]}{NV_0 + \sigma^2} \tag{11.19}$$

where W_2 is the nonresponse rate for the population, $W_1 = 1 - W_2$, and σ^2 and σ_2^2 are the variances for the entire population and the nonresponse group, respectively. The variance of $\bar{y}^*$ can be estimated by

$$\hat{V}(\bar{y}^*) = \frac{k - 1}{n}W_2 s_2^2 + \left(\frac{N - n}{Nn}\right)s^2$$

where s_2^2 estimates the variance of the nonresponse group and s^2 estimates the overall population variance.

EXAMPLE 11.6

A mailed questionnaire is to be used to collect data for estimating the average amount per week that a certain group of 1000 college males spends on entertainment. From past experience the response rate is anticipated to be about 60%. It is thought that $\sigma^2 \approx 120$ and $\sigma_2^2 \approx 80$. (The nonresponse group tends to be those not interested in entertainment and hence spend less and have less variation in spending habits.) Suppose $c_0 = 0$, $c_1 = 1$, and $c_2 = 4$ and a simple random sample is to be used initially. Find n and k so that the variance of the resulting estimator is approximately 5 units.

SOLUTION

Observe that $W_2 = 1 - W_1 = .4$. Then from Equations (11.18) and (11.19)

$$k = \sqrt{\frac{4[120 - .4(80)]}{80(1)(.6)}} = 2.71$$

$$n = \frac{1000[120 + 1.71(.4)(80)]}{1000(5) + 120} = 34.1 \text{ or } 35$$

Since $E(n_2) = nW_2 = 35(.4) = 14$, we can expect that approximately 21 persons will respond initially, and

$$r = \frac{n_2}{k} \approx \frac{14}{2.71} = 5.2 \text{ or } 6$$

callbacks will have to be made.

11.6
SUMMARY

This chapter presented three useful techniques for estimating population parameters when the assumptions underlying the elementary sample survey designs are not valid.

The effect of interviewer bias can be reduced by using interpenetrating subsamples. The estimator of the population mean in this case is given by Equation (11.2), and the estimated variance of this estimator is given by Equation (11.3).

An inadequate frame generates the problem of estimating means and totals over subpopulations. The estimator of the subpopulation mean is given by Equation (11.5), and estimators of the subpopulation total are given by Equations (11.8) and (11.11).

When persons being interviewed will not give correct answers to sensitive questions, a random–response technique can sometimes be used.

The method for estimating a population proportion p by using this procedure is explained in Section 11.4.

Sometimes, one can treat the nonrespondents as a separate stratum, for purposes of choosing an optimal number of callbacks, as shown in Section 11.5.

EXERCISES

11.1 A researcher is interested in estimating the average yearly medical expense per family in a community of 545 families. The researcher has eight assistants available to do the fieldwork. Skill is required to obtain accurate information on medical expenses because some respondents are reluctant to give detailed information on their health. Since the assistants differ in their interviewing abilities, the researcher decides to use 8 interpenetrating subsamples of 5 families each, with one assistant assigned to each subsample. Hence a simple random sample of 40 families is selected and divided into 8 random subsamples. The interviews are conducted and yield the results shown in the accompanying table. Estimate the average medical expense per family for the past year, and place a bound on the error of estimation.

Subsample	Amount (in dollars) of Medical Expenses for Past Year				
1	101	95	310	427	680
2	157	192	108	960	312
3	689	432	187	512	649
4	322	48	93	162	495
5	837	649	152	175	210
6	1015	864	325	470	295
7	837	249	1127	493	218
8	327	419	291	114	287

11.2 An experiment is designed to gauge the emotional reaction to a city's decision on school desegregation. A simple random sample of 50 people is interviewed, and the emotional reactions are given a score from 1 to 10. The scale on which scores are

Subsample	Scores				
1	5	4	6	1	8
2	4	6	5	2	7
3	9	8	9	7	5
4	8	5	4	6	3
5	6	4	5	7	9
6	1	5	6	4	7
7	6	4	3	5	2
8	5	6	7	3	4
9	2	4	4	5	3
10	9	7	8	6	4

assigned runs from extreme anger to extreme joy. Ten interviewers do the questioning and scoring, with each interviewer working on a random subsample (interpenetrating subsample) of 5 people. Interpenetrating subsamples are used because of the flexible nature of the scoring. The results are given in the accompanying table. Estimate the average score for people in the city, and place a bound on the error of estimation.

11.3   A retail store wants to estimate the average amount of all past-due accounts. The available list of past-due accounts is outdated because some accounts have since been paid. Because drawing up a new list would be expensive, the store uses the outdated list. A simple random sample of 20 accounts is selected from the list, which contains 95 accounts. Of the 20 sampled accounts, 4 have been paid. The 16 past-due accounts contain the following amounts (in dollars): 3.65, 15.98, 40.70, 2.98, 50.00, 60.31, 67.21, 14.98, 10.20, 14.32, 1.87, 32.60, 19.80, 15.98, 12.20, 15.00. Estimate the average amount of past-due accounts for the store, and place a bound on the error of estimation.

11.4   For Exercise 11.3, estimate the total amount of past-due accounts for the store, and place a bound on the error of estimation.

11.5   An employee of the store in the Exercise 11.3 decides to look through the list of past-due accounts and mark those that have been paid. He finds that only 83 of the 95 accounts are past due. Estimate the total amount of past-due accounts by using this additional information and the data of Exercise 11.3. Place a bound on the error of estimation.

11.6   A study is conducted to estimate the average number of miles from home to place of employment for household heads living in a certain suburban area. A simple random sample of 30 people is selected from the 493 heads of households in the area. While conducting interviews, the experimenter finds some household heads are not appropriate for the study because they are retired or do not go to a place of employment for various reasons. Of the 30 sampled household heads, 24 are appropriate for the study, and the data on miles to place of employment are as follows:

8.5	10.2	25.1	5.0	6.3	7.9	15.8	2.1
9.2	4.2	8.3	4.2	6.7	10.1	15.6	22.1
10.0	6.1	7.9	1.5	8.0	11.0	20.2	9.3

Estimate the average distance between home and place of employment for household heads who commute to a place of employment. Place a bound on the error of estimation.

11.7   For the data of Exercise 11.6, estimate the total travel distance between home and place of employment for all household heads in the suburban area. Place a bound on the error of estimation.

11.8   Suppose you know that 420 out of the 493 household heads (Exercise 11.6) commute to a place of employment. Estimate the total travel distance for all household heads in the suburban area, making use of this additional information. Place a bound on the error of estimation.

11.9   A public health official wants to estimate the proportion of dog owners in a city who have had their dogs vaccinated against rabies. She knows that a dog owner often gives incorrect information about rabies shots out of fear that something might happen to his dog if it has not had the shots. Thus the official decides to use a randomized-response technique. She has a stack of cards with .8 of the cards marked A for the group having the shots and .2 marked B for the group not having the shots. A simple random sample of 200 dog owners is selected. Each sampled owner is interviewed and asked to draw a card and to respond with "yes" if the letter on the card agrees with the group he is in. The official obtained 145 yes responses. Estimate the proportion of dog owners who have had their dogs vaccinated, and place a bound on the error of estimation. Assume that the number of dog owners in the city is very large.

11.10 A corporation executive wants to estimate the proportion of corporation employees who have been convicted of a misdemeanor. Since the employees would not want to answer the question directly, the executive uses a randomized-response technique. A simple random sample of 300 people is selected from a large number of corporation employees. In separate interviews each employee draws a card from a deck that has .7 of the cards marked "convicted" and .3 marked "not convicted." The employee responds "yes" if the card agrees with his or her category and "no" otherwise. The executive obtains 105 yes responses. Estimate the proportion of employees who have been convicted of a misdemeanor, and place a bound on the error of estimation.

EXPERIENCES WITH REAL DATA

Select a simple random sample from the appropriate population in at least one of the situations outlined below. Estimate the indicated proportion or average, and place a bound on the error, by using the appropriate results from Section 11.3 on subpopulations. In each case, assume that the items in the subpopulation cannot be classified as such until after they are observed.

11.1 Estimate the proportion of voters favoring a certain local governmental proposal from among those who voted in the most recent election.

11.2 Estimate the proportion of students on your campus favoring the quarter system from among those who have been college students on the quarter system and at least one other system.

11.3 Estimate the average amount spent for utilities in the past month for *homeowners* in a certain neighborhood.

11.4 Estimate the average number of words per page among pages that contain no boxed formulas or tables in this book.

12
SUMMARY

12.1
SUMMARY OF THE DESIGNS AND METHODS

You will recall that the objective of statistics is to make inferences about a population from information contained in a sample. This text discusses the design of sample surveys and associated methods of inference for populations containing a finite number of elements. Practical examples have been selected primarily from the fields of business and the social sciences where finite populations of human responses are frequently the target of surveys. Natural resource management examples are also included.

The method of inference employed for most sample surveys is estimation. Thus we consider appropriate estimators for population parameters and the associated two-standard-deviation bound on the error of estimation. In repeated sampling the error of estimation will be less than its bound, with probability approximately equal to .95. Equivalently, we construct confidence intervals that, in repeated sampling, enclose the true population parameter approximately 95 times out of 100. The quantity of information pertinent to a given parameter is measured by the bound on the error of estimation.

The material in this text falls naturally into five segments. The first is a review of elementary concepts, the second contains useful sample survey designs, the third considers an estimator that utilizes information obtained on an auxiliary variable, the fourth gives methods of estimating the size of wildlife populations, and the fifth considers methods for making inferences when one or more of the basic assumptions with the standard techniques are not satisfied.

The first segment, presented in chapters 1, 2, and 3, reviews the objective of statistics and points to the peculiarities of problems arising in the social sciences, business, and natural resource management that make them different from the traditional type of experiment conducted in the laboratory. These peculiarities primarily involve sampling from finite populations along with a number of difficulties that occur in drawing samples from human populations. The former requires modification of the formulas for the bounds on the error of estimation that are encountered in an introductory course in statistics. The difficulties associated with sampling from human populations suggest specific sample survey designs that reduce the cost of acquiring a specified quantity of information.

In Chapters 4, 5, 7, 8, and 9 we consider specific sample survey designs and their associated methods of estimation. The basic sample survey design, simple random sampling, is presented in Chapter 4. For this design the sample is selected so that every sample of size n in the population has an equal chance of being chosen. The design does not make a specific attempt to reduce the cost of the desired quantity of information. It is the most basic type of sample survey design, and all other designs are compared with it.

The second type of design, stratified random sampling (Chapter 5), divides the population into homogeneous groups called strata. This procedure usually produces an estimator that possesses a smaller variance than can be acquired by simple random sampling. Thus the cost of the survey can be reduced by selecting fewer elements to achieve an equivalent bound on the error of estimation.

The third type of experimental design is systematic sampling (Chapter 7), which is usually applied to population elements that are available in a list or line, such as names on file cards in a drawer or people coming out of a factory. A random starting point is selected and then every kth element thereafter is sampled. Systematic sampling is frequently conducted when collecting a simple random or a stratified random sample is extremely costly or impossible. Once again, the reduction in survey cost is primarily associated with the cost of collecting the sample.

The fourth type of sample survey design is cluster sampling, which is presented in Chapters 8 and 9. Cluster sampling may reduce cost because each sampling unit is a collection of elements usually selected so as to be physically close together. Cluster sampling is most often used when a frame that lists all population elements is not available or when travel costs from element to element are considerable. Cluster sampling reduces the cost of the survey primarily by reducing the cost of collecting the data.

A discussion of ratio, regression, and difference estimators, which utilize information on an auxiliary variable, is covered in the third segment of material, Chapter 6. The ratio estimator illustrates how additional information, frequently acquired at little cost, can be used to reduce the variance of the estimator and, consequently, reduce the overall cost of a survey. It also suggests the possibility of acquiring more sophisticated estimators by using information on more than one auxiliary variable. This chapter on ratio estimation follows naturally the discussion on simple random sampling in Chapter 4. That is, you can take a measurement of y, the response of interest, for each element of the simple random sample and utilize the traditional estimators of Chapter 4. Or as suggested in Chapter 6, you might take a measurement on both y and an auxiliary variable x for each element and utilize the additional information contributed by the auxiliary variable to acquire a better estimator of the parameter. Thus although it was not particularly stressed, ratio estimators can be employed with any of the designs discussed in the text.

Chapter 10 deals with the specific problems of estimating the size of populations. Two estimators employed use recapture data, which requires that the sampling be done in at least two stages.

The fifth and final segment of material is contained in Chapter 11, which deals with four situations in which some of the basic assumptions of the standard procedures cannot be satisfied. The situations are (1) interviewer biases, which can sometimes be minimized by using interpenetrating sub-samples, (2) an inadequate frame, which can sometimes be accounted for by using an estimator for subpopulations of the sampled population, (3) information on sensitive questions, which can be obtained by using a randomized-response model, and (4) nonresponse, which can be planned for and designed into the survey by treating nonrespondents as a separate stratum.

To summarize, we have presented various elementary sample survey designs along with their associated methods of inference. Treatment of the topics has been directed toward practical applications so that you can see how sample survey design can be employed to make inferences at minimum cost when sampling from finite social, business, or natural resource populations.

12.2
COMPARISONS AMONG THE DESIGNS AND METHODS

With an array of sampling designs and methods of analysis available, we now summarize earlier discussions on how one chooses an appropriate design for a particular problem.

Simple random sampling is the basic building block and point of reference for all other designs discussed in this text. However, few large-scale surveys use only simple random sampling, because other designs often provide greater accuracy or efficiency or both.

Stratified random sampling produces estimators with smaller variance than those from simple random sampling, for the same sample size, when the measurements under study are homogeneous within strata but the stratum means vary among themselves. The ideal situation for stratified random sampling is to have all measurements within any one stratum equal but have differences occurring as we move from stratum to stratum.

Systematic sampling is used most often simply as a convenience. It is relatively easy to carry out. But this form of sampling may actually be better than simple random sampling, in terms of bounds on the error of estimation, if the correlation between pairs of elements within the same systematic sample is negative. This situation will occur, for example, in periodic data if the systematic sample hits both the high points and the low points of the periodicities. If, in contrast, the systematic sample hits only the high points, the results are very poor. Populations that have a linear trend in the data or that have a periodic structure that is not completely understood may be better sampled by using a stratified design. Economic time series, for example, can be stratified by quarter or month, with a random sample selected from each stratum. The stratified and the systematic sample both force the sampling to be carried out along the whole set of data, but the stratified design offers more random selection and often produces a smaller bound on the error of estimation.

Cluster sampling is generally employed because of cost-effectiveness or because no adequate frame for elements is available. However, cluster sampling may be better than either simple or stratified random sampling if the measurements within clusters are heterogeneous and the cluster means are nearly equal. The ideal situation for cluster sampling is, then, to have each cluster contain measurements as different as possible but to have the cluster means equal. This condition is in contrast to that for stratified random sampling, in which strata are to be homogenous but stratum means are to differ.

Another way to contrast the last three designs is as follows. Suppose a population consists of $N = nk$ elements, which can be thought of as k systematic samples each of size n. The nk elements can be thought of as n clusters of size k, and the systematic sample merely selects one such cluster. In this case the clusters should be heterogeneous for optimal systematic sampling. By contrast, the nk elements can also be thought of as n strata of k elements each, and the systematic sample selects one element from each stratum. In this case the strata should be as homogeneous as possible, but the stratum means should differ as much as possible. This design is consistent with the cluster formulation of the problem and once again produces an optimal situation for systematic sampling. So we see that the three sampling designs are different, and yet they are consistent with one another with regard to basic principles.

Some final comments are in order on how to make use of an auxiliary variable x to gain more information on our variable of interest y. Ratio estimation is optimal if the regression of y on x produces a straight line through the origin and if the variation in the y's increases with increasing x. Regression estimation is better than ratio estimation if the regression of y on x does not go through the origin and if the variation in the y's remains relatively constant as x varies. Difference estimation is as good as regression estimation if the regression coefficient is nearly equal to unity.

We now provide some exercises for which you can decide the appropriate method of analysis.

EXERCISES

12.1   A shipment of 6000 automobile batteries is to contain, according to the manufacturer's specifications, batteries weighing approximately 69 pounds each and having positive-plate thicknesses of 120 thousandths of an inch. Thirty batteries were randomly selected from this shipment and tested. The data are recorded in Table 12.1. Do you think either of the manufacturer's specifications is met for this shipment? (Each battery contains 24 positive plates.)

TABLE 12.1 Battery specifications

Battery	Weight (in pounds)	Number of Positive Plates Sampled	Average Plate Thickness (in thousandths of an inch)	Standard Deviation of Thicknesses
1	61.5	8	109.6	0.74
2	63.5	16	110.0	1.22
3	63.5	16	107.0	1.83
4	63.8	16	111.6	2.55
5	63.8	17	110.7	1.65
6	64.0	16	108.7	1.40
7	64.0	16	111.4	2.63
8	65.0	13	112.8	2.06
9	64.2	16	107.8	3.35
10	64.5	8	109.9	1.25
11	66.5	16	107.8	3.19
12	63.5	16	110.2	1.22
13	63.8	12	112.0	1.81
14	63.5	12	108.5	1.57
15	64.0	12	110.4	1.68
16	64.0	12	111.8	1.64
17	63.2	12	111.9	1.68
18	66.5	12	112.5	1.00
19	63.0	12	109.2	2.44
20	62.0	12	106.1	2.23
21	63.0	12	112.0	0.95
22	63.5	12	112.8	1.75
23	64.0	12	110.2	2.05
24	63.5	12	108.0	2.37
25	66.5	7	112.4	0.79
26	67.0	12	106.6	2.47
27	66.5	12	110.5	1.62
28	65.5	12	113.3	1.23
29	66.5	12	112.7	1.23
30	66.0	12	110.6	1.68

12.2   The Department of Revenue in a state carefully audits sales tax returns from retail stores. If the department thinks a firm is understating its taxable sales, it can order an audit of the the firm's accounts. Just such an audit was ordered for a firm with many retail outlets across the state. Records on taxable sales were kept by each retail store. Hence the auditors decided to randomly sample sales by store-months. That is, sales records were obtained for randomly selected months at randomly selected stores. The auditors then recorded total taxable sales for comparison with the taxable sales reported by the store. The Department of Revenue wants to estimate the proportional increase in audited taxable sales over reported taxable sales. How can you make this estimate, with a bound on the error, from the data given in the accompanying table for 15 store-months? (Figures are in thousands of dollars.)

Store-Month	Audited Taxable Sales	Reported Taxable Sales
1	31.5	23.2
2	31.8	22.9
3	21.1	17.6
4	34.7	29.8
5	21.0	16.8
6	40.8	35.1
7	21.3	23.3
8	31.3	26.1
9	19.9	18.8
10	30.9	25.7
11	32.2	29.6
12	32.4	27.1
13	31.7	29.9
14	28.8	31.5
15	30.7	28.4

12.3   The U.S. Department of Interior, Geological Survey, monitors water flow in United States rivers. The data given in Table 12.2 show the mean daily flow rates, in cubic feet per second, for a specific monitoring station in a certain Florida river for a two-year period, 1977–1979.
(a)   Take a sample of 20 measurements from the data in order to compute a quick estimate of the average daily flow rate for the two-year period. Place a bound on the error of estimation.
(b)   Estimate the ratio of the average April flow rate to the average September flow rate, and place a bound on the error of estimation. Do you think you need data for more years to make a good estimate here?

12.4   Foresters estimate the net volume of standing trees by measuring the diameter at breast height and the tree height and then observing visible defects and other characteristics of the tree. The actual volume of usable timber can only be found after the tree is felled and processed into boards. For a sample of 20 trees data on both estimated and actual volume is recorded, along with the species of the tree, in the accompanying table. The total estimated volume for all 180 trees is 60,000 board feet. Use the data in the table to solve the following problems.

TABLE 12.2 Discharge water (in cubic feet per second), October 1977 to September 1979 (mean values)

1977–1978

Day	Oct.	Nov.	Dec.	Jan.	Feb.	Mar.	Apr.	May	June	July	Aug.	Sept.
1	51	11	3.6	6.1	13	37	9.8	16	3.6	9.3	26	127
2	45	11	3.6	6.1	14	26	8.6	18	4.1	3.4	47	122
3	42	11	4.0	5.7	17	26	8.4	18	3.6	2.5	48	126
4	38	11	3.3	4.8	16	54	8.0	25	9.1	1.6	40	132
5	35	9.5	2.9	4.4	14	55	8.7	27	12	1.3	34	136
6	32	7.4	3.5	5.3	13	45	8.4	19	12	1.9	33	139
7	29	5.4	2.6	5.9	12	37	9.2	50	12	2.6	36	143
8	28	5.5	2.6	6.6	12	34	9.5	49	9.9	2.2	38	133
9	28	5.5	2.8	9.0	31	111	9.9	28	9.1	4.1	46	126
10	27	6.3	3.1	8.8	47	216	10	22	8.4	5.3	63	123
11	26	6.0	3.1	7.4	44	134	11	20	6.8	6.6	78	120
12	27	4.8	2.8	6.1	35	91	11	18	6.8	11	85	116
13	26	4.7	2.6	11	28	77	12	18	6.8	5.8	88	114
14	24	5.0	2.6	15	25	68	13	17	6.1	6.8	89	114
15	23	5.5	2.8	12	24	64	13	15	6.1	17	91	98
16	23	5.5	5.4	9.3	35	62	12	14	6.8	3.8	96	86
17	23	6.0	6.2	11	112	50	13	13	6.8	7.7	98	81
18	22	6.0	5.6	17	234	42	13	14	5.4	19	103	77
19	21	5.7	4.7	18	243	36	14	20	4.7	48	108	78
20	21	4.8	4.0	29	138	30	17	14	4.7	65	127	75
21	20	5.4	4.1	29	125	27	17	13	4.1	35	119	60
22	19	5.6	4.2	29	113	21	16	12	3.3	20	104	49
23	16	7.5	4.9	26	98	16	14	9.6	4.1	17	96	40
24	17	4.8	5.1	24	82	14	14	7.7	4.1	14	93	31
25	17	4.8	5.0	23	69	13	14	7.1	3.7	13	91	24

Continues

TABLE 12.2 *Continued*

1977–1978

Day	Oct.	Nov.	Dec.	Jan.	Feb.	Mar.	Apr.	May	June	July	Aug.	Sept.
26	16	4.4	6.1	23	58	12	16	6.9	3.0	22	90	24
27	15	4.3	5.9	23	49	12	17	11	2.3	29	91	19
28	14	3.6	5.0	19	41	11	17	12	2.7	24	98	13
29	13	3.8	5.2	18	—	11	17	9.5	2.8	22	95	9.6
30	13	3.8	5.4	16	—	9.5	16	4.6	7.9	19	100	8.5
31	12	—	6.8	14	—	9.6	—	2.6	—	20	114	—
Total	763	185.6	129.5	442.3	1742	1451.1	377.5	531.2	182.8	459.9	2465	2544.1
Mean	24.6	6.19	4.18	14.3	62.2	46.8	12.6	17.1	6.09	14.8	79.5	84.8
Max.	51	11	6.8	29	243	216	17	50	12	65	127	143
Min.	12	3.6	2.6	4.4	12	9.5	8.0	2.6	2.3	1.3	26	8.5

1978–1979

Day	Oct.	Nov.	Dec.	Jan.	Feb.	Mar.	Apr.	May	June	July	Aug.	Sept.
1	10	1.2	2.6	7.3	23	29	11	3.5	38	6.1	13	521
2	8.4	.77	1.8	9.5	21	23	11	2.3	36	4.3	13	508
3	9.1	.63	2.6	11	19	21	9.2	1.5	34	3.3	14	401
4	8.1	.60	2.5	8.3	18	19	6.7	1.4	39	4.0	16	345
5	7.1	.50	4.1	6.5	19	18	6.8	1.4	40	8.2	26	284
6	9.4	.65	4.1	4.7	19	40	8.0	2.1	32	11	27	237
7	8.7	.56	3.7	6.0	19	101	7.6	2.6	26	6.8	58	205
8	6.1	.64	3.3	7.6	20	85	7.9	504	22	17	84	183
9	6.1	.57	3.0	8.5	19	65	8.7	1300	19	14	133	165
10	5.4	.56	3.0	8.0	17	50	9.0	489	16	8.5	114	155

11	4.6	.70	3.0	7.5	16	44	9.4	275	14	6.2	150	150
12	5.6	.76	3.4	27	16	37	9.4	220	12	6.8	490	145
13	6.8	.74	3.1	59	15	33	9.9	185	11	11	379	226
14	9.3	.80	4.3	49	15	31	8.5	185	9.6	14	296	314
15	5.7	.87	3.5	30	15	27	8.5	160	8.6	18	296	287
16	5.0	.96	3.1	22	15	25	7.8	140	7.8	12	264	333
17	4.7	1.0	3.0	19	15	21	6.4	126	7.3	14	281	324
18	4.5	1.1	2.6	17	14	21	6.1	116	6.4	11	261	290
19	4.4	1.1	3.1	14	14	19	6.3	95	6.4	10	228	276
20	5.8	1.1	4.7	13	13	16	6.1	78	5.9	10	337	254
21	6.6	1.1	5.6	17	12	16	6.8	76	5.8	16	331	235
22	7.9	1.3	7.7	19	12	15	5.9	74	5.6	14	281	472
23	9.6	1.6	8.6	17	12	17	4.8	62	5.2	14	288	472
24	6.6	1.8	10	33	13	20	5.0	74	5.1	19	321	897
25	5.6	2.2	11	40	56	18	5.9	82	4.7	25	305	764
26	5.2	2.0	12	33	106	16	4.8	70	3.7	17	296	878
27	5.8	2.0	15	26	62	13	4.7	60	3.9	12	328	786
28	7.6	2.1	28	24	36	12	4.5	54	7.2	11	305	680
29	3.8	2.6	19	23	—	12	3.7	49	14	10	287	580
30	2.7	.93	12	23	—	11	3.6	45	9.1	11	465	500
31	1.4	—	8.8	24	—	12	—	41	—	12	492	—
Total	197.6	33.44	202.2	613.9	651	887	214.0	4574.8	455.3	357.2	7189	11867
Mean	6.37	1.11	6.52	19.8	23.3	28.6	7.13	148	15.2	11.5	232	396
Max.	10	2.6	28	59	106	101	11	1300	40	25	492	897
Min.	1.4	.50	1.8	4.7	12	11	3.6	1.4	3.7	3.3	13	145

Source: U.S. Department of Interior, Geological Survey.

(a)   Estimate the total actual board feet for the 180 trees.
(b)   Estimate the proportion of balsam fir trees in the entire stand.
(c)   Estimate the total actual board feet of balsam fir in the stand.
(d)   Estimate the total actual board feet of balsam fir if there are 110 balsam fir trees
in the stand.
Place bounds on the error of estimation in all four cases.

Species (S = black spruce; F = balsam fir)	Estimated Net volume (in board feet)	Actual Net Volume (in board feet)
F	130	141
S	450	474
S	268	301
F	227	215
F	190	210
F	432	400
S	501	487
F	397	368
F	248	262
S	184	195
S	230	280
F	287	243
F	312	255
F	260	282
S	410	375
S	325	280
F	422	490
S	268	325
F	250	210
F	195	236

12.5   The Environmental Protection Agency and the University of Florida recently cooper-
ated in a large study of the possible effects of drinking water on kidney stone disease.

	Carolinas		Rockies	
	New Stone	Recurrent Stone	New Stone	Recurrent Stone
Sample size	363	467	259	191
Age	42.2 (10.9)	45.1 (10.2)	42.5 (10.8)	46.4 (9.8)
Calcium (in parts per million)	11.0 (15.1)	11.3 (16.6)	42.4 (31.8)	40.1 (28.4)
Proportion now smoking	.73	.78	.57	.61

Kidney stone patients were sampled in the Carolinas and in the Rocky Mountain
States. The patients were divided into "new stones" (the current episode being their
first encounter with stone disease) and "recurrent stones." Measurements on three
variables of interest, age of patient, amount of calcium in their home drinking water,

and smoking activity, are recorded in the accompanying table. (Measurements are averages or proportions; standard deviations are given in parentheses.)

(a) Estimate the average age of all stone patients in the population, and place a bound on the error of estimation.

(b) Estimate the average calcium concentration in drinking water supplies for stone patients in the Carolinas. Place a bound on the error of estimation.

(c) Estimate the average calcium concentration in drinking water supplies for stone patients in the Rockies. Place a bound on the error of estimation. Does the answer here differ considerably from that in part (b)?

(d) Estimate the proportion of smokers among new stones, and place a bound on the error.

12.6 In Exercise 12.5 the data were actually collected by first sampling hospitals from the two regions and then sampling stone patients from within hospitals. Explain how you would conduct the analysis asked for in Exercise 12.5 with the data supplied by the hospitals. What additional data would you need?

12.7 Suppose in Exercise 12.6 that the hospitals within regions vary greatly in size. How can you use the information on hospital size advantageously in your sampling design?

12.8 The toxic effects of chemicals on fish are measured in the laboratory by subjecting a certain species of fish to various concentrations of a chemical added to the water. The concentration of chemical that is lethal to 50% of the fish, over the test period, is called the LC 50. Tests in a tank in which water is not renewed during the test process are called static. If new water is constantly coming into the tank, the test is called flow-through. Static tests are cheaper and easier to run, but flow-through tests better approximate reality. Thus experimenters often estimate a static–to–flow-through conversion factor. From the data given in the accompanying table on 12 static and flow-through tests (the measurements are in milligrams per liter), estimate a factor by which a static-test result should be multiplied to make it comparable to a flow-through–test result. Place a bound on the error of estimation.

Toxicant	LC 50, Flow-through	LC 50 Static
Malathion	0.5	0.9
DDT	0.8	1.8
Parathion	4.5	2.1
Endrin	5.5	1.3
Azinphosmethyl	1.2	0.2
DDT	3.5	2.3
Parathion	5.0	1.5
Endrin	0.5	3.2
Zectran	83.0	12.0
Chlordane	4.0	10.0
Fenthion	5.8	12.0
Malathion	12.0	90.0

Source: *Federal Register*, vol. 43, no. 97, May 18, 1978.

12.9 Refer to Exercise 12.8. Can you suggest some improvements in the sampling so as to obtain a better estimate of the conversion factor?

12.10 Refer to Exercise 12.2. Can you suggest a better design for sampling the retail stores? Keep in mind that sales vary from store to store and from month to month.

12.11 Raw sugar is delivered from a grower to a refining mill in bulk form, transported in large tank trucks. The amount paid by the mill for a truckload of sugar depends on the pure-sugar content of the load. This pure-sugar content is determined by laboratory analysis of small test samples, each test sample containing enough raw sugar to fill a test tube. Discuss possible sampling designs to obtain these test samples. (Only a few test samples can be run per truckload of raw sugar.)

12.12 Baled wool from Australia is inspected as it comes into the United States, and an import duty is paid on the basis of pure-wool content. Core samples are taken from bales and analyzed to determine the proportion of the bale that is pure wool. Discuss possible sampling designs for the estimation of pure-wool content in a shipload of bales.

12.13 The Florida Public Service Commission requires companies that sell natural gas to make sure that the meters attached to houses and commerical buildings are operating correctly. However, they will allow a sampling inspection plan rather than a detailed annual examination of every meter. Suppose 20% of the meters owned by a certain company must be checked each year, and the proportion of the company's meters operating correctly must be estimated. (If this proportion is low, the company will be forced to check more meters.) Suggest a sampling plan for this meter inspection policy, keeping the following points in mind: (1) Meters are of varying ages; (2) gas use varies greatly from user to user; and (3) meters are being connected and disconnected continuously.

APPENDIX

REFERENCES _____

Bailey, A. D. 1981. *Statistical Auditing*. New York: Harcourt Brace Jovanovich.

Bailey, N. T. J. 1951. "On Estimating the Size of Mobile Populations from Recaptive Data." *Biometrika*, 38:292–306.

Bergsten, J. W. 1979. "Some Methodological Results from Four Statewide Telephone Surveys Using Random Digit Dialing." *American Statistical Association Proceedings of the Section on Survey Research Methods*. pp. 239–243.

Bryson, M. C. 1976. "The Literary Digest Poll: Making of a Statistical Myth." *American Statistician*, 30 (4):184–185.

Bureau of Labor Statistics, *Handbook of Methods*, vols. I and II. 1982. Washington, D.C.: U.S. Department of Labor.

Campbell, C., and Joiner, B. 1973. "How to Get the Answer Without Being Sure You Asked the Question." *American Statistician*, 27:229–231.

Careers in Statistics. 1980. Washington, D.C.: American Statistical Association.

Chapman, D. G. 1952. "Inverse, Multiple and Sequential Sample Censuses." *Biometrics*, 8:286–306.

Cochran, W. G. 1950. "Estimation of Bacterial Densities by Means of the 'Most Probable Number.'" *Biometrics*, 6:105.

———. 1977. *Sampling Techniques*. 3rd ed. New York: Wiley.

Deming, W. E. 1960. *Sample Design in Business Research*. New York: Wiley.

Frankel, L. R. 1976. "Statisticians and People—The Statistician's Responsibility." *Journal of the American Statistical Association*, 7:9–16.

Gallup, George. 1972. *The Sophisticated Poll Watcher's Guide*. Princeton, N.J.: Princeton Opinion Press.

Greenberg, B. G.; Kuebler, R. R.; Abernathy, J. R.; and Horvitz, D. G. 1971. "Application of Randomized Response Technique in Obtaining Quantitative Data." *Journal of the American Statistical Association*, 66:245–250.

Hansen, M. H.; Hurwitz, W. N.; and Madow, W. G. 1953. *Sample Survey Methods and Theory*, vol. 1. New York: Wiley.

Harper, W. B.; Westfall, R.; and Stasch, S. F. 1977. *Marketing Research*. Homewood, Ill.: Irwin.

Jessen, Raymond T. 1978. *Statistical Survey Techniques*. New York: Wiley.

Jones, H. L. 1956. "Investigation of the Properties of a Sample Mean by Employing Random Subsample Means." *Journal of the American Statistical Association*, 51:54–83.

Kinnear, T. C., and Taylor, J. R. 1983. *Marketing Research, An Applied Approach*. New York: McGraw-Hill.

Kish, L. 1965. *Survey Sampling*. New York: Wiley.

Levy, P. S., and Lemeshow, S. 1980. *Sampling for Health Professionals*. Belmont, Calif.: Lifetime Learning.

Leysieffer, F., and Warner, S. 1976. "Respondent Jeopardy and Optimal Designs in Randomized Response Models." *Journal of the American Statistical Association*, 71:649–656.

Mendenhall, W. 1983. *Introduction to Probability and Statistics*. 6th ed. Boston: Duxbury Press.

Mosby, H. S., ed. 1969. *Wildlife Investigational Techniques*. 3rd ed. Washington, D.C.: Wildlife Society.

Raj, Des. 1968. *Sampling Theory*. New York: McGraw-Hill.

Ray, A. A., ed. 1982. *SAS Users Guide: Statistics*. Cary, N.C.: SAS Institute.

Roberts, D. 1978. *Statistical Auditing*. New York: American Institute of Certified Accountants.

Ryan, T. A.; Joiner, B. L.; and Ryan, B. F. 1976. *Minitab Student Handbook*. Boston: Duxbury Press.

Schuman, Howard, and Presser, Stanley. 1981. *Questions and Answers in Attitude Surveys*. New York: Academic Press.

Stephan, Frederick F., and McCarthy, Philip M. 1958. *Sampling Opinions, An Analysis of Survey Procedure*. New York: Wiley.

Sudman, Seymour. 1976. *Applied Sampling*. New York: Academic Press.

Swindel, B. F. 1983. "Choice of Size and Number of Quadrats to Estimate Density from Frequency in Poisson and Binomially Dispersed Populations." *Biometrics*, 39:455.

Tanur, J. M.; Mosteller, F.; Kruskal, W. H.: Pieters, R. S.; and Rising, G. R., eds. 1972. *Statistics: A Guide to the Unknown*. San Francisco: Holden-Day.

Warner, S. L. 1965. "Randomized Response: A Survey Technique for Eliminating Evasive Answer Bias." *Journal of the American Statistical Association*, 60:63–69.

Weeks, M. F.; Jones, B. L.: Folsom, R. E.; and Benrud, C. H. 1980. "Optimal Times to Contact Sample Households." *Public Opinion Quaterly*, 44:101–114.

Williams, B. 1978. *A Sampler on Sampling*. New York: Wiley.

TABLES

TABLE 1 Normal curve areas

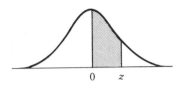

z	.00	.01	.02	.03	.04	.05	.06	.07	.08	.09
0.0	.0000	.0040	.0080	.0120	.0160	.0199	.0239	.0279	.0319	.0359
0.1	.0398	.0438	.0478	.0517	.0557	.0596	.0636	.0675	.0714	.0753
0.2	.0793	.0832	.0871	.0910	.0948	.0987	.1026	.1064	.1103	.1141
0.3	.1179	.1217	.1255	.1293	.1331	.1368	.1406	.1443	.1480	.1517
0.4	.1554	.1591	.1628	.1664	.1700	.1736	.1772	.1808	.1844	.1879
0.5	.1915	.1950	.1985	.2019	.2054	.2088	.2123	.2157	.2190	.2224
0.6	.2257	.2291	.2324	.2357	.2389	.2422	.2454	.2486	.2517	.2549
0.7	.2580	.2611	.2642	.2673	.2704	.2734	.2764	.2794	.2823	.2852
0.8	.2881	.2910	.2939	.2967	.2995	.3023	.3051	.3078	.3106	.3133
0.9	.3159	.3186	.3212	.3238	.3264	.3289	.3315	.3340	.3365	.3389
1.0	.3413	.3438	.3461	.3485	.3508	.3531	.3554	.3577	.3599	.3621
1.1	.3643	.3665	.3686	.3708	.3729	.3749	.3770	.3790	.3810	.3830
1.2	.3849	.3869	.3888	.3907	.3925	.3944	.3962	.3980	.3997	.4015
1.3	.4032	.4049	.4066	.4082	.4099	.4115	.4131	.4147	.4162	.4177
1.4	.4192	.4207	.4222	.4236	.4251	.4265	.4279	.4292	.4306	.4319
1.5	.4332	.4345	.4357	.4370	.4382	.4394	.4406	.4418	.4429	.4441
1.6	.4452	.4463	.4474	.4484	.4495	.4505	.4515	.4525	.4535	.4545
1.7	.4554	.4564	.4573	.4582	.4591	.4599	.4608	.4616	.4625	.4633
1.8	.4641	.4649	.4656	.4664	.4671	.4678	.4686	.4693	.4699	.4706
1.9	.4713	.4719	.4726	.4732	.4738	.4744	.4750	.4756	.4761	.4767
2.0	.4772	.4778	.4783	.4788	.4793	.4798	.4803	.4808	.4812	.4817
2.1	.4821	.4826	.4830	.4834	.4838	.4842	.4846	.4850	.4854	.4857
2.2	.4861	.4864	.4868	.4871	.4875	.4878	.4881	.4884	.4887	.4890
2.3	.4893	.4896	.4898	.4901	.4904	.4906	.4909	.4911	.4913	.4916
2.4	.4918	.4920	.4922	.4925	.4927	.4929	.4931	.4932	.4934	.4936
2.5	.4938	.4940	.4941	.4943	.4945	.4946	.4948	.4949	.4951	.4952
2.6	.4953	.4955	.4956	.4957	.4959	.4960	.4961	.4962	.4963	.4964
2.7	.4965	.4966	.4967	.4968	.4969	.4970	.4971	.4972	.4973	.4974
2.8	.4974	.4975	.4976	.4977	.4977	.4978	.4979	.4979	.4980	.4981
2.9	.4981	.4982	.4982	.4982	.4984	.4984	.4985	.4985	.4986	.4986
3.0	.4987	.4987	.4987	.4988	.4988	.4989	.4989	.4989	.4990	.4990

Abridged from Table I of *Statistical Tables and Formulas* by A. Hald (New York: John Wiley & Sons, Inc., 1952). Reproduced by permission of A. Hald and the publishers, John Wiley & Sons, Inc.

TABLE 2 Random numbers

Line/Col.	(1)	(2)	(3)	(4)	(5)	(6)	(7)	(8)	(9)	(10)	(11)	(12)	(13)	(14)
1	10480	15011	01536	02011	81647	91646	69179	14194	62590	36207	20969	99570	91291	90700
2	22368	46573	25595	85393	30995	89198	27982	53402	93965	34095	52666	19174	39615	99505
3	24130	48360	22527	97265	76393	64809	15179	24830	49340	32081	30680	19655	63348	58629
4	42167	93093	06243	61680	07856	16376	39440	53537	71341	57004	00849	74917	97758	16379
5	37570	39975	81837	16656	06121	91782	60468	81305	49684	60672	14110	06927	01263	54613
6	77921	06907	11008	42751	27756	53498	18602	70659	90655	15053	21916	81825	44394	42880
7	99562	72905	56420	69994	98872	31016	71194	18738	44013	48840	63213	21069	10634	12952
8	96301	91977	05463	07972	18876	20922	94595	56869	69014	60045	18425	84903	42508	32307
9	89579	14342	63661	10281	17453	18103	57740	84378	25331	12566	58678	44947	05585	56941
10	85475	36857	53342	53988	53060	59533	38867	62300	08158	17983	16439	11458	18593	64952
11	28918	69578	88231	33276	70997	79936	56865	05859	90106	31595	01547	85590	91610	78188
12	63553	40961	48235	03427	49626	69445	18663	72695	52180	20847	12234	90511	33703	90322
13	09429	93969	52636	92737	88974	33488	36320	17617	30015	08272	84115	27156	30613	74952
14	10365	61129	87529	85689	48237	52267	67689	93394	01511	26358	85104	20285	29975	89868
15	07119	97336	71048	08178	77233	13916	47564	81056	97735	85977	29372	74461	28551	90707
16	51085	12765	51821	51259	77452	16308	60756	92144	49442	53900	70960	63990	75601	40719
17	02368	21382	52404	60268	89368	19885	55322	44819	01188	65255	64835	44919	05944	55157
18	01011	54092	33362	94904	31273	04146	18594	29852	71585	85030	51132	01915	92747	64951
19	52162	53916	46369	58586	23216	14513	83149	98736	23495	64350	94738	17752	35156	35749
20	07056	97628	33787	09998	42698	06691	76988	13602	51851	46104	88916	19509	25625	58104
21	48663	91245	85828	14346	09172	30168	90229	04734	59193	22178	30421	61666	99904	32812
22	54164	58492	22421	74103	47070	25306	76468	26384	58151	06646	21524	15227	96909	44592
23	32639	32363	05597	24200	13363	38005	94342	28728	35806	06912	17012	64161	18296	22851
24	29334	27001	87637	87308	58731	00256	45834	15398	46557	41135	10367	07684	36188	18510
25	02488	33062	28834	07351	19731	92420	60952	61280	50001	67658	32586	86679	50720	94953
26	81525	72295	04839	96423	24878	82651	66566	14778	76797	14780	13300	87074	79666	95725
27	29676	20591	68086	26432	46901	20849	89768	81536	86645	12659	92259	57102	80428	25280
28	00742	57392	39064	66432	84673	40027	32832	61362	98947	96067	64760	64584	96096	98253
29	05366	04213	25669	26422	44407	44048	37937	63904	45766	66134	75470	66520	34693	90449
30	91921	26418	64117	94305	26766	25940	39972	22209	71500	64568	91402	42416	07844	69618

31	00582	04711	87917	77341	42206	35126	74087	99547	81817	42607	43808	76655	62028	76630
32	00725	69884	62797	56170	86324	88072	76222	36086	84637	93161	76038	65855	77919	88006
33	69011	65795	95876	55293	18988	27354	26575	08625	40801	59920	29841	80150	12777	48501
34	25976	57948	29888	88604	67917	43708	18912	82271	65424	69774	33611	54262	85963	03547
35	09763	83473	73577	12908	30883	18317	28290	35797	05998	41688	34952	37888	38917	88050
36	91567	42595	27958	30134	04024	86385	29880	99730	55536	84855	29080	09250	79656	73211
37	17955	56349	90999	49127	20044	59931	06115	20542	18059	02008	73708	83517	36103	42791
38	46503	18584	18845	49618	02304	51038	20655	58727	28168	15475	56942	53389	20562	87338
39	92157	89634	94824	78171	84610	82834	09922	25417	44137	48413	25555	21246	35509	20468
40	14577	62765	35605	81263	39667	47358	56873	56307	61607	49518	89656	20103	77490	18062
41	98427	07523	33362	64270	01638	92477	66969	98420	04880	45585	46565	04102	46880	45709
42	34914	63976	88720	82765	34476	17032	87589	40836	32427	70002	70663	88863	77775	69348
43	70060	28277	39475	46473	23219	53416	94970	25832	69975	94884	19661	72828	00102	66794
44	53976	54914	06990	67245	68350	82948	11398	42878	80287	88267	47363	46634	06541	97809
45	76072	29515	40980	07391	58745	25774	22987	80059	39911	96189	41151	14222	60697	59583
46	90725	52210	83974	29992	65831	38857	50490	83765	55657	14361	31720	57375	56228	41546
47	64364	67412	33339	31926	14883	24413	59744	92351	97473	89286	35931	04110	23726	51900
48	08962	00358	31662	25388	61642	34072	81249	35648	56891	69352	48373	45578	78547	81788
49	95012	68379	93526	70765	10592	04542	76463	54328	02349	17247	28865	14777	62730	92277
50	15664	10493	20492	38391	91132	21999	59516	81652	27195	48223	46751	22923	32261	85653
51	16408	81899	04153	53381	79401	21438	83035	92350	36693	31238	59649	91754	72772	02338
52	18629	81953	05520	91962	04739	13092	97662	24822	94730	06496	35090	04822	86774	98289
53	73115	35101	47498	87637	99016	71060	88824	71013	18735	20286	23153	72924	35165	43040
54	57491	16703	23167	49323	45021	33132	12544	41035	80780	45393	44812	12515	98931	91202
55	30405	83946	23792	14422	15059	45799	22716	19792	09983	74353	68668	30429	70735	25499
56	16631	35006	85900	98275	32388	52390	16815	69298	82732	38480	73817	32523	41961	44437
57	96773	20206	42559	78985	05300	22164	24369	54224	35083	19687	11052	91491	60383	19746
58	38935	64202	14349	82674	66523	44133	00697	35552	35570	19124	63318	29686	03387	59846
59	31624	76384	17403	53363	44167	64486	64758	75366	76554	31601	12614	33072	60332	92325
60	78919	19474	23632	27889	47914	02584	37680	20801	72152	39339	34806	08930	85001	87820
61	03931	33309	57047	74211	63445	17361	62825	39908	05607	91284	68833	25570	38818	46920
62	74426	33278	43972	10119	89917	15665	52872	73823	73144	88662	88970	74492	51805	99378
63	09066	00903	20795	95452	92648	45454	09552	88815	16553	51125	79375	97596	16296	66092
64	42238	12426	87025	14267	20979	04508	64535	31355	86064	29472	47689	05974	52468	16834
65	16153	08002	26504	41744	81959	65642	74240	56302	00033	67107	77510	70625	28725	34191

Continues

TABLE 2 Continued

Line/Col.	(1)	(2)	(3)	(4)	(5)	(6)	(7)	(8)	(9)	(10)	(11)	(12)	(13)	(14)
66	21457	40742	29820	96783	29400	21840	15035	34537	33310	06116	95240	15957	16572	06004
67	21581	57802	02050	89728	17937	37621	47075	42080	97403	48626	68995	43805	33386	21597
68	55612	78095	83197	33732	05810	24813	86902	60397	16489	03264	88525	42786	05269	92532
69	44657	66999	99324	51281	84463	60563	79312	93454	68876	25471	93911	25650	12682	73572
70	91340	84979	46949	81973	37949	61023	43997	15263	80644	43942	89203	71795	99533	50501
71	91227	21199	31935	27022	84067	05462	35216	14486	29891	68607	41867	14951	91696	85065
72	50001	38140	66321	19924	72163	09538	12151	06878	91903	18749	34405	56087	82790	70925
73	65390	05224	72958	28609	81406	39147	25549	48542	42627	45233	57202	94617	23772	07896
74	27504	96131	83944	41575	10573	08619	64482	73923	36152	05184	94142	25299	84387	34925
75	37169	94851	39117	89632	00959	16487	65536	49071	39782	17095	02330	74301	00275	48280
76	11508	70225	51111	38351	19444	66499	71945	05422	13442	78675	84081	66938	93654	59894
77	37449	30362	06694	54690	04052	53115	62757	95348	78662	11163	81651	50245	34971	52924
78	46515	70331	85922	38329	57015	15765	97161	17869	45349	61796	66345	81073	49106	79860
79	30986	81223	42416	58353	21532	30502	32305	86482	05174	07901	54339	58861	74818	46942
80	63798	64995	46583	09785	44160	78128	83991	42865	92520	83531	80377	35909	81250	54238
81	82486	84846	99254	67632	43218	50076	21361	64816	51202	88124	41870	52689	51275	83556
82	21885	32906	92431	09060	64297	51674	64126	62570	26123	05155	59194	52799	28225	85762
83	60336	98782	07408	53458	13564	59089	26445	29789	85205	41001	12535	12133	14645	23541
84	43937	46891	24010	25560	86355	33941	25786	54990	71899	15475	95434	98227	21824	19585
85	97656	63175	89303	16275	07100	92063	21942	18611	47348	20203	18534	03862	78095	50136
86	03299	01221	05418	38982	55758	92237	26759	86367	21216	98442	08303	56613	91511	75928
87	79626	06486	03574	17668	07785	76020	79924	25651	83325	88428	85076	72811	22717	50585
88	85636	68335	47539	03129	65651	11977	02510	26113	99447	68645	34327	15152	55230	93448
89	18039	14367	61337	06177	12143	46609	32989	74014	64708	00533	35398	58408	13261	47908
90	08362	15656	60627	36478	65648	16764	53412	09013	07832	41574	17639	82163	60859	75567

91	79556	29068	04142	16268	15387	12856	66227	38358	22478	73373	88732	09443	82558	05250
92	92608	82674	27072	32534	17075	27698	98204	63863	11951	34648	88022	56148	34925	57031
93	23982	25835	40055	67006	12293	02753	14827	23235	35071	99704	37543	11601	35503	85171
94	09915	96306	05908	97901	28395	14186	00821	80703	70426	75647	76310	88717	37890	40129
95	59037	33300	26695	62247	69927	76123	50842	43834	86654	70959	79725	93872	28117	19233
96	42488	78077	69882	61657	34136	79180	97526	43092	04098	73571	80799	76536	71255	64239
97	46764	86273	63003	93017	31204	36692	40202	35275	57306	55543	53203	18098	47625	88684
98	03237	45430	55417	63282	90816	17349	88298	90183	36600	78406	06216	95787	42579	90730
99	86591	81482	52667	61582	14972	90053	89534	76036	49199	43716	97548	04379	46370	28672
100	38534	01715	94964	87288	65680	43772	39560	12918	86537	62738	19636	51132	25739	56947

Abridged from *Handbook of Tables for Probability and Statistics*, Second Edition, edited by William H. Beyer (Cleveland: The Chemical Rubber Publishing Company, 1968). Reprinted by permission. Copyright CRC Press, Inc., Boca Raton, FL.

TABLE 3 United States Population

Regions, Divisions, and States	Resident Population (4/1/1980) (thousands)	Resident Population (4/1/1970) (thousands)	Percent Change, 1970–1980	Crude Birth Rate, 1978	Crude Death Rate, 1978	Infant Mortality Rate, 1978	Net Migration of 1970 Population	Percent of Population 65 and Over	Percent Population in Metropolitan Area, 1978	Black Population Percent of Total, 1976	Per Capita Income, 1977
United States	226,505	203,302	11.4	15.3	8.8	13.8	2.0	11.2	73	11.5	$5,751
Northeast	49,137	49,061	0.2	12.9	9.4	13.1	-3.7	12.1	85	9.2	5,882
New England	12,348	11,847	4.2	12.5	8.9	11.4	0.0	12.0	82	8.6	5,814
Maine	1,125	994	13.2	14.5	9.3	10.4	5.2	12.3	30	0.2	4,627
New Hampshire	921	738	24.8	14.3	8.5	10.4	14.3	11.1	51	0.2	5,365
Vermont	511	445	15.0	14.6	9.1	13.6	5.0	11.3	0	0.2	4,770
Massachusetts	5,737	5,689	0.8	11.9	9.1	11.1	-1.6	12.3	96	3.3	5,826
Rhode Island	947	950	-0.3	12.4	9.5	13.6	-5.2	13.2	91	2.7	5,589
Connecticut	3,108	3,032	2.5	12.0	8.4	11.6	-1.2	11.4	92	6.9	6,564
Middle Atlantic	36,788	37,213	-1.1	13.0	9.6	13.7	-4.9	12.1	87	11.0	5,904
New York	17,557	18,241	-3.8	13.2	9.4	14.0	-7.1	12.0	88	12.5	5,849
New Jersey	7,364	7,171	2.7	12.8	9.0	13.0	-1.8	11.5	92	10.8	6,492
Pennsylvania	11,867	11,801	0.6	12.9	10.2	13.7	-3.4	12.7	80	8.7	5,622
North Central	58,854	56,500	4.0	15.3	8.9	13.6	-2.8	11.2	70	8.8	5,868
East N. Central	41,670	40,263	3.5	15.2	8.8	13.8	-3.7	10.6	77	10.4	6,003
Ohio	10,797	10,657	1.3	15.0	9.0	13.3	-5.3	10.6	80	9.4	5,796
Indiana	5,490	5,195	5.7	15.5	8.8	13.1	-2.9	10.6	70	6.7	5,751
Illinois	11,418	11,110	2.8	15.5	9.2	15.7	-4.9	10.9	81	15.4	6,358
Michigan	9,258	8,882	4.2	15.2	8.2	13.8	-3.4	9.6	81	11.2	6,130
Wisconsin	4,705	4,418	6.5	14.7	8.6	11.2	1.2	11.8	63	3.2	5,660
West N. Central	17,184	16,328	5.2	15.5	9.3	13.2	-0.6	12.6	53	5.0	5,523
Minnesota	4,077	3,806	7.1	15.4	8.3	12.0	0.4	11.6	64	1.3	5,778
Iowa	2,913	2,825	3.1	15.3	9.4	12.6	-2.0	13.1	37	1.2	5,439
Missouri	4,917	4,678	5.1	15.0	10.2	14.8	-0.3	13.0	64	11.8	5,493
North Dakota	653	618	5.6	17.3	8.4	13.5	-1.3	12.1	35	0.3	4,856
South Dakota	690	666	3.6	17.7	9.4	13.5	-3.5	13.1	28	0.1	4,529
Nebraska	1,570	1,485	5.7	16.0	9.2	13.0	0.0	13.0	45	3.1	5,326
Kansas	2,353	2,249	5.1	15.7	9.2	12.5	-0.2	12.7	46	5.6	5,861
South	75,349	62,893	20.0	16.0	9.0	15.3	6.6	11.3	64	18.8	5,289
South Atlantic	36,943	30,679	20.4	14.6	9.1	15.5	7.8	11.9	66	20.8	5,516

Delaware	595	548	8.6	14.9	8.5	13.2	-0.4	9.7	68	13.4	5,883
Maryland	4,216	3,924	7.5	13.3	8.0	14.7	0.1	9.2	85	20.7	6,561
D. of Columbia	638	757	-15.7	14.0	10.5	27.3	-17.4	11.1	100	71.7	7,074
Virginia	5,346	4,651	14.9	14.2	7.9	13.8	4.9	9.3	66	15.9	5,883
West Virginia	1,950	1,744	11.8	15.7	10.6	15.1	2.7	12.0	36	2.1	4,851
N. Carolina	5,874	5,084	15.5	14.8	8.5	16.6	3.0	10.2	45	22.4	4,876
S. Carolina	3,119	2,591	20.4	17.1	8.3	18.6	4.0	9.2	48	31.6	4,628
Georgia	5,464	4,588	19.1	16.6	8.5	15.4	2.8	9.5	57	27.2	5,071
Florida	9,740	6,791	43.3	13.1	11.0	14.1	27.2	18.1	86	15.5	5,761
East S. Central	14,663	12,808	14.5	16.4	9.3	15.4	2.9	11.9	52	20.6	4,686
Kentucky	3,661	3,221	13.7	16.4	9.6	12.7	2.8	11.2	45	8.8	4,851
Tennessee	4,591	3,926	16.9	15.4	8.9	14.8	5.1	11.2	63	16.4	4,845
Alabama	3,890	3,444	12.9	16.2	9.3	16.1	2.2	11.2	62	26.7	4,712
Mississippi	2,521	2,217	13.7	18.4	9.7	18.7	0.3	11.4	27	35.6	4,120
West S. Central	23,743	19,326	22.3	17.8	8.6	15.0	7.2	10.4	69	14.6	5,313
Arkansas	2,286	1,923	18.8	16.1	10.1	16.4	7.1	13.7	39	17.2	4,443
Louisiana	4,204	3,645	15.3	18.8	8.9	17.3	0.9	9.4	63	28.6	4,790
Oklahoma	3,025	2,559	18.2	16.1	9.8	14.3	6.9	12.5	56	6.7	5,245
Texas	14,228	11,199	27.1	18.2	8.0	14.3	9.3	9.7	80	11.6	5,633
West	43,165	34,838	23.5	17.0	7.7	12.1	9.5	10.0	79	5.3	6,238
Mountain	11,368	8,290	37.1	19.5	7.2	12.2	16.6	9.5	61	2.3	5,600
Montana	787	694	13.3	17.4	8.3	11.6	5.4	10.6	24	0.4	5,288
Idaho	944	713	32.4	22.0	7.4	11.7	14.0	10.0	17	0.0	3,072
Wyoming	471	332	41.6	20.4	7.3	13.0	24.2	8.1	0	0.8	6,454
Colorado	2,889	2,210	30.7	16.3	6.8	11.2	15.8	8.6	81	3.5	6,118
New Mexico	1,300	1,017	27.8	19.8	7.0	14.1	9.3	8.8	34	1.7	4,837
Arizona	2,718	1,775	53.1	18.2	8.1	13.1	26.1	11.8	75	2.6	5,545
Utah	1,461	1,059	37.9	29.5	6.0	11.4	7.6	7.7	79	0.7	5,135
Nevada	799	489	63.5	16.5	7.6	12.5	33.8	8.6	81	6.3	6,533
Pacific	31,797	26,548	19.8	16.1	7.8	12.0	7.3	10.2	86	6.3	6,459
Washington	4,130	3,413	21.0	15.5	8.0	12.5	8.6	10.6	71	1.9	6,394
Oregon	2,633	2,092	25.9	15.9	8.5	12.9	14.6	11.6	59	1.3	6,018
California	23,669	19,971	18.5	16.0	7.9	11.8	6.2	10.2	92	7.8	6,487
Alaska	400	303	32.4	21.6	4.1	14.4	15.6	2.6	45	4.3	9,170
Hawaii	965	770	25.3	18.6	5.1	11.1	4.6	7.7	80	0.7	6,005

Source: Reproduced by permission of Population Reference Bureau, Inc., Washington, D.C.

DERIVATION OF SOME MAIN RESULTS _____

In this section we present the mathematical derivation of some of the main formulas used throughout the text. We assume the reader has some knowledge of probability theory, so that expectations, variances, and covariances can be manipulated with little explanation.

Let y_i denote a random variable with probability distribution $p(y)$. Then we have the following definitions from elementary probability theory:

$$E(y) = \sum_y yp(y) = \mu$$

$$E[g(y)] = \sum_y g(y)p(y)$$

$$V(y) = E(y - \mu)^2 = \sum_y (y - \mu)^2 p(y) = \sigma^2$$

where E denotes expected value, V denotes variance, and $g(y)$ is a function of y.

Suppose $y_1, y_2, \ldots, y_n$ denotes a sample of size n and $a_1, a_2, \ldots, a_n$ are constants. If

$$U = \sum_{i=1}^n a_i y_i$$

then

$$E(U) = \sum_{i=1}^n a_i E(y_i) \tag{A.1}$$

and

$$V(U) = \sum_{i=1}^n a_i^2 V(y_i) + 2 \sum\sum_{i<j} a_i a_j \, \text{Cov}(y_i, y_j) \tag{A.2}$$

where Cov denotes covariances. If the y_i's are uncorrelated, then

$$V(U) = \sum_{i=1}^n a_i^2 V(y_i) \tag{A.3}$$

SIMPLE RANDOM SAMPLING

Suppose $y_1, y_2, \ldots, y_n$ denotes a simple random sample from a population of values $\{u_1, u_2, \ldots, u_N\}$. Considering y_i by itself (a simple random sample of size one), we have

$$E(y_i) = \sum_{i=1}^N u_i \left(\frac{1}{N}\right) = \mu$$

and

$$V(y_i) = \sum_{i=1}^N (u_i - \mu)^2 \left(\frac{1}{N}\right) = \sigma^2$$

By Equation (A.1),

$$E(\bar{y}) = \frac{1}{n} \sum_{i=1}^n E(y_i) = \frac{1}{n} \sum_{i=1}^n \mu = \mu$$

Also, $\text{Cov}(y_i, y_j) = E[(y_i - \mu)(y_j - \mu)] = E(y_iy_j) - \mu^2$

$$= \sum_{i \neq j}^{N} u_iu_j \left[\frac{1}{N(N-1)} \right] - \frac{1}{N^2} \left(\sum_{i=1}^{N} u_i \right)^2$$

$$= \frac{1}{N} \left[\sum_{i \neq j}^{N} \frac{u_iu_j}{N-1} - \frac{1}{N} \left(\sum_{i=1}^{N} u_i \right)^2 \right]$$

$$= \frac{1}{N} \left[\frac{\left(\sum_{i=1}^{N} u_i \right)^2 - \sum_{i=1}^{N} u_i^2}{N-1} - \frac{1}{N} \left(\sum_{i=1}^{N} u_i \right)^2 \right]$$

$$= -\frac{1}{N} \left[\frac{1}{N-1} \sum_{i=1}^{N} u_i^2 - \frac{1}{N(N-1)} \left(\sum_{i=1}^{N} u_i \right)^2 \right]$$

$$= -\frac{1}{N(N-1)} \sum_{i=1}^{N} (u_i - \mu)^2 = -\frac{1}{N-1} \sigma^2$$

Using this fact and Equation (A.2), we can find the variance of $\bar{y}$. We have

$$V(\bar{y}) = V\left(\frac{1}{n} \sum_{i=1}^{n} y_i \right) = \frac{1}{n^2} \left[\sum_{i=1}^{n} \sigma^2 + 2 \sum \sum_{i<j} \text{Cov}(y_i, y_j) \right]$$

$$= \frac{1}{n^2} \left[\sum_{i=1}^{n} \sigma^2 + 2 \sum \sum_{i<j} \frac{-\sigma^2}{N-1} \right]$$

$$= \frac{1}{n^2} \left\{ n\sigma^2 - \frac{2\sigma^2}{N-1} \left[\frac{n(n-1)}{2} \right] \right\}$$

since there are $n(n-1)/2$ pairs (i, j) selected from the integers $1, 2, \ldots, n$ so that $i < j$. Therefore

$$V(\bar{y}) = \frac{\sigma^2}{n} \left(\frac{N-n}{N-1} \right)$$

We now show that $[(N-n)/N]/(s^2/n)$ is an unbiased estimator of $V(\bar{y})$. We have

$$E(s^2) = E\left[\left(\frac{1}{n-1} \right) \sum_{i=1}^{n} (y_i - \bar{y})^2 \right] = \left(\frac{1}{n-1} \right) E\left\{ \sum_{i=1}^{n} [(y_i - \mu) - (\bar{y} - \mu)]^2 \right\}$$

$$= \left(\frac{1}{n-1} \right) E\left[\sum_{i=1}^{n} (y_i - \mu)^2 - n(\bar{y} - \mu)^2 \right]$$

$$= \frac{1}{n-1} \left[\sum_{i=1}^{n} E(y_i - \mu)^2 - nE(\bar{y} - \mu)^2 \right]$$

$$= \frac{1}{n-1} [n\sigma^2 - nV(\bar{y})] = \frac{1}{n-1} \left[n\sigma^2 - n\left(\frac{N-n}{N-1} \right)\left(\frac{\sigma^2}{n} \right) \right]$$

$$= \frac{\sigma^2}{n-1} \left(n - \frac{N-n}{N-1} \right) = \frac{N}{N-1} \sigma^2 .$$

Therefore

$$E\left[\left(\frac{N-n}{N}\right)\left(\frac{s^2}{n}\right)\right] = \left(\frac{N-n}{N}\right)\left(\frac{1}{n}\right)\left(\frac{N}{N-1}\right)\sigma^2$$

$$= \left(\frac{N-n}{N-1}\right)\left(\frac{\sigma^2}{n}\right) = V(\bar{y})$$

which was to be shown.

This derivation results in Equations (4.2), (4.3), and (4.4). Now $\hat{\tau} = N\bar{y}$ is an unbiased estimator of τ by Equation (A.1), and Equation (4.6) follows from Equation (A.3) and results already shown.

Since $\hat{p}$ is actually a $\bar{y}$ for $\{0, 1\}$ data, $\hat{p}$ is an unbiased estimator of p, and Equation (4.16) follows directly, after observing that

$$\frac{s^2}{n} = \frac{\hat{p}(1-\hat{p})}{n-1}$$

for the $\{0, 1\}$ data.

STRATIFIED RANDOM SAMPLING

In stratified random sampling

$$\bar{y}_{st} = \sum_{i=1}^{L} \left(\frac{N_i}{N}\right)\bar{y}_i$$

is of the same form as U, and the $\bar{y}_i$'s are independently selected through simple random sampling. Thus

$$E(\bar{y}_{st}) = \sum_{i=1}^{L}\left(\frac{N_i}{N}\right)E(\bar{y}_i) = \sum_{i=1}^{L}\left(\frac{N_i}{N}\right)\mu_i = \mu$$

by Equation (A.1), and

$$V(\bar{y}_{st}) = \sum_{i=1}^{L}\left(\frac{N_i}{N}\right)^2 V(\bar{y}_i)$$

by Equation (A.3), and Equation (5.2) follows.

In the sample size and allocation formulas of Chapter 5, we set $N/(N-1)$ equal to 1 for convenience.

RATIO ESTIMATION

The ratio estimator r is approximately an unbiased estimator of $R = (\mu_y/\mu_x)$ if n is reasonably large. That is, $E(\bar{y}/\bar{x})$ is approximately R. Hence

$$V(r) = V\left(\frac{\bar{y}}{\bar{x}}\right) \doteq E\left(\frac{\bar{y}}{\bar{x}} - R\right)^2 = E\left(\frac{\bar{y} - R\bar{x}}{\bar{x}}\right)^2$$

$$\doteq E\left(\frac{\bar{y} - R\bar{x}}{\mu_x}\right)^2 = \frac{1}{\mu_x^2}E(\bar{y} - R\bar{x})^2 = \frac{1}{\mu_x^2}V(\bar{y} - R\bar{x})$$

since $E(\bar{y} - R\bar{x}) = 0$. Since $\bar{y} - R\bar{x}$ is the sample mean of quantities $y_i - Rx_i$, with $E(y_i - Rx_i) = 0$, then $V(\bar{y} - R\bar{x})$ can be estimated by

$$\left(\frac{N - n}{Nn}\right)\left(\frac{1}{n - 1}\right) \sum_{i=1}^{n} (y_i - Rx_i)^2$$

If R is replaced by r in the latter expression, Equation (6.2) follows. Variance expressions (6.6) and (6.9) follow by using Equation (A.3).

SINGLE-STAGE CLUSTER SAMPLING

The estimator of Equation (8.1) from cluster sampling is a ratio estimator, and its variance (8.2) follows from results derived above. The variance expression (8.5) then comes about by applying Equation (A.3). The estimator of τ given in Equation (8.8) is simply based on a sample mean of cluster totals, and Equation (8.9) follows from basic principles used above.

TWO-STAGE CLUSTER SAMPLING

Since this situation requires careful manipulation of between–cluster variances and within–cluster variances, we illustrate the derivations only for the case in which all clusters are of the same size. That is, we assume

$$m_1 = m_2 = \cdots = m_N = m$$

and

$$M_1 = M_2 = \cdots = \bar{M} = \frac{M}{N}$$

In this case

$$\hat{\mu} = \left(\frac{N}{M}\right)\left(\frac{1}{n}\right) \sum_{i=1}^{n} M_i \bar{y}_i = \sum_{i=1}^{n} \bar{y}_i$$

We find the mean and variance of $\hat{\mu}$ by first fixing the n clusters in the sample and then averaging over all possible samples of n clusters. Expectation and variance operations when the n clusters are fixed will be denoted by E_2 and V_2, respectively. Similarly, expectations and variances over all possible samples of n clusters will be denoted by E_1 and V_1. (When the n clusters are fixed, the cluster sample looks like a stratified random sample.)

Now

$$E(\hat{\mu}) = E_1[E_2(\hat{\mu})] = E_1\left[\frac{1}{n} \sum_{i=1}^{n} E_2(\bar{y}_i)\right]$$

$$= E_1\left(\frac{1}{n} \sum_{i=1}^{n} \mu_i\right)$$

where μ_i is the mean of cluster i. Since the expected value of a sample mean is the corresponding population mean in simple random sampling,

$$E_1\left(\frac{1}{n}\sum_{i=1}^{n}\mu_i\right) = \frac{1}{N}\sum_{i=1}^{N}\mu_i = \frac{1}{N\bar{M}}\sum_{i=1}^{N}\bar{M}\mu_i$$

$$= \frac{1}{M}\sum_{i=1}^{N}\tau_i = \frac{\tau}{M} = \mu$$

where τ_i is the total for cluster i. Thus $\hat{\mu}$ is an unbiased estimator of μ.

From a basic result in probability theory,

$$V(\hat{\mu}) = V_1[E_2(\hat{\mu})] + E_1[V_2(\hat{\mu})]$$

Now

$$V_1[E_2(\hat{\mu})] = V_1\left(\frac{1}{n}\sum_{i=1}^{n}\mu_i\right) = \left(\frac{N-n}{N}\right)\left(\frac{1}{n}\right)\left(\frac{1}{N-1}\right)\sum_{i=1}^{N}(\mu_i - \bar{\mu})^2 \quad \text{(A.4)}$$

where $\bar{\mu} = (1/N)\sum_{i=1}^{N}\mu_i$. This expression follows from basic results explored earlier. Also,

$$E_1[V_2(\hat{\mu})] = E_1\left[V_2\left(\frac{1}{n}\sum_{i=1}^{n}\bar{y}_i\right)\right] = E_1\left[\frac{1}{n^2}\sum_{i=1}^{n}V_2(\bar{y}_i)\right]$$

$$= \left(\frac{1}{n}\right)\left(\frac{1}{N}\right)\sum_{i=1}^{N}V(\bar{y}_i) \quad \text{(A.5)}$$

where $V(\bar{y}_i)$ is the usual variance of a sample mean for a simple random sample of m elements from $\bar{M}$ elements.

We must now estimate the two parts of $V(\hat{\mu})$. For the first part we might start with

$$\left(\frac{1}{\bar{M}}\right)^2 s_b^2 = \frac{1}{n-1}\sum_{i=1}^{n}(\bar{y}_i - \hat{\mu})^2$$

We have, as its expected value,

$$E_1\left\{E_2\left[\left(\frac{1}{n-1}\right)\sum_{i=1}^{n}(\bar{y}_i - \hat{\mu})^2\right]\right\}$$

$$= E_1\left\{\left(\frac{1}{n-1}\right)E_2\left[\sum_{i=1}^{n}\bar{y}_i^2 - \frac{1}{n}\left(\sum_{i=1}^{n}\bar{y}_i\right)^2\right]\right\}$$

$$= E_1\left\{\left(\frac{1}{n-1}\right)E_2\left[\sum_{i=1}^{n}\bar{y}_i^2 - \frac{1}{n}\left(\sum_{i=1}^{n}\bar{y}_i^2 + 2\sum\sum_{i<j}\bar{y}_i\bar{y}_j\right)\right]\right\}$$

$$= E_1\left\{\left(\frac{1}{n-1}\right)E_2\left[\left(1-\frac{1}{n}\right)\sum_{i=1}^{n}\bar{y}_i^2 - \frac{2}{n}\sum\sum_{i<j}\bar{y}_i\bar{y}_j\right]\right\}$$

$$= E_1\left\{\left(\frac{1}{n-1}\right)\left[\left(1-\frac{1}{n}\right)\sum_{i=1}^{n}(V_2(\bar{y}_i) + \mu_i^2) - \frac{2}{n}\sum\sum_{i<j}\mu_i\mu_j\right]\right\}$$

$$= E_1\left\{\left(\frac{1}{n-1}\right)\left[\left(1-\frac{1}{n}\right)\sum_{i=1}^{n}V_2(\bar{y}_i) + \sum_{i=1}^{n}\mu_i^2 - \frac{1}{n}\left(\sum_{i=1}^{n}\mu_i\right)^2\right]\right\}$$

$$= E_1\left\{\frac{1}{n}\sum_{i=1}^{n}V_2(\bar{y}_i) + \left(\frac{1}{n-1}\right)\left[\sum_{i=1}^{n}\mu_i^2 - \frac{1}{n}\left(\sum_{i=1}^{n}\mu_i\right)^2\right]\right\}$$

$$= \frac{1}{N}\sum_{i=1}^{N}V(\bar{y}_i) + \frac{1}{N-1}\sum_{i=1}^{N}(\mu_i - \bar{\mu})^2$$

Thus
$$\left(\frac{N-n}{N}\right)\left(\frac{1}{n\bar{M}^2}\right)s_b^2$$

estimates Equation (A.4) plus a term

$$\left(\frac{N-n}{N}\right)\left(\frac{1}{nN}\right)\sum_{i=1}^{N}V(\bar{y}_i) \qquad\qquad (A.6)$$

and we need to find an estimate of (A.5) − (A.6) in order to estimate $V(\hat{\mu})$. But

$$(A.5)-(A.6) = \frac{1}{nN}\sum_{i=1}^{N}V(\bar{y}_i)\left(1-\frac{N-n}{N}\right)$$

$$= \left(\frac{1}{N}\right)\left(\frac{1}{N}\right)\sum_{i=1}^{N}V(\bar{y}_i)$$

which can be estimated unbiasedly by

$$\left(\frac{1}{N}\right)\left(\frac{1}{n}\right)\sum_{i=1}^{n}\left(\frac{\bar{M}-m}{\bar{M}}\right)\left(\frac{s_i^2}{m}\right)$$

where
$$s_i^2 = \frac{1}{m-1}\sum_{i=1}^{m}(y_{ij}-\bar{y}_i)^2$$

The estimator of $V(\hat{\mu})$ is, then,

$$\hat{V}(\hat{\mu}) = \left(\frac{N-n}{N}\right)\left(\frac{1}{n\bar{M}^2}\right)s_b^2 + \frac{1}{nN}\sum_{i=1}^{n}\left(\frac{\bar{M}-m}{\bar{M}}\right)\left(\frac{1}{m-1}\right)s_i^2$$

This equation is equivalent to Equation (9.2) in the case of equal cluster sizes. The case for unequal cluster sizes is derived analogously.

SELECTED ANSWERS

Chapter 4 Exercises

4.1. $\sigma^2 = 12$, $V(\bar{y}) = .75$

4.5. $\hat{p} = \frac{5}{6}$, $B = .1313$

4.6. $n = 128$

4.7. $\bar{y} = 12.5$, $B = 7.04$

4.8. $\hat{\tau} = 125,000$, $B = 70,412.50$

4.9. $\hat{\mu}_1 = 2.30$, $\hat{\mu}_2 = 4.52$, $B = .0703$,
       $B = .0858$

4.10. $\hat{p} = .625$, $B = .1535$

4.11. $\hat{\mu} = 2.0$, $B = .9381$

4.12. $\hat{p} = .43$, $B = .0312$

4.13. $n = 2392$

4.14. $\hat{\tau} = 100$, $B = 31.29$

4.15. $\hat{\mu} = 2.1$, $B = .170$

4.16. $n = 4$

4.17. $\hat{\mu} = 5.01$, $B = .8711$

4.18. $\hat{p} = \frac{11}{60}$, $B = .0958$

4.19. $n = 87$

4.20. $\hat{\tau} = 37,800$, $B = 3379.94$

4.21. $n = 400$

4.22. $\hat{\tau} = \$3,898,000$, $B = \$263,918.17$

4.29. $\hat{\tau} = \$17,333.33$, $B = \$4,479.23$

4.30. $\hat{\tau} = \$98,550.00$, $B = \$19,905.83$

4.31. $\hat{p} = .3$, $b = .2060$

4.37. (a) $\hat{p} = .22$, $B = .0893$
       (b) $\hat{p} = .63$, $B = .1041$
       (c) $\hat{p} = .10$, $B = .0880$
       (d) $\hat{p} = .90$, $B = .0880$

Chapter 5 Exercises

5.1. $\hat{p}_{st} = .30$, $B = .117$

5.2. $n_1 = 18$, $n_2 = 10$, $n_3 = 2$

5.3. $\hat{\tau} = 1903.90$, $B = 676.80$

5.4. $\bar{y}_{st} = 53,208.63$, $B = 560.48$

5.5. $n = 26$, $n_1 = 16$, $n_2 = 7$, $n_3 = 3$

5.6. $\bar{y}_{st} = 59.99$, $B = 3.032$

5.7. $n_1 = 12$, $n_2 = 20$, $n_3 = 18$

5.8. $n = 33$

5.9. $n = 32$

5.10. $\hat{\tau} = 50,505.60$, $B = 8,663.12$

5.11. $n = 60$

5.12. $n = 29$

5.13. $n = 158$, $n_1 = 39$, $n_2 = 17$, $n_3 = 69$,
       $n_4 = 33$

5.14. $\hat{p}_{st} = .701$, $B = .0503$

5.15. $n = 62$, $n_1 = 17$, $n_2 = 6$, $n_3 = 26$, $n_4 = 13$

5.16. (a) $\bar{y}_{st} = 250.20$, $\hat{V}(\bar{y}_{st}) = 137.90$
       (b) $\bar{y}_{st} = 249.21$, $\hat{V}(\bar{y}_{st}) = 191.23$

5.17. dividing points: 40, 70, 90

5.18. stratum 1: 0–\$200;
       stratum 2: \$200–\$350

5.19. $\bar{y}_{st} = 63.88$, $B = .628$

5.21. (a) $\hat{p} = .16$, $B = .074$
       (b) $\hat{p}_{st} = .159$, $B = .081$

Chapter 6 Exercises

6.1. $\hat{\tau}_y = 1578.36$, $B = 192.56$

6.2. $\hat{\tau}_y = 2937.5$, $B = 713.38$

6.3. $r = .1467$, $B = .0102$

6.4. $\hat{\tau}_y = 145,943.78$, $B = 7353.67$

6.5. $\hat{\mu}_y = 1186.53$, $B = 59.79$

6.6. $\hat{\mu}_y = 17.59$, $B = .2710$

6.7. $\hat{\mu}_y = 4.1646$, $B = .0847$

6.8. $r = .283$, $B = .0616$

6.9. $\hat{\tau}_y = 5492.31$, $B = 428.44$

6.10. $r = 1.072$, $B = .00643$

6.11. $\hat{\mu}_y = 1061.04$, $B = 139.95$

6.12. $\hat{\tau}_y = 231,611,86$, $B = 3073.83$

6.13. $n = 14$

6.14. $\hat{\mu}_{yL} = 1186.5457$, $B = 61.35$

6.16. $\hat{\tau}_y = 5515.50$, $B = 448.61$

6.17. $r = .835$, $B = .012$

6.20. $r = .401$, $B = .128$

6.21. (a) $r = 1.043$, $B = .0733$
       (b) $r = .870$, $B = .176$

6.22. $r = 1.128$, $B = .0271$

6.23. (a) $\hat{\tau}_y = 2433.30$, $B = 45.95$
       (b) $\hat{\tau}_{yL} = 2432.91$, $B = 48.64$
       (c) $\hat{\tau}_{yD} = 2455.90$, $B = 180.07$

6.25. $\hat{\tau}_{yRc} = 48,209.84$,
       $\hat{V}(\hat{\tau}_{yRc}) = 557,095.07$

Chapter 7 Exercises

7.3.   (a) $V(\hat{p}) = .1275$
       (b) $V(\hat{p}) = .00875$

7.4. $\hat{p}_{sy} = .66$, $B = .0637$

7.5. $n = 1636$

7.6. $\hat{\mu}_{sy} = 11.94$, $B = .026$

7.7. $n = 28$

7.8. $\hat{\mu}_{sy} = 2007.11$, $B = 74.505$

7.9. $\hat{p}_{sy} = .81$, $B = .036$

7.10. $n = 1432$

7.11. $\hat{t}_{sy} = 127,500$, $B = 30,137.06$

7.12. $n = 259$

7.13. $\hat{\mu}_{sy} = 3.54$, $B = .406$

7.14. $\hat{\mu}_{sy} = 225.47$, $B = 6.75$

7.15. $\hat{\tau}_{sy} = 48,680$, $B = 1370.34$

7.16. $\hat{\mu}_{sy} = 7038.10$, $B = 108.74$

7.17. $\hat{p}_{sy} = .738$, $B = .104$

7.18. $\hat{\tau}_{sy} = 4400$, $B = 784.08$

7.20. (a) $\hat{\mu}_{sy} = 1,926,935$, $B = 139,437.35$

   (b) $\hat{\mu}_{sy} = 19.67$, $B = 3.17$

7.21. $\hat{\mu}_{sy} = 2.26$, $B = .576$

Chapter 8 Exercises

8.2. $\hat{\mu} = 19.73$, $B = 1.78$

8.3. $\hat{\tau} = 12,312$, $B = 3175.07$

8.4. $\hat{\tau} = 14,008.85$, $B = 1110.78$

8.5. $n = 14$

8.6. $\hat{\mu} = 51.56$, $B = 1.344$

8.7. $n = 13$

8.8 $\hat{p} = .709$, $B = .048$

8.9. $n = 7$

8.10. $\hat{\mu} = 40.17$, $B = .640$

8.11. $\hat{\tau} = 157,020$, $B = 6927.88$

8.12. $n = 30$

8.13. $\hat{\mu} = 16.005$, $B = .0215$

8.14. $\hat{p} = .5701$, $B = .0307$

8.15. $n = 21$

8.16. $\hat{\mu} = 5.91$, $B = .322$

8.17. $\hat{p} = .40$, $B = .116$

8.18. $\hat{\tau} = 3532.8$, $B = 539.50$

8.19. $\hat{\mu} = 2.685$, $\hat{V}(\bar{y}^*) = .056$

8.20. (a) $\hat{\mu} = 3.153$, $B = .460$

   (b) $\hat{\mu} = 5.99$, $B = .929$

8.21. $\hat{p} = .133$, $B = .075$

8.22. $\hat{\tau} = 80$, $B = 40.44$

8.23. $\hat{p} = .0918$, $B = .0390$

8.26. $\hat{\tau} = 600$, $B = 308.22$

8.27. $\hat{\mu} = .25$, $B = .128$

Chapter 9 Exercises

9.2. $\hat{\mu}_r = 9.3789$, $B = 1.4536$

9.3. $\hat{\mu} = 9.5593$, $B = 1.3672$

9.4. $\hat{p} = .2865$, $B = .1116$

9.5. $\hat{p} = .351$, $B = .1767$

9.6. $\hat{\tau} = 3980.7$, $B = 274.7317$

9.7. $\hat{p} = .1200$, $B = .0067$

9.8. $\hat{\tau} = 1276.2425$, $B = 333.4435$

9.9. $\hat{\mu} = 7.9333$, $B = .0924$

9.11. $\hat{\mu}_r = 97.97$, $B = 10.996$

9.12. $\hat{\tau} = 57,608$, $B = 6465.37$

9.14. $\hat{\tau} = 3900$, $B = 896.10$

9.15. $\hat{\mu} = .9811$, $B = .225$

Chapter 10 Exercises

10.4. $N = 444.444 \approx 445$, $B = 150.596$

10.5. $\hat{N} = 1811$, $B = 344.512$

10.6. $\hat{N} = 10,868$, $B = 715.82086$

10.7. $\hat{N} = 3348.2143$, $B = 445.10$

10.8. $\hat{N} = 200$, $B = 78.88$

10.9. $\hat{V}(\hat{N})/N = 12.67$ or $t \approx 625$, $n \approx 625$

10.10. $\hat{N} = 1067$, $B = 507.7182$

10.11. $\hat{N} = 750$, $B = 441.588$

10.13. $\hat{N} = 250$, $B = 52.04$

10.14. $\hat{\lambda} = 2.1$, $B = .0748$

10.15. $\hat{\lambda} = .0171$, $B = .00191$

10.16. $\hat{M} = 1920$, $B = 135.76$

10.19. $\hat{\lambda} = 2.792$, $B = .216$

Chapter 11 Exercises

11.1. $\bar{y} = 407.125$, $B = 93.703$

11.2. $\bar{y} = 5.26$, $B = .7889$

11.3. $\bar{y} = 23.6113$, $B = 9.0972$

11.4. $\hat{\tau} = 1794.455$, $B = 778.1539$

11.5. $\hat{\tau}_1 = 1959.7338$, $B = 763.5104$

11.6. $\bar{y} = 9.8042$, $B = 2.3758$

11.7. $\hat{\tau} = 3866.7633$, $B = 1171.2750$

11.8. $\hat{\tau}_1 = 4117.764$, $B = 999.8094$

11.9. $\hat{p} = .875$, $B = .1052$

11.10. $\hat{p} = .125$, $B = .1377$

INDEX